# Bacteria to AI

# *Bacteria to AI*

## Human Futures with Our Nonhuman Symbionts

N. KATHERINE HAYLES

The University of Chicago Press
Chicago and London

The University of Chicago Press, Chicago 60637
The University of Chicago Press, Ltd., London
© 2025 by The University of Chicago
All rights reserved. No part of this book may be used or reproduced in any manner whatsoever without written permission, except in the case of brief quotations in critical articles and reviews. For more information, contact the University of Chicago Press, 1427 E. 60th St., Chicago, IL 60637.
Published 2025
Printed in the United States of America

34  33  32  31  30  29  28  27  26  25       1  2  3  4  5

ISBN-13: 978-0-226-83598-3 (cloth)
ISBN-13: 978-0-226-83747-5 (paper)
ISBN-13: 978-0-226-83746-8 (e-book)
DOI: https://doi.org/10.7208/chicago/9780226837468.001.0001

Library of Congress Cataloging-in-Publication Data

Names: Hayles, N. Katherine, 1943– author.
Title: Bacteria to AI : human futures with our nonhuman symbionts / N. Katherine Hayles.
Description: Chicago ; London : The University of Chicago Press, 2025. | Includes bibliographical references and index.
Identifiers: LCCN 2024020844 | ISBN 9780226835983 (cloth) | ISBN 9780226837475 (paperback) | ISBN 9780226837468 (ebook)
Subjects: LCSH: Cognition—Philosophy. | Posthumanism. | Cognition—Social aspects. | Subliminal perception—Social aspects. | Artificial intelligence—Social aspects. | Technology—Social aspects. | Evolution.
Classification: LCC BF311 .H3948 2025 | DDC 153—dc23/eng/20240605
LC record available at https://lccn.loc.gov/2024020844

♾ This paper meets the requirements of ANSI/NISO Z39.48-1992 (Permanence of Paper).

## Contents

1 An Integrated Cognitive Framework  1
2 Can Computers Create Meanings? A Technosymbiotic Perspective  51
3 The Emergence of Technosymbiosis and Gaia Theory  74
4 Cellular Cognition: Mimetic Bacteria and Xenobot Creativity  101
5 Rocks and Microbes: The Two Different Temporal Regimes of Biological and Mineral Evolution  119
6 Inside the Mind of an AI: Materiality and the Crisis of Representation  138
7 GPT-4: The Leap from Correlation to Causality and Its Implications  166
8 Subversion of the Human Aura: Three Fictions of Conscious Robots  180
9 Collective Intelligences: Assessing the Roles of Humans and AIs  204
10 Planetary Reversal: Ecological Relationality versus Political Liberalism  228

*Acknowledgments*  249
*Notes*  253
*Bibliography*  265
*Index*  281

# 1

# An Integrated Cognitive Framework

A central purpose of this book is to combat anthropocentrism and suggest other perspectives more conducive to survival and flourishing. While we cannot avoid anthropomorphism—we see with human eyes and think with human brains—anthropo*centrism* can and should be contested. Anthropocentrism positions humans "at the center" conceptually, politically, socially, and economically, declaring that humans are the most important species on Earth and the most entitled to exploit Earth's resources for their/our own benefit. While there are many belief systems and practices driving our present ecological crises—capitalism, neoliberalism, political corruption, blind ignorance, and greed, for example—anthropocentrism is certainly among them.

The linchpin of anthropocentrism is cognition. Humans are not the fastest or strongest species on Earth, not the one with the most acute senses or most important to Earth's flourishing (that honor would undoubtedly go to microorganisms). It is, however, the one most cognitively advanced (so the anthropocentric argument goes). This presumed superiority makes up for everything else, enabling humans to invent cars faster than cheetahs, deploy machines stronger than elephants, create sensors more sensitive than grizzly noses, and so forth. Therefore, to attack anthropocentrism it is essential to rethink and reposition human cognitive capabilities.

Many books could be (and have been) written on this topic. From my viewpoint, one of the most effective approaches is to decouple cognition and consciousness, for millennia seen as virtually synonymous with each other. In *Unthought: The Power of the Cognitive Nonconscious* (Hayles 2017), I argued that human *nonconscious* cognitive abilities, which operate on a neuronal level below consciousness, are in fact essential for consciousness to function. Moreover, nonconscious cognitive organisms are far more numerous in the biosphere than are conscious ones (which include primarily mammals, cephalopods such as octopi, and some avion species, for example parrots and ravens). Nonconscious organisms, by contrast, include unicellular organisms such as bacteria, all plant species, fungi, insects, and everything else. By biomass, about 90 percent by weight of Earth species are nonconscious. Only human hubris would have ignored nonconscious cognition for so long and have failed so spectacularly in exploring its capacities, including recognizing its importance in conscious functioning, in the biosphere, and in computational media.

This book aspires to combat anthropocentrism by urging a transition from the liberal philosophy we inherited from the Enlightenment to a different conceptual orientation that I call the integrated cognitive framework (ICF). One of the purposes of ICF is to create a platform in which human conscious cognition can be put *in relation* to nonconscious cognitions both within humans and within the greater-than-human world. "Integrated" here does not mean flattening out but precisely the opposite; it is a framework in which overlaps and divergences between different kinds of cognitive capacities and origin stories can be compared and contrasted. It bears resemblance to Cary Wolfe's "jagged ontologies" (Wolfe 2024), creating a densely textured conceptual landscape of high dimensionality that never resolves to a smooth homogeneous surface.

The integrated cognitive framework compares and contrasts tribraided origin stories in quantum mechanics, evolutionary biology, and technics that insist, in different ways and with differently situated dynamics, on the coconstitutive power of relationalities. "Integrated" here means an approach that keeps all three strands in view. Juxtaposing origin stories of quantum mechanics (origin of phenomena) with evolutionary biology (origin of multicellular bodies) and technics (origin of the human) reveals the structural similarities and also the accelerating pace with which they are becoming entangled with each other through such developments as quantum computers, gene editing, and artificial intelligence. I call this configuration the micro/evo/techno relationality.

This synthetic approach is well suited to deal with the exponentially accelerating trajectory of AI. In this context, ICF compares and contrasts human cognition with large language models (LLMs) such as OpenAI's GPTs that, for the first time in history, produce human-competitive texts. Although LLMs have enormous implications for virtually all aspects of the economy and human communicative practices, they pose especially urgent challenges to literary studies, theory, and criticism, fields focused on the interpretations of verbal texts.

Before progressing to the unfolding implications of ICF, it may be useful to consider anthropocentrism in more detail, and why it should be avoided.

Anthropocentrism implies human supremacy, but more than that, it assumes that the human is and should be the measure by which all other organisms are judged. The Greek philosopher Protagoras (490–420 BCE) proclaimed, "Of all things the measure is man: of those that are, that they are, and of those that are not, that they are not" (Bonazzi 2023). Whether "man" is interpreted as referring to specific individuals or to the human race (there are philosophical debates on the question), the anthropocentric orientation is clear. The implication is that only humans have the right to judge what is in the world, and what is not. One argument that will run through this book is that all creatures have world horizons, or *umwelten*, specific for their species, including humans, and that these world horizons may overlap, but they never entirely coincide. The ethical implication of this viewpoint is that no umwelt has the right to dominate the others and declare that it alone is the proper and right way to see the world. Rabbits see as rabbits, octopi as octopi, humans as humans: none is inherently superior, although each has specific ways in which its world horizon bestows functionality within a given ecological niche. Another obnoxious implication of anthropocentrism is that humans are literally at the center of the action, overseeing and dictating what happens. This is absurdly far from the truth; increasingly, evolutionary biologists and others realize that human life is interpenetrated at many scales and in many ways by other lifeforms, on which it is entirely dependent for its continued existence. Moreover, the coconstitution of multicellular bodies by unicellular ones implies that relationality is not peripheral or after the fact, but deeply involved in the origin of multicellularity from the beginning.

Finally, anthropocentrism implies that humans alone are the deciders, the ones who make decisions for everyone else. Yet our increasing reliance on computational media means that decisions, like agency, are distributed throughout the collectivities of humans, nonhumans, and computational media that I call cognitive assemblages. Even when humans appear to be in control, their assumptions have been formed and mediated by prior decisions and interpretations made by computational media, so that they can scarcely be considered autonomous or self-determining. To sum up, then, anthropocentrism spreads its tentacles through a mass of mistaken and dangerous beliefs, including human supremacy and dominance, human viewpoints as the right and meaningful ones, human decisions as the only ones that matter, and the subsequent denigration of all other species and cognitive entities.

To return now to the main line of this book's argument, it begins by recasting cognition as a much broader and more capacious capacity than consciousness, including in humans. In *Unthought: The Power of the Cognitive Nonconscious* (2017) mentioned earlier, I referenced work in neuroscience, cognitive science, cognitive psychology, and other fields demonstrating that much of human action and behavior is influenced by neuronal processes that happen below the threshold of consciousness and yet are critical for consciousness to function. I called these processes *nonconscious cognition*. Their existence demonstrates that consciousness is not necessary for cognitive acts to take place. If so, then cognition may be extended to nonconscious organisms, even those without brains such as clams and date palms.[1] Moreover, it may also include computational media, such as computers, cell phones, and many other devices. Indeed, I argue that computational media, working together with humans in cognitive assemblages, are responsible for most of the world's work today.

(Did you notice? Twice in the above paragraph I used "humans" as nouns. I once had a copy editor tell me that "human" could only be an adjective, and that to form a noun, the correct formulation is "human being."[2] Why do we not speak about "whale beings," "dog beings," or "octopus beings"? The answer is obvious: because we do not recognize these creatures as worthy of the title "beings." As a feminist, I try to avoid masculinist language; for similar reasons, I try to avoid language that reinscribes anthropocentric assumptions. So, humans.)

I want now to explore further the claim that cognitive assemblages are responsible for most of the world's work. Isn't work making widgets in a factory, unloading cargo ships, taking food orders in a diner? Well, yes and no. Increasingly, such activities include computational media as controllers, processors, transmitters, and storers of information. Cognitive media have deeply penetrated into the infrastructures of contemporary societies because they control the flow of information that allows large, complicated, and massive electromechanical equipment to be directed and controlled by much smaller and more efficient electronic mediators. It's as if a flea sitting in an elephant's ear could control its movements and direct what it does.

To help visualize this, I will focus on a typically American technology that has been featured in innumerable movies and TV shows: the starter motor for an automobile engine. The differences between 1950 and 2023 models will tell the tale of the remarkable transformations that occurred throughout industrial infrastructures during this historically tumultuous period, which saw the invention of the transistor, the spread of electronics, and the miniaturization and increased memory that made possible the desktop computer, along with the invention of the internet and subsequently the web.

Around 1950, cars began to incorporate solenoids into the circuit connecting the key to the starter motor.[3] Previously, the circuit was typically closed by pushing a starter button (sometimes located in the floorboard), which brought two heavy copper wires into contact. This took quite a lot of pressure. The solenoid, by contrast, was activated by a much smaller current; its function was to convert that small current to the large output current needed for the starter motor. The solenoid was electric, but in its design philosophy, it was edging toward electronics: the idea was to control something large and heavy with a much smaller component that took a lot less work to activate. Here is a description of what happens when the starter motor engages:

> On the engine, a flywheel, with a ring gear attached around the edge, is fitted to the end of the crankshaft. On the starter, the pinion is designed to fit into the grooves of the ring gear. When you turn the ignition switch, the electromagnet inside the body engages and pushes out a rod to which the pinion is attached. The pinion meets the flywheel and the starter motor turns. This spins the engine over, sucking in air (as well as fuel). As the engine turns over, the starter motor disengages, and the electromagnet stops. The rod retracts into the starter motor once more, taking the pinion out of contact with the flywheel and preventing potential damage. ("Starter Motors," n.d.)

The technology described here is a large, complex, and heavy electromechanical system that was the result of many decades of innovations and design improvements. These improvements culminated in an innovation that changed the distribution of work (understood as mass moved through distance with the application of force). With a solenoid, the entire mechanism could be started with a current as small as a few amps and a delicate turn of the ignition key.

The process of minimizing the work needed to start a car does not end there, of course. Our 2023 model does not have a key that fits into the ignition switch at all. The car is started proximally, with a fob that you can keep in your pocket, purse, or briefcase. "The fob transmits a unique low-frequency signal to the car's computer system, which then validates that the correct signal has been sent and allows you to push a button on the dashboard or console to start the engine. Shutting off the motor is just as hassle-free: simply press the start/stop button. And, in addition to keyless ignition, most systems also include keyless entry, allowing you to enter the car without inserting a key or hitting a button on the fob" ("What You Need to Know," n.d.).

Thus into the electromechanical system has been inserted a computer system capable of receiving, processing, transmitting, and storing information. It is now the computer system, and its corresponding fob equipped with an RFID (radio frequency identification), that controls the electromechanical process, so that even the slight pressure needed to turn the ignition key has been eliminated. Although the computer made large, heavy machinery easier for humans to operate, it also introduced new vulnerabilities. If you fry the computer system, the car simply cannot be started at all. If you lose the fob, you may not even be able to open the car door.

I drive a late-model Toyota Camry, and this car, like many contemporary models, has a variety of sensors that convey information to the car's computer system, including laser range finders, image detectors, and so forth. If I swerve over a white line, there is a slight resistance as the car gives me a "nudge" to return to the lane. Similarly, if I come too close to another car, a dash display screams "BRAKE" in red lights. The self-driving cars now on the horizon go much further in their ability to sense information, process it, and make choices about a large variety of actions.

This example can be multiplied a million times over, as computational media are introduced into complex systems as controllers, processors, and transmitters of information. The crane that unloads the container ship incorporates computerized components that control the movements and introduce safety protocols; the factory that makes widgets transitioned last year from humans on an assembly line to industrial robots fabricating the widgets; the diner introduced a computerized order system, bypassing the waitperson and connecting the customer directly with the short-order cook.

Of course, the computational components are not all that is required. The crane computer communicates with both the human operator and the electromechanical systems that function as sensors and actuators for the huge steel tower and lifting apparatus; the factory robots must be programmed, maintained, and overseen by human technicians; the computer ordering system interfaces both with a database containing the menu selections and with the screen at which the cook peers. Each of these collectivities qualifies as a cognitive assemblage, comprising humans, computational media, and electromechanical systems, through which information, interpretations, and meanings circulate. It is these assemblages that now drive huge cargo ships, fly airplanes, run trains, and control traffic on urban streets.

Cognitive assemblages also include nonhuman lifeforms, all of which have cognitive capabilities. Increasingly, computational media are used to harness the power of microorganisms' cognitive abilities. Gene editing is a spectacular example, as explained in chapter 4. In addition, computational media played crucial roles in developing the Pfizer and Moderna vaccines for Covid-19, which rely on messenger RNA to send instructions to cells to build the spike proteins on outer walls that catalyze the development of antibodies. The development of these vaccines in record time would not have been possible without computational power to analyze results and assist in gene sequencing.

There are also instances where nonhuman animals have joined with computational media to provide data for humans. For example, Jennifer Gabrys in *Program Earth: Environmental Sensing Technology and the Making of a Computational Planet* describes a project that attached computational devices to dolphins to monitor their movements and gain information about their ocean environments (Gabrys 2016, 126–36). Karen Bakker in *The Sounds of Life* discusses research such as the Interspecies Internet project that aims to decode and translate animal vocalizations so that humans can understand them (Bakker 2022, 158–70). In the Cetacean Translation Initiative (CETI), launched by researchers from Harvard and Berkeley, the songs of sperm whales are collected in a data set large enough so that, using neural net algorithms, it may soon be possible to construct a dictionary of "Sperm Whalish" (Bakker 2022, 159). In addition to studying animal language, researchers are also inventing ways to let the animals speak for themselves. The idea, Bakker writes, is to "design devices that can communicate with nonhumans using their own modalities of communication" (163), such as smart phones and tablets adapted so dolphins can use them. Here the cognitive assemblage consists of the humans who receive the information, the computational components that process and transmit it, and the dolphins whose behaviors create it.

There have also been numerous art projects that incorporate lifeforms into cognitive assemblages. For example, Kenneth Goldberg's *Telegarden* created a garden with living plants watered and tended by robotic mechanisms that could be controlled by people logging into the Telegarden site (Goldberg 1995–2004). Eduardo Kac's *Eighth Day, a Transgenic Artwork* used robotic mechanisms that tended fluorescent transgenic organisms including plants, amoeba, fish, and mice (Kac 2001; see also Britton and Collins 2003).

Perhaps the most crucial way in which cognitive assemblages function in developed societies is through infrastructures that rely on computational media to sense, process, direct, communicate, and store information flows. As a result, if computational media were to malfunction on a massive scale, the toll in human lives and in social chaos would be enormous. As early as July 1997, George Ullrich, deputy director of the Defense Special Weapons Agency in the Department of Defense, testified to a House subcommittee investigating electromagnetic pulses (EMPs) about the fragility of unshielded electronic systems (such pulses do not affect electrical systems, so the pervasiveness of electronics introduced new vulnerabilities. Virtually all contemporary digital computational media rely on electronics).[4] "A high-altitude burst, detonated a few hundred kilometers above the surface of the Earth, has as its salient featured effect the ability to simultaneously bathe an entire continent in EMPs. The ability of EMPs to induce potentially damaging voltages and currents in unprotected electronic circuits and components is well-known. The immense footprint of EMPs can therefore simultaneously place at risk unhardened military systems, as well as critical infrastructure systems to include power grids, telecommunication networks, transportation systems, banking systems, medical services, civil emergency systems and so forth" (Ullrich 1997).

Nearly three decades further along, computational systems have penetrated even more deeply into the infrastructures of US society. Let us suppose that a high-altitude EMP burst happened. What would be the result? Cars and trucks would not start; railroad command centers would not function; airline towers could not communicate with or control the airplanes, which would be flying blind; ATMs would not work; Wall Street trading would collapse in chaos; the electrical grid, controlled by electronic components, would be disabled; cell phones would be useless; water would not flow from taps; food and supply networks would be disrupted; and so forth. I list this depressing litany not to forecast an apocalypse (which of course I fervently hope never happens), but to vivify how deeply we have entered into alliances with cognitive assemblages and how dependent we have become on their proper functioning.

In 1950, it may have been possible to choose other paths. In the third decade of the new millennium, however, our reliance of cognitive assemblages and computational media has progressed so far that there is no going back. The only feasible options are to go forward from where we are now. There are, however, multiple paths that we might choose, and our choices within the next few years are likely to be decisive for the kinds of futures we and our children will have. It behooves us, then, to investigate more fully the nature of the cognitive assemblages so crucial to our societies. If we are serious about finding solutions to the global problems that confront us, one path forward is through the ICF framework.

## Cognitive Assemblages and the Integrated Cognitive Framework

I chose "cognition" and not "thought" for this framework because, for my purposes here, "thought" (like one of those inedible fruitcakes soaked in bad whiskey) has an off-putting smell, in this case of anthropocentrism. The present book develops further the line of argument I pursued in the previously mentioned book, *Unthought: The Power of the Cognitive Nonconscious*. There I was looking for a definition of cognition that would have a low threshold for something to count as cognitive, yet was capable of scaling up to the kind of high-level cognitions evident in humans. Here is the definition I offered: "cognition is a process that interprets information within contexts that connect it with meaning" (Hayles 2017, 22). The definition emphasizes cognition's processual nature; it is always in the process of becoming. Moreover, it specifies an integral connection between information and cognition; information is the material on which cognition operates. In this sense, it is compatible with the theories in physics postulating that information, rather than matter or energy, may be the primordial entity in the universe, with matter and energy emerging from it.[5]

The integrated cognitive framework is built on the premise that all lifeforms have ways to sense their environments, absorb information from them, and interpret that information through their sensory and organismic capacities. Thus all lifeforms, including those without central nervous systems such as nematode worms and maidenhair ferns, have cognitive capabilities. Moreover, the interpretations that organisms perform on environmental information are crucial to their survival and reproduction and, in this sense, have meanings relevant to organisms and their milieux. The research area that has most extensively developed this concept of meaning is biosemiotics. Working through the triadic framework of Peircean semiotics consisting of sign vehicle, interpretant, and representamen, biosemiotics understands "meaning" not as an abstract concept but rather as a response to environmental stimuli. As Wendy Wheeler put it, "meaning is always a kind of doing" (2016b, 7).

This conceptual move has enormous implications, for it opens the entire realm of biota to the creation, exchange, and interpretation of signs, and thus to meaning-making practices. Humans in this view are not unique or special in their emphasis on meaning creation; all living beings create, process, and interpret meanings. The uniqueness of humans rather lies in their ability to extend meaning making into abstract symbols, as Terrence W. Deacon argues (1998). Moreover, the signs that living creatures create are performed in environments that include all the other signs. Like the birdsongs and creaturely cries that greet the dawn in an Amazonian rainforest, the signs intermingle, each influencing and being influenced by the others. Jesper Hoffmeyer calls this realm of interacting signs the semiosphere, a grand symphony across scales audible and inaudible produced by creatures as they go about the business of living (1997, vii).

## Actors and Agents

The emphasis on cognition has another important implication, for it provides a principled way to distinguish between cognitive acts and material processes. By material processes I mean the physicochemical processes that are the foundations on which all life is built, from chemical reactions in the soil and oceans, on up to the nuclear fission reactions that fuel the Sun. Make no mistake: these processes have agency. Indeed, they often release powerful forces that dwarf anything humans can do. What they do not have is the ability to interpret information and make choices (or selections) based on those interpretations. A tornado cannot choose to plow through a field rather than a populous town; an avalanche will not hold back its pounding cascades because climbers are on the slope. It took the spark of life to create those possibilities, and thus to unleash the power of signs and meaning creation. To encode this distinction, I use the term "agents" to refer to material processes, reserving the term "actors" for living organisms and other entities that embody cognitive capabilities.

The agent/actor distinction is further authorized by the different temporal regimes occupied by material processes and the evolution of biological lifeforms. Physicochemical processes follow trajectories that can be predicted by mapping them into phase spaces; in chapter 5, this is explained by mapping the trajectory of a swinging pendulum into a phase space showing momenta and position. In a seminal article, Giuseppe Longo, Maël Montévil, and Stuart Kauffman (2012) show how this is accomplished even for chaotic nonlinear systems such as a double-jointed pendulum. In contrast, they argue, biological evolution follows no such predictable trajectories. Rather, it leaps from one niche to another in a pattern that Kauffman elsewhere calls "adjacent possibles" (Kauffman 2000, 22).

The unpredictable nature of biological evolution points to another systemic difference between biological lifeforms and physicochemical process: living creatures *have stakes* in what happens to them, whereas material processes do not, a fact that Michel Levin and Daniel Dennett (2020) emphasize in their subtitle, which refers to "agents with agendas." Every living being, even a unicellular organism, will act so as to continue its existence; as Darwin noted a century and a half ago, the need to survive and reproduce is universal throughout the animal and plant kingdoms. I call this desire to survive the biological mandate. Through the billions of years of the evolution of life on Earth, all creatures incorporated the biological mandate into their ways of being in world, for if a hypothetical species did not, it would cease to exist and go extinct. Rocks decay, mountains erode, lakes evaporate, but they neither know nor care that these events happen; they simply are.

This position partially coincides with those taken by many theorists writing under the banner of neomaterialism, including Jane Bennett (2010), Elizabeth Grosz (2011), and Rosi Braidotti (2013), among others. Like them, I agree that material processes have agency that we ignore at our peril, and like them I agree that these processes are intimately involved with the processes that take place in living organisms, including humans. Where we differ is in the distinction I make between the agential powers of material processes and the interpretive/cognitive capabilities of living organisms.

This difference is consequential, for it entails a number of other important differences, including an emphasis on information and interpretation, on cognition as a primary capacity of lifeforms, and on the creation, dissemination, and interpretation of signs as pervasive for living organisms. I can only speculate about why neomaterialists do not choose to move in this direction; I suspect that many of them may feel that emphasizing these qualities reinforces anthropocentrism rather than overcomes it. To this I would respond that much of anthropocentrism's sting is removed by the insistence that these are capacities that all lifeforms share. Moreover, I think it is important not to throw the baby out with the bathwater. In our desire to overcome anthropocentrism (an enterprise that I believe is urgently necessary), it is also crucial to account for the ways in which humans have capabilities that other species do not. These include (but are not limited to) the ability to anticipate events far into the future, coordinate with strangers across far-flung networks, engage in symbolic reasonings of many kinds, use language extensively, write symphonies, develop mathematical theorems, create computer science and computer languages, and all the other achievements unique to the human species. A framework built on cognition is able to make such an accounting, while at the same time broadening the realm of the cognitive so that it includes all living creatures and some computational media.

## The Cognitive Revolution

The Industrial Revolution discovered ways to harness the powers of material processes, from early steam engines to water turbines, fossil fuel engines to electrical power. Only in the twentieth century, however, did the subsequent Cognitive Revolution discover ways to create cognitive capabilities in artificial media and, more recently still, to harness the cognitive powers of microorganisms. The development of computational media, beginning in the 1930s with Konrad Zuse and other researchers, blossomed in post–World War II research into the development of von Neumann machines that now dominate the media landscape of cognitive machines.[6] With intense miniaturization and exponentially increasing memory, the 1980s saw the field move from mainframes to powerful desktop machines. Then, by the late 1990s, the transition from stand-alone devices to networked computers was accomplished. The frontiers of computation have now been opened to neural nets and deep learning algorithms, moving away from the von Neumann architectures into systems such as generative adversarial networks (GANs) that take their inspiration from the way that cognition works in biological organisms. With the development of neuromorphic chips based on artificial "neurons" and "synapses," such as the SyNAPSE chip under development at IBM, the biological approach is carried further toward an implementation of an artificial "brain" that has flexibility and fault tolerance comparable to biological brains (Modha 2015; Edwards 2011).

Early debates about a computer's capabilities focused on whether a computer simply manipulated symbols with no understanding of what its operations implied, or whether its capabilities could accurately be described as cognitive, a term that implies the ability to interpret and make choices or selections.[7] The ICF argues that even stand-alone computers do indeed have cognitive powers, including the ability to perform interpretations and thus to create meanings. The key point here is to reject an anthropocentric approach that insists meanings must be those that a human would produce; rather, the ICF approach situates a computer's operations within its internal and external milieu and locates the specificity of its meaning making within these contexts. As stand-alone computers give way to networked machines with sensors and actuators, the case for meaning making is even stronger, as computational media become increasingly environment aware and capable of interpreting sensory inputs in flexible ways.

Thus the integrated cognitive framework is extended *to form bridges between biological cognition and computational media*. This is one of its principal contributions: to provide integrated ways to talk about cognition in general, and especially about the processes that enable biological and computational cognitions to work together in cognitive assemblages. In visualizing cognitive assemblages, we may think first about our interactions with the computational devices in our personal sphere, from desktop computers to cell phones. Important as these are, they are the tip of the iceberg. Far more extensive are the communications *between* computational devices, particularly all the infrastructural interfaces that keep flowing the information on which contemporary developed societies absolutely depend.

In addition to providing a bridge, the ICF also fully acknowledges the profound differences of embodiment between lifeforms and computational media. Among the consequences emerging from those differences are their inverse evolutionary trajectories. Whereas biological organisms from day one were immersed in their environments and therefore had to have ways to cope with fluctuating and uncertain events, computers began by being rigid and deterministic, only gradually developing the flexibility to deal with uncertainties and ambiguities. Conversely, it took eons of evolution for biological organisms to develop the capability to deal with abstract symbols as *Homo sapiens* can, whereas computers possessed these abilities in their earliest days. In this sense their evolutionary trajectories are the inverse of each other, with biological organisms going from immersion to symbol manipulation, while computers go from symbol manipulation to immersion. The figure in classical rhetoric that describes this kind of pattern is a chiasmus. The figure works by inverting the grammar or concept (or both) of a first phrase in the following second phrase. "She went to church, but to the bar went he" shows grammatical as well as a conceptual inversion (she went/went he, church/bar). By analogy, the First Great Inversion can be figured as a kind of temporal chiasmus, in which two trajectories mirror each other in inverse order.

The Second Great Inversion concerns the fact that all organisms follow the biological mandate to survive and reproduce; as noted above, they otherwise go extinct. As *Homo sapiens* emerged along with other mammals, they supplemented this mandate by conceiving of purposes and designing ways to accomplish them. Computers, conversely, began with mandates of design and purpose, assigned to them by their human creators. As computational media evolve, they increasingly self-design their programs; our science fiction writers are already anticipating the day when computational media also desire to survive and reproduce. Again, the evolutionary trajectories form a chiastic structure, with organisms going from survival and reproduction to design and purpose, whereas computational media began with designs and purposes and may, through advanced artificial intelligence, progress to survival and reproduction.

Biological adaptations are generally understood as opening new functionalities for organisms, new ecological niches that adaptations evolve to exploit and fill. With this understanding, we can ask what kind of functionalities cultural adaptations make available, and what kinds of niches they open up. On this basis, I judge that computational media, and especially AI, is the most important cultural adaptation since the invention of language. Consider the new functionalities created by computational media. Here is a partial list, which only reveals the tip of the iceberg: (1) the ability to create, process, and automate large data sets too expansive for human-enacted arithmetic procedures; (2) the ability to automate sensors and actuators too extensive, remote, or sensitive for human operators alone to access and manipulate; (3) simulation modeling of large-scale phenomena such as the weather and climate change too vast and rapid for human-alone calculation; (4) real-time monitoring and analysis of global financial changes.

### Nonhuman Powers and Human Hubris

*Homo sapiens* have a long history of utilizing the powers of nonhuman forces, from the prehistoric domestication of fire through fire-driven steam engines to the fiery inferno of the atomic bomb. Very often we humans think of these achievements as our accomplishments. As Timothy James LeCain has observed, this interpretation reflects our hubris: "we are neither particularly powerful nor especially intelligent and creative—at least not on our own." He credits neomaterialism with pointing out that "we humans derive much of what we like to think of as *our* power, intelligence, and creativity, from the material things around us. Indeed, in many ways these things should be understood as constituting who we are. . . . Neo-materialism suggests that [humans] accomplished these things only at the price of throwing their lot in with a lot of other things, like coal and oil, whose powers they only vaguely understood and certainly did not control" (LeCain 2015, 4).

With the dawn of the computer era, the "things" became even more powerful in their partnerships with humans, because they had increasingly potent cognitive abilities. It is no exaggeration to say that developed societies have entered into a deep symbiosis with computational media.[8] Symbiosis of course is a term deployed in a biological context to denote interactions between two species living in close proximity to one another, like the cattle egret that follows cows and eats the bugs that torment the beasts. The closeness of the symbiotic relation varies from independent organisms that occasionally interact, like the bee that fertilizes an orchid shaped like a female bee, to the codependent relationship between humans and computational media, to the mitochondria inside a eukaryotic cell that have lost their autonomy and are now fully absorbed into the cellular dynamics. Although in common usage the primary connotation is positive, implying that the interactions benefit both species, symbiosis as a technical term in biology also can denote negative results, for example with parasitism, which hurts one species while benefiting the other. So too with our symbiosis with computational media. While the partnerships have brought enormous benefits in everything from international commerce to advanced medical devices, there are also new risks, including cyberwarfare, malicious hacking, computer viruses and malware, and a host of other vulnerabilities. What is undeniable, in my view, is that the symbiosis has advanced so far that developed societies cannot afford *not* to use computational media.

The pervasiveness, centrality, and criticality of computational media mean that we need to think seriously about what it implies to live in a society where cognitive assemblages do most of the informational and physical work. With customary hubris, many humans like to think that these assemblages are designed, implemented, and controlled by humans. While it may still be true that some are designed by humans, increasingly software programs take over more and more of the design work, because it has become so complex and interwoven that no single person, or even large teams of people, can keep it all in mind. Similarly with implementation: software programs are used to develop and implement still more complicated programs. Google's project AutoML, for example, uses machine-learning software to write more machine-learning software (Simonite 2017). As for control, even if a human has the final decision (for example with nuclear launch codes), the information on which such a decision is based has been so completely processed by computational media that the "final" decision is in name only. The true state of affairs is that control has been massively disseminated through all the participants in the cognitive assemblage, including the computational components, which often perform the majority of the interpretations, anticipations, estimations, and intermediary decisions on which final outcomes depend.

In such circumstances, anthropocentrism becomes a dangerous illusion, for it implies, as we have seen, that humans are at the center of the action. We *Homo sapiens* like to see ourselves as the active ones dominating our partnerships, and we tend exaggerate our own powers relative to the nonhuman forces that we deploy, activate, or direct. Such fantasies are already deeply wrong when they concern material processes. When the nonhuman forces become cognitive and, in the case of computational media, powerfully so, the fantasies of control become risky in the extreme, because they tempt people into experiments and arrangements that they think they can control—until they can't. The historical trajectory of *Homo sapiens*' interactions with their environments progress from material processes (domestication of fire) to simple mechanical devices (inclined plane, lever) to complex machines (steam engines) to electrical/electromechanical machines (cars, planes) to computational media. At each stage, the interactions produced profound cultural and biological changes in humans as agency became increasingly distributed through the partnerships. With the last stage, however, it is not only agency that is distributed but cognition itself. This introduces a host of new issues, among them the question of who or what makes the decisions. It a dangerous illusion to assume that only humans make the decisions, and even more so to think that humans are the only ones in control.

Consider, for example, gene editing, discussed in chapter 4. For the first time in history, through the cognitive assemblages that humans have formed with bacteria, we have the power to direct evolution—our own as well as that of other species. But who is in control of this technology, and what may be the consequences when it goes out of control? A case in point is the Covid-19 pandemic. According to an unclassified summary of a report by the US Intelligence Agencies released in August 2021, there is a "plausible" possibility that the Wuhan Virology Laboratory created the chimeric virus from a bat virus, making it more contagious for humans (National Intelligence Council 2021). While the truth of this situation may never be definitively determined (largely because the Chinese government refuses access to crucial information), it illustrates, with a global death toll approaching seven million people, the incredible dangers of doing gene editing with inadequate regulations to control the experiments.

Nor do the dangers end there. Much of the research on general artificial intelligence (GAI) aims to develop computational devices with capabilities, flexibilities, and world horizons that are human equivalent or more than human. If this research is successful, it is likely, as Nick Bostrom has argued, that it would be impossible to control such a superintelligence, because any controls humans devise could be circumvented by the much greater intelligence of the GAI (Bostrom 2016, 115–19, 175–76). A few years ago it would have been easier to dismiss such concerns as so far into the future that only techno-enthusiasts like Bostrom would worry about them. However, with the advent of Transformer architecture described in chapter 6 and large language models demonstrating human-level language abilities, the chorus of concern has grown much louder. From Noam Chomsky to Geoffrey Hinton (the so-called godfather of AI; see Metz 2023),[9] esteemed researchers and thinkers have been issuing warnings about advanced AI. They warn us, and I agree, that we simply do not know how such experiments will turn out. There is a nonnegligible possibility that they could be as disastrous in their own way as the Covid-19 pandemic has been in its deadly impact.

In short, human hubris, and the anthropocentrism that underlies it, are now not merely mistaken beliefs but planet-level risks to humans and other species. Since anthropocentrism has its origins in the liberal political philosophies that grew out of Enlightenment visions of the human, it is crucial that we rethink and reenvision those philosophies for more planet-friendly alternatives—which is to say, there are compelling reasons to embrace the integrated cognitive framework and similar approaches if we want our species and our ecologies to continue into the future. As one ecologist of my acquaintance likes to say, "If we plan on saving the planet . . . ," leaving the quiet part hanging in the air; "if" we do so plan, we urgently need to rethink basic assumptions about agency, decision making, control, and our relations to our biological brethren and artificial cognizers.

None of the above exempts the human species from taking positive actions to avoid environmental collapse. The cruel conundrum of our time is that humans are the ones driving the planet toward catastrophe, and yet humans are the only species capable of taking positive actions on a global scale to prevent collapse. The path out of the conundrum runs through a concept of responsibility that recognizes cognition is a basic property of all living beings as well as computational media, that decision making and agency are massively distributed among humans and nonhumans, that control is never perfect, and that humility and respect for others (human and nonhuman) are essential components of a responsible approach toward the continuation of life (human and nonhuman) on Earth.

### Integration: Micro/Evo/Techno Relationalities

Feminist theory, along with critical race theory and postcolonial/decolonial theory, are among the areas that have explored relationality at length. Rosi Braidotti, Elizabeth Grosz, Zakiyyah Iman Jackson, and others have interrogated the dynamics of relationality in a variety of contexts and explored its potential for serving as a model for interactions that are mutually constructive and biophilic rather than exploitive and destructive of Earth's flourishing. Relationality is crucial to the story I want to tell in this book, and consequently feminist theory is also central to its aims. To explore the resources that feminist theory offers for combating anthropocentrism, I focus on two prominent voices in this effort: Karen Barad and Donna Haraway.

I draw on Barad's pioneering work in quantum mechanics to analyze and explain how microphenomena at atomic scale produce the phenomena revealed in quantum experiments. This is the first strand in the tribraided complex (attentive readers may have noticed the shifting foci of this chapter, which is designed to simulate the braiding process). Barad argues these must be understood as constitutively coproduced by the choice of apparatus and measurement for a given experiment in a theory she calls agential realism. For Barad, a crucial source of inspiration is the epistemological argument of Niels Bohr, which she extends to ontological claims. Reality as it manifests to us does not preexist the experiments; rather, it is brought into existence as observable phenomena precisely through the process of measurement. Therefore, the specificities of the experimental apparatus are crucial.

A countervailing force to this argument is the issue of scale. The coconstitution for which Barad argues occurs at the quantum level, but these effects disappear for macrophenomena, primarily because quantum fluctuations are averaged or "smoothed out" at macroscale levels. Nevertheless, her account amounts to a kind of origin story for phenomena, which I will compare and contrast with Gilbert Simondon's theory of individuation. I will also compare it to my modest effort in a similar vein that I called "constrained constructivism" (Hayles 1993).

Donna Haraway's richly layered, turbo-charged writing and thinking have long been a source of admiration and inspiration for me. Her work in evolutionary biology points toward the second strand of the tribraided complex. I focus on her claims in *Staying with the Trouble: Making Kin in the Chthulucene* (2016) for symbiosis, symbiogenesis, and sympoiesis in evolutionary biology. Drawing on two seminal articles in biological evolution, she argues that multicellular bodies were interpenetrated by unicellular organisms from the beginning, and that in fact unicellular organisms were responsible for their emergence in the first place. "Nothing makes itself," she argues; "nothing is really autopoietic or self-organizing" (Haraway 2016, 58).

In alluding to autopoiesis, she gestures toward the countervailing forces that limit her claim. In chapter 3 I explore the complexities of autopoiesis, a theory developed by Humberto Maturana and Francisco Varela, in relation to evolutionary biology. Suffice it here simply to identify it as a vision that partially contradicts Haraway's claim that no multicellular organism is an individual. In strong tension with this claim is Maturana and Varela's concept of autopoiesis as the defining characteristic of living systems. They emphasized an organism's separation from the environment (while also acknowledging its embeddedness within an environment). An autopoietic entity interprets all information solely through its own processes; in this sense it creates its own self-enclosed world. It follows that a living system not only acts as an individual but moreover can act *only* in this way. Haraway, in insisting that symbiosis precedes and partially negates autopoiesis (or at least changes its meaning in critical ways), can be understood as crafting an origin story for multicellular bodies distinct from, and in some ways contradictory to, autopoiesis.

The third strand in this tribraided complex is technics. My own work lies mostly within this area, particularly with the idea of technogenesis. My focus on technogenesis was inspired by the work of Bernard Stiegler, who in *Technics and Time, 1* (1998) argues that the "human" has been involved with technics from the very beginning of the species, and that technics coconstitutes how the "human" has evolved. As Derek Woods (2022) usefully clarifies, technics is distinguished from technology because it does not assume that technologies are passive tools used in an instrumental way. On the contrary, technics emphasizes the feedback between the artifact and the human, which coevolve together. Like the origin story of microphenomena in quantum mechanics and the origin story of multicellularity in evolutionary biology, this amounts to an origin story about how the human is coconstituted and coevolved from the beginning through its interactions with technics, from the domestication of fire to the fiery inferno of atomic energy.

I had a small part to play in how this dynamic played out in contemporary theory through my interactions with Stiegler, who for a brief time was catalyzed by my argument for how contemporary computational technologies are affecting the attention spans of modern humans, especially children. Our thoughts took different paths, however, as I went on to argue for a nuanced view of the effects of computational media on youth, whereas he became more and more alarmed by the growing dominance of screens in contemporary cultures. Preoccupied with these issues, he was late in realizing the implications of the momentous developments in computer processing strategies and exterior memory storage as von Neumann architectures were succeeded by neural nets, choosing instead to argue passionately for the necessity of regulation for computational media. The choice has its ironic overtones, for it was his work in the trilogy *Time and Technics* that initially drew attention to the importance of exterior memory in the hominization process.

The larger narrative about micro/evo/techno relationalities and their associated origin stories (along with countervailing forces) is that these three strands, which up until now have proceeded along parallel but clearly demarcated disciplinary divides, have now burst through their disciplinary boundaries and are in active interplay with one another. The name I use to denote this condition, which is at least as inadequate as most names, is ecological relationality. "Ecological" in this sense means that each of the three domains—micro, evo, and techno—has now become the environment for, and hence interacts with, the others, opening the way to more intense and faster dynamics.

At first "ecological relationality" might strike us as soothingly pleasant: we should all be more ecological, more attuned to relationality, right? In actuality, this seemingly benign, seemingly placid name is anything but: it should scare the hell out of us.

In my view, here is our present planetary condition. Each of the three terms, *micro*, *evo*, *techno*, designates emergent dynamics within its own domain. Through their interactions, they are now creating second-order emergences, characterized by highly volatile, highly interactive cultural/scientific/physical fields in which changes ripple through the systems at speed, causing cascading effects whose consequences are increasingly unpredictable. Borrowing a term from Gilbert Simondon (2020 82), I characterize these emerging fields as metastable. Bursting with potential energy, they are poised for transformative becoming. It is possible they can accelerate human evolution to science-fictional heights, but they can also quickly go south through complex dynamics that we likely will not be able to recognize in time, much less control.

How did we get to this point? I view technics as a powerful accelerant acting on the two other dynamics of micro and evo through technological change. For anthropologists, a tool is defined as an artifact used to create other artifacts (remembering here also its feedback on what counts as human). Imagine how tedious it must have been to fashion the first tool, because by definition there were no other tools to help in its fabrication. Tool by tool, and then ensembles of tools by ensembles, technical infrastructures expanded and deepened. Their trajectories map as exponential curves, at first slowly, slowly creeping up over the millennia and centuries from nearly horizontal slopes. By the 1800s, they began to accelerate, and by 1900, they were growing faster. In the contemporary moment, the trajectories have begun their upward explosion, with change happening at accelerating and now dizzying velocities.

Contributing to this exponential explosion are the second-order emergences now in play as the emergent dynamics of the three domains begin to interact with each other. Unwieldy as the neologism micro/evo/techno is, it has the virtue of indicating that the three domains are no longer separate but are becoming proximate with each other. More subtly, it suggests through the quasi divisions of the connecting/dividing slash that their entangled effects may, in a theoretical sense, be unbraided to make them available for interrogation and analysis. The theoretical project of narrating the stories of micro/evo/techno ecological relationalities, then, is first to position the three domains as parallel origin stories of emergence, then to separate out the three stands for analysis and interrogation through case studies, and finally to rebraid them with deeper insight into their combined emergent dynamics. That basically is the plan of this book.

Ecological relationality is the (mis)leading phrase that gestures toward the complexities of this project. It names the highly interactive dynamics that provide the broader contexts for the ICF. The *integrated* framework traces the intersections, convergences, and divergences as the parallel origin stories of micro/evo/techno interact in complex ways. It connotes the braiding process, which recognizes the accelerating interconnections between the three strands; the unbraiding necessary to explore their histories and analyze their implications; and the rebraiding together again, which takes advantage of the preceding interrogations to arrive at more comprehensive understandings.

Here is a brief indication of how the interconnections are working at present, achieved through the rebraiding process mentioned above. With gene editing, biological evolution is now open to (possibly world-altering) mutations through intentional interventions of humans into genetic processes. Quantum effects are being harnessed to expand and accelerate the capacities of computational media through quantum computers, where research efforts are responding to the necessity of handling already vast and ever-increasing amounts of calculations and data storage. Neuromorphic chips such as SyNAPSE are being developed to leap over the fast-approaching physical limits of the silicon atom's diameter to enable further miniaturization and computational power. Computational power is essential to carry out the calculations, data storage, and processing necessary for gene editing. Artificial intelligences, especially neural nets, are achieving human-competitive and more-than-human results in staggeringly diverse fields of competence. AIs are now also interacting with gene editing to construct protein foldings that form the basis for new drugs and further interventions in genetic processes. In addition, AIs are now communicating with humans through natural language, increasing AIs' penetration not only into contemporary infrastructures but into people's daily lives, with accompanying influences on human behaviors and psyches. Through all this, the second-order emergences continue to accelerate and expand.

How does one begin to untangle these entanglements, separate out the strands for analysis and interrogation? My approach is to explore the strands through case studies ranging from gene editing to autopoiesis and Gaia theory, bacteria and xenobots to artificial intelligence. None of these captures the *full* implications, of course, but I think that aspiration is probably beyond human capacity. My hope is that the chapters that follow will in some measure contribute to the collective endeavor to investigate and understand how second-order emergences are rapidly transforming our contemporary world. But first, let us return to trace in more detail how the origin stories of micro/evo/techno took shape in the twentieth and twenty-first centuries.

## Micro Relationalities: Karen Barad

Barad bases her theory of agential realism in experimental results from quantum mechanics and their interpretation by Neils Bohr. In contrast to Heisenberg's well-known interpretation of the experiments showing that light could manifest as either a particle or a wave, which posited that the apparatus was interfering with the result, Bohr took a more radical approach, arguing that reality enters the realm of knowledge only once reality is measured. Barad's project takes Bohr's ideas from epistemology into ontology, arguing that quantum reality has no determinate qualities until it is measured with a specific apparatus (Barad 2007, 31). Therefore, the apparatus and measuring procedure coconstitute the phenomena, understood as what is perceived at the experiment's conclusion.

Barad's *Meeting the Universe Halfway: Quantum Mechanics and the Entanglement of Matter and Meaning* (2007) masterfully draws out the implications of this insight. She argues, for example, that representationalism fundamentally distorts our situation, for it assumes that there is a stable object to be represented that precedes its existence as a perceived object. Notwithstanding its subversive effects, reflexivity is also suspect because it assumes that there is a representation that can be subverted.[10] Instead Barad argues for performative approaches, which "call into question representationalism's claim that there are representations, on the one hand, and ontologically separate entities awaiting representation, on the other" (Barad 2007, 49). "Knowing does not come from standing at a distance and representing but rather from a *direct material engagement with the world*" (49, emphasis in original). Elsewhere she articulates a similar idea in her succinct summary—"We are a part of that nature we seek to understand" (29)—using this insight as a foundation from which to launch her theory of agential realism.[11]

The adjective "agential" underscores the performative nature of knowledge practices. "According to agential realism, knowing, thinking, measuring, theorizing and observing are material practices of intra-acting within and as part of the world. . . . We do not uncover preexisting facts about independently existing things. . . . Rather we learn about phenomena—about specific material configurations of the world's becoming" (Barad 2007, 91). It is important to note that agential realism should not be understood as saying that reality has no underlying regularities. Just as "agential" is important, so is "realism." As Barad points out, if the same apparatus and measuring procedure are used repeatedly, each time they will give the same result. Therefore, underlying regularities (formerly known through the misleading term "laws of nature") must exist, but the problem comes in disentangling our ways of knowing them from what they are in themselves—an inherently impossible task, since it supposes they can somehow be grasped "objectively," that is, from a God's-eye viewpoint impossible to achieve.

I came to similar conclusions in the early 1990s, at the height of the so-called science wars. The tensions of that tumultuous period are captured in George Levine's 1993 collection *Realism and Representation: Essays on the Problem of Realism in Relation to Science, Literature, and Culture*. This edited volume was preceded by a conference in which speakers lined up like players at a baseball game, with the constructivists in one dugout and the realists in the other, jeering at each other from their respective positions (as such activities are performed through the decorous protocols of academia). I had read enough of STS (science and technology studies) to think that the constructivists had powerful arguments, but my scientific training also had instilled in me a rock-solid belief that scientific methodologies are the best options we have to achieve durable, robust, and reliable knowledge about the world.

The problem, it seemed to me, was precisely what Barad would later address through her theory of agential realism. Scientific experiments cannot tell us about the "truth of the world"; they can only tell us about how our experiments, measurements, theories (*and* cognitive capacities) experience the world. Thus they can never say, "reality is so-and-so"; they can only say, "this measurement/perception/model is *consistent* with what reality is." Since it is convenient to have a term to refer to reality as such, I took to calling it the "unmediated flux."

Crucially, scientific experiments can also give negative answers: "whatever reality is, this model/perception/theory is *not* consistent with it." This power of disconfirmation is what to my mind separates science from religion. Religion has no such intersubjective disconfirmation process; if someone says he or she has seen the face of God, who could contest it, and on what basis? Without disconfirmation, religions also have no intersubjective confirmation process either (other, perhaps, than collective prayer), which is why virtually all religions emphasize the importance of faith, the willingness to believe regardless of the absence or presence of reliable proof. Scientific inquiry, in addition to serving as a counterweight to religious faith, also serves as a bulwark against misconceptions, conspiracy theories, and other cultural fantasies, popular though they may be. It has ways to test, through interpersonal protocols that can be replicated, whether a given theory or idea is consistent with reality (the unmediated flux) or is not.

Because of this constraint, I called my own version of the problems addressed in Barad's agential realism "constrained constructivism" (Levine 1993, 27–43). Yes, our theories are constructed and thus subject to all the vagaries of cultural, political, and economic influences, just as STS work argues, but they are not *simply* constructed, because they are constrained to models consistent with the unmediated flux. As Karl Popper ([1934] 1959) argued nearly a century ago, it is this property that has enabled scientific theories and methodologies to construct reliable and robust knowledges about the world.[12] Sandra Harding (2005) has famously called this property "strong objectivity," to distinguish it from the unattainable God's-eye objectivity. It is strong precisely *because* it is culturally, politically, economically, and socially constructed. It fits with our evolutionary umwelten and all the other culturally derived and culturally influenced perceptions of who and what we are. What would we do with "objective" knowledge if we somehow were able to attain it? It would be as alien to us as are the gods themselves. We are humans, and we know *as* humans, fashioned through eons of evolving biological, cultural, and technical heritages.

Barad, in her exposition of agential realism, argues that quantum mechanics principles apply at all scale levels. "Significantly, quantum mechanics is not a theory that applies only to small objects; rather, quantum mechanics is thought to be the correct theory of nature that applies to all scales. As far as we know, the universe is not broken up into two separate domains (i.e., the microscopic and the macroscopic) identified with different length scales with different sets of physical laws for each" (Barad 2007, 85). This is technically correct, and quantum effects such as entanglement have recently been observed in two mechanical oscillators with masses of seventy picograms—very small, but still much larger than quantum scales (Kotler et al. 2021). However, it also remains true that no one would use quantum mechanics methods to calculate a rocket trajectory to the Moon. At macroscopic scales, quantum effects tend to average together and thus cancel each other out, much like random small ripples in a pond interfere with each other and thus don't cohere into larger waves. Thus for practical purposes, such as the rocket trajectories, classical mechanics works fine.[13]

How important is this constraint on Barad's larger claims about the performative nature of what we perceive as reality? Since she bases much of her argument on detailed expositions of specific quantum mechanics experiments, there remains a large gap between the scope of her ontological claim and evidence that would support it in macroscopic terms. Similarly, she spends a good deal of ink writing about her "diffractive methodology" (Barad 2007, esp. 91), but this sits uneasily with her other claim that she is not dealing in analogies or homologies but literal facts. Diffraction gratings use grooves in a substrate to break light waves into different wave lengths, thus spreading out the white light that hits them into different wave lengths and therefore colors. Barad proposes "diffractive methodology" that reads different viewpoints and disciplinary practices through each other, with the idea that this practice will cause their implicit assumptions to be spread out and thus be revealed, much as white light is broken into colors by diffraction gratings. "Diffractive methodology," Barad writes, "is a commitment to understanding which differences matter, how they matter, and for whom. It is a critical practice of engagement, not a distance-learning practice of reflecting from afar" (2007, 91). This could be a powerful methodology, but it nevertheless relies on analogy rather than literal fact. The argument for diffractive methodology would be more persuasive if Barad's book contained case studies of how it would work in practice beyond the quantum scale, which are largely lacking.[14]

These limitations notwithstanding, there is a fierce insistence on rigor in Barad's work that I find refreshing. We can see this if we compare Barad's arguments to Simondon's, which they somewhat resemble, although Simondon is far more speculative than Barad is willing to be.[15] Just as Barad focuses on phenomena and contests the idea that objects to be represented preexist their representations, so Simondon also contests the idea that individuals preexist their individuation. Rather, he sees individuation as a *process* that brings individuals into existence. Barad's thought is saturated with and emerges from quantum physics, similarly to the way in which thermodynamics is crucial to Simondon's vocabulary. Thus being as such (the unmediated flux) is discussed by Simondon in terms of potential energy, polyphasal transitions, and so forth. This highlights a crucial difference between Barad and Simondon: Barad never attempts to characterize being as such, other than to acknowledge the underlying regularities that make scientific work possible—and even here, she gestures toward it rather than making it the center of her inquiry. Simondon, by contrast, emphasizes that "we have chosen to take the different regimes of individuation as the basis of various domains such as matter, life, mind, and society. . . . The more fundamental notions of first information, metastability, internal resonance, energetic potential, and orders of magnitude are substituted for the notions of substance, form, and matter" (Simondon 2020, 12). He characterizes being as such as having a "*transductive unity*, i.e. it can overflow itself on both sides from *its center*"; individuation is then labeled as dephasing, resulting simultaneously in both an individual and its milieu (12). Perhaps one way to think about this contrast is to say that for Barad, the focus is squarely on the processes that bring phenomena into existence, whereas for Simondon, the focus is not only on the processes but also on the nature of that within and on which the processes operate. Barad is a physicist-turned-philosopher, whereas Simondon is a philosopher first and an experimentalist second (to the extent that he shows a keen interest in material reality, for example, in his discussion of the distinction between abstract and concrete mechanisms in On the Mode of Existence of Technical Objects [2017]).

Returning to the tribraided entanglement of micro/evo/techno, I note that micro up to now has designated the quantum realm. However, as we turn to the evo dynamic, micro assumes another meaning as the microscopic as opposed to the macroscopic, as the unseen (though not unseeable) realm of bacteria, viruses, and archaea. Of course, these tiny lifeforms are orders of magnitude larger than quantum particles, but they nevertheless have been found to use quantum effects, specifically vibronic mixing, to move energy between different pathways (see, for example, Higgins et al. 2021). Thus the "micro" element of the term may be considered a transition zone in which both microorganisms and quantum effects are in play.

### Evo Relationalities: Donna Haraway

In *Staying with the Trouble*, Donna Haraway writes, "Nothing makes itself; nothing is really autopoietic or self-organizing" (2016, 58). Thus she declares herself firmly on the side of symbiosis rather than the older view of biological "individuals," an argument proposed in two seminal articles she cites advocating the symbiotic view. Following this trail, I found the evidence for nearly universal symbiosis compelling indeed. The articles' titles tell the story: "A Symbiotic View of Life: We Have Never Been Individuals" (Gilbert et al. 2012) and "Animals in a Bacterial World, a New Imperative for the Life Sciences" (McFall-Ngai et al. 2013).[16] Instancing symbiotic mechanisms present in almost every process characteristic of animals, from metabolism to reproduction and development, both articles argue that the traditional view of biological "individuals" should be jettisoned in favor of symbiotic views that see organisms as collectivities (holobionts) rather than single entities. "Animals are composites of many species living, developing, and evolving together," Gilbert et al. write (2012, 326). They continue, "The discovery of symbiosis throughout the animal kingdom is fundamentally transforming the classical conception of an insular individuality into one in which interactive relationships among species blurs the boundaries of the organism and obscures the notion of essential identity" (326).

Although I am far from expert in this field, the evidence they present seems to me overwhelming. Nevertheless, I wonder about the other side of the story—the conventional narrative, perhaps fueled by the Western ideology of possessive individualism, that animals make decisions as individuals, struggle to survive as individuals, and reproduce as pairs of individuals (the exception here is that they don't die as individuals, for their demise means the death of their symbionts as well).[17] The evidence for symbiosis is very strong, but isn't the evidence for individuals equally overwhelming? I understand that from the viewpoint of the authors (and Haraway), the mainstream narrative is badly misleading and needs to be corrected. Granting this point, I wonder if it is possible to hold both views in balance: the view that all animals are holobionts, and the opposing view that they have the capacity to act as individuals, distinct from the influence of their symbionts.

Without doubt, scale and advanced technologies are both at issue here. Gilbert et al. acknowledge that crucial evidence was collected through "nucleic acid analysis, especially genomic sequencing and high-throughput RNA techniques" that provided a strong basis for the claim of pervasive symbiotic activities in multicellular organisms (2012, 325). As a consequence of this evo/techno intervention, the Neo-Darwinian view of the gene's supreme role has been disrupted. Symbio*genesis* (origins created through symbiosis) constitute, Gilbert et al. summarize, "a second mode of genetic inheritance, providing selectable genetic variation for natural selection" (2012, 325). In addition, McFall-Ngai et al. also point to evidence suggesting "an ancient involvement of bacteria in the initiation of multicellularity" (2013, 3230). They acknowledge that "the origin of multicellularity has been a topic of intense debate in biology," but they point out that "a microbial role in animal origins does not obviate other perspectives on the evolution of complex multicellularity but adds a necessary functional and ecological dimension to these considerations" (3230). Thus in their view symbiosis is necessary, although not sufficient, to account for the origin story of multicellularity.

Within this debate, autopoiesis is a crucial term. As Maturana and Varela envisioned it, autopoiesis (*poiesis* from the Greek root meaning "to make," thus self-making) is the defining characteristic of living systems, in contrast to allopoietic systems, such as automobiles for example, which neither make nor repair themselves. In contrast, living systems produce and reproduce the components that make up their structures (for example, cell renewal), and their organizations both result from, and are the result of, their structures. This circularity of operations is emphasized in autopoietic theory, especially in the context of reflexivity or self-reference. Chapter 3 deals with this topic in depth, but here I note that autopoiesis as envisioned by Maturana and Varela also claimed that a living organism, although embedded in an environment, receives no information from the environment as such; rather, stimuli coming from an environment merely act as "triggers" to the organism's own sensors and nervous system, which convert the triggers into information the organism can use Maturana and Varela [1972] 1980, 95).

When Niklas Luhmann took inspiration from their work to describe social systems, he crucially changed the idea of *informational* closure to *operational* closure (Luhmann 1995). A social system is in communication with the environment, but it does not interact with its environment directly. Rather, it only operates through its own codes, programs, and memory; no operations can enter the system from the outside. Like living systems, social systems produce and self-reproduce themselves, thus creating the circularities of self-reference intrinsic to their operations.

Moreover, according to Luhmann this operational closure is essential to a system's ability to exist, because it protects the system against the overwhelming complexities of the environments in which it is embedded, which are always orders of magnitude more complex than the system itself. Were a system to be open to this exterior complexity, it would collapse from its inability to handle all the information. Because it is operationally closed, however, it can respond to the exterior complexity by (re)creating interior complexities in its own terms, which remain at a manageable level and allow the system to continue to operate. Hence another crucial attribute for a system to exist is a clear distinction (drawing on *The Laws of Form* by George Spenser-Brown) between the system as such and its environment.

In 1994, I had an opportunity to sit with Luhmann in an interview/debate at the University of Indiana, arranged by Cary Wolfe and William Rasch. From my point of view then, systems theory had a major blind spot. In its excision of individuals and causal connections with the environment (which Luhmann adopted from Maturana and Varela), systems theory was setting itself up as the "other" to narrative, which focuses precisely on these qualities. Luhmann acknowledged in his work that every distinction creates a blind spot, something obscured or rendered invisible by the given distinction's choice of where the cut is made. That narrative is so darkened by systems theory means that all the insights conveyed by narratives from time immemorial to the present are minimized by systems theory, which seemed to me a heavy price to pay for the insights that systems theory admittedly provides, such as the necessity for operational closure. I also wanted to indicate by this argument that systems theory alone cannot tell the full story of human and environmental complexities.

By that time Luhmann was in his late sixties, and only four years from his death. I am not sure he grasped the point I was trying to make (or perhaps did not want to grasp it), so the "debate" happened more in print than in person (Hayles 1995). Rereading my article now, three decades after it was published, I am struck by the continuing tension between the autopoietic/systems theory's insistence on closure (informational and operational) and the mounting evidence instanced by Haraway, among others, on all the ways in which symbiotic interactions render that idea unfeasible.

This brief account should make clear why autopoiesis and associated ideas are incompatible with sympoiesis as Haraway uses it. The very idea that there is a clear boundary between a system and its environment is precisely what is in question in symbiosis; as Gilbert et al. write, symbiosis "blurs the boundaries" between self and other, inside and outside (2012, 326). Haraway, crediting M. Beth Dempster, a Canadian graduate student, with inventing the term "sympoiesis," quotes Dempster's definition: "collectively producing systems that do not have self-defined spatial or temporal boundaries. Information and control are distributed among components. The systems are evolutionary and have the potential for surprising change" (Haraway 2016, 61). However, some animals cited as holobionts in McFall-Ngai et al. (2013) do have clear spatial and temporal boundaries (at least on a macro-level), for example, cows and humans, so clearly sympoiesis and symbiosis are not simply synonymous. Haraway's own characterization of sympoiesis is more poetic than Dempster's, but perhaps less clear: "Sympoiesis is a simple word: it means 'making with.' . . . Sympoiesis is a word proper to complex, dynamic, responsive, situated historical systems. It is a word for worlding-with, in company" (2016, 58). And then comes a crucial qualification: "Sympoiesis enfolds autopoiesis and generatively unfurls and extends it" (58).

What are we to make of her implication that sympoiesis is able to engulf autopoiesis, apparently without indigestion? Elsewhere she seeks to clarify her position: "As long as autopoiesis does not mean self-sufficient 'self-making,' autopoiesis and sympoiesis, foregrounding and backgrounding different aspects of systemic complexity, are in generative friction or generative enfolding, rather than opposition" (61). Except, of course, self-sufficient self-making are precisely what autopoiesis *has* meant, from Maturana and Varela's *Autopoiesis and Cognition* ([1972] 1980) to the present. Perhaps the reason that Haraway wants to hang onto the word, rather than jettison it altogether, is because her hero (and mine) Lynn Margulis used it (as noted in chapter 3). Calling Margulis "a student of interlocked and multileveled systemic processes of nonreductionist organization and maintenance that make earth itself and earth's living beings unique," Haraway notes that "Margulis called these processes autopoietic." She adds, somewhat wistfully, "Perhaps she would have chosen the term *sympoietic*, but the word and concept had not yet surfaced" (2016, 61).

Elsewhere, she makes her preference crystal clear. "The earth of the ongoing Chthulucene is sympoietic, *not autopoietic*. . . . Autopoietic systems are hugely interesting—witness the history of cybernetics and information sciences; but they are not good models for living and dying worlds and their critters. Autopoietic systems are not closed, spherical, deterministic, or teleological; but they are not quite good enough models for the mortal SF world. Poiesis is symchthonic, sympoietic, always partnered all the way down, with no starting and subsequently interacting 'units'" (Haraway 2016, 33, emphasis added).[18]

Derek Woods, in his article "Prosthetic Symbiosis" (2022), identifies the problem with this vision. Picking up on the tension between autopoiesis and symbiogenesis (the idea that symbiosis was the genesis of speciation and perhaps of multicellularity), he writes, "Haraway sees them as contradictory. For her, there can be no self-enclosed units of life. The recursive closure demanded by autopoiesis is incompatible with a biosphere composed of relations all the way down, a biosphere in which life forms are constantly enfolded and merged with one another—in other words, a biosphere in which symbiosis is so constitutive that it would be impossible to spot the chimeras and distinguish them from basic singular beings. Haraway offers 'sympoiesis' as a way to move past the impasse of autopoiesis and symbiogenesis" (Woods 2022, 159). He continues, "the problem with sympoiesis is that it provides no replacement for the bootstrapping abilities of life—abilities that are not about some kind of anti-ecological isolation from the environment, but rather present a compelling answer to the question of what makes life distinct and irreducible to the chemistry of replicators" (159–60).

Haraway would likely object that symbiogenesis provides exactly that kind of. bootstrapping capacity, since it focuses on endosymbiotic relationships that lead to new kinds of cells, metabolisms, and even species. But what Woods evidently is objecting to here is a vision of symbiosis so complete that the distinction between organism and environment simply collapses, along with such entailed activities as self-organization. He writes, "One downside [of such a vision] is that symbiosis ends up applied widely across scales and registers" so that it becomes the dominant force "at a planetary scale. Some of the fascinating specificity of symbiosis as something life offers to ecotheory is lost or vitiated" (Woods 2022, 161).

As we have seen, fundamental to every kind of systems theory is the idea that a cut can be made distinguishing organism from environment. If such a cut is ruled out of court, all the accomplishments of systems theory would dissipate as well. Just as organisms do act as individuals, so the autopoietic capacity for self-organization, self-maintenance and homeostasis has been documented so thoroughly that no reasonable person could doubt it exists. Although the Neo-Darwinist focus on genes alone obviously needs correction to account for genetic transfer through symbiosis, in my view it is a question of finding a reasonable balance in which organisms exist as individuals as well as symbionts, genetic transfer happens through multiple pathways, and lifeforms are recognized as having autopoietic capacities as well as symbiotic connections.

As one pathway toward this kind of vision, Woods offers his idea of prosthetic symbiosis. Crucially from my point of view, it includes attention to scale, which as we have seen is an issue not only with symbiosis but also with microquantum effects. "There may be scales where symbiosis matters and scales where it does not," he observes (2022, 166). He also notes that "even if it is difficult to demarcate the boundaries of individuals, they remain individuals to the extent that they can die. Even if species are really multispecies assemblages, fuzzily bounded 'species' continue to exist insofar as they can go extinct" (166). The novelty of his proposal is to regard symbiosis as a "primordial form of technics" in which "species use one another as tools to open new relationships to their milieus. They extend their functions, or add new ones, through their relationships with other species" (168). With this move, he proposes to explore what I have called a second-order emergence, this time in ecotheory, in which evo-and-techno relationalities come together to create a new narrative about the origin of the human *and* the origin of biological symbiosis.

## Techno Relationalities: Bernard Stiegler

Bernard Stiegler is a towering figure arguing for the importance of technics and its role in the coevolution of humans. In *Technics and Time, 1: The Fault of Epimetheus* (1998), he does a deep dive into the work of French anthropologist André Leroi-Gourhan, especially his book *Gesture and Speech*, to show that technics did not just accompany the rise of humans but in fact played a central role in coconstituting what the human is, including gait, upright posture, jaw and speech-organ formation, opposable thumbs, and other physiocerebral changes, along of course with all the associated cultural transformations. Stiegler focused specifically on the role of technics, as organized inorganic material, in creating exteriorized forms of memory, which he argued are constitutive of human temporality. In a crucial passage in *Time and Technics, 1*, he writes:

> If molecular biology is correct in claiming that the sexual being is defined by the somatic memory of the epigenetic and the germinal memory of the genetic, which in principle do not communicate with each other (to which Darwin devoted himself, contra Lamarck), exteriorization is a rupture in the history of life resulting from the appearance of a third—tertiary—memory I have called epiphylogenetic. Epiphylogenetic memory, essential to the living human being, is technics: inscribed in the non-living body. It is a break with the "law of life" in that, considering the hermetic separation between the somatic and the germinal, the epigenetic experience of an animal is lost to the species when the animal dies, while in a life proceeding by means other than life, the being's experience, registered in the tool (in the object), becomes transmissible and cumulative: thus arises the possibility of a heritage. (1998, 4)

Calling technics "life proceeding by means other than life," he weaves an origin story for the human in which tertiary memory, exteriorized first through cave paintings, quipu knots, and other prehistoric devices, accelerates and accumulates until it becomes, in his view, the defining characteristic of the human.

I was inspired by this vision to carry the line of thought into the contemporary moment, which culminated in my book *How We Think: Digital Media and Contemporary Technogenesis* (Haykes 2015). In *Time and Technics*, Stiegler is mostly concerned with prehistoric humans, but it seemed to me that his argument is stronger than ever when considered in the context of the twentieth and twenty-first centuries. Memory storage, transmission, and processing have been enormously accelerated by the invention of the digital computer; moreover, the impact of digital devices on human brains and nervous systems has also been growing exponentially as people spend more and more time looking at screens of all kinds—desktop computers, cell phones, digital notebooks, and so on.

Some time before the publication of my book *How We Think* (2015), I had spent a year as a Phi Beta Kappa Visiting Scholar, visiting colleges and universities across the country. Everywhere I went, I heard teachers, professors, and grad students complaining about a lack of attention in their students. Whereas a standard teaching technique in business schools, for example, was to ask students to read thirty-page case histories, a business professor said she had been receiving so many complaints that she had begun condensing them into two-to-three-page abstracts. A literature professor said that he could not get his undergraduates to read long Victorian novels, so he had taken to assigning short stories. Subsequently I published a widely cited article that argued digital technologies were causing a cortical (and generational) shift in cognitive modes of attention, shifting from what I called "deep attention" to "hyper attention" (Hayles 2007a).

The article attracted Stiegler's notice, and for a time he cited my work in his lectures. However, I went on to argue that hyper attention had adaptive advantages as well as disadvantages, for example the ability to rapidly shift focus from one task to another (useful for air traffic controllers, stock traders watching multiple screens, etc.), while at the same time he grew increasingly alarmed at the effects of digital technologies (or what he called the "programming industries" [Stiegler 2010, 16, 181]), including TV, on the populace in general, and specifically on young people, whose brain circuitry is still relatively malleable.[19]

Our differences came sharply into focus in 2011, when Marcel O'Gorman at the University of Waterloo arranged a debate between us called "Attention!" at the annual conference of the Society for Literature, Science and the Arts, where Stiegler was the keynote speaker. In his keynote address, he urged "adoption, not adaptation," which is to say, he wanted to ban some programs and to limit or eliminate screen times for young children, as well as to initiate strict regulatory controls on digital technologies. As I listened to his talk, I was struck by the irony that this towering intellect, who had led the way in arguing for the adaptive powers of technics in forming what counts as the human, was now attempting to marshal the forces of social consent and regulation against the very adaptive coevolution he was noted for recognizing. A line from *Blade Runner 2049* (Villeneuve 2017) kept running through my mind as I listened: "You can't hold back the tide with a broom."

And the tide was coming in strongly. Stiegler's untimely death by suicide on August 5, 2020, left the world bereft of a great mind, but even before his death, he had not completely grasped the full implications of the epiphylogenetic—or tertiary, as he called it—memory in the contemporary moment. As the new millennium dawned and more and more data were collected, compiled, organized, and processed, it was inevitable that algorithms, designed to extract useful information from the resulting relational databases, would explode into prominence, affecting almost every aspect of culture in developed societies.[20] The need not only to process correlations between database attributes but also to interpret the *content* of this data hoard led directly to the invention of advanced AI, including the GPTs of OpenAI, such as GPT-3, GPT-4, and Chat GPT, discussed in chapters 6, 7 and 8. In my view, the most likely stimulus for the next important coevolution of humans and technics lies precisely at this juncture between humans and advanced AIs, difficult as it may now be to envision exactly what forms that will take.

Let us now return to Derek Woods's article, "Prosthetic Symbiosis," mentioned in the previous section. In a theoretical move as bold in conception as it is subtle in its implications, Woods proposes that symbiosis can be considered to be a nonhuman, nonconscious form of technics and therefore prosthetic (strictly speaking, a kind of technics) in its origins. For this claim, he relies on Stiegler's view that "technics constitutes a system to the extent that it cannot be understood as a means—as in Saussure the evolution of language, which forms a system of extreme complexity, escapes the will of those who speak it" (Stiegler 1998, 24; quoted in Woods 2022, 169). In this view, technics assumes its own evolutionary trajectory, which only partly depends on humans; rather, it has its own selective pressures, variabilities, and fitness criteria that operate by themselves, "escap[ing] the will of those [humans]" who invent it (Stiegler 1998, 24; quoted in Woods 2022, 169).

There may be some truth in this, but it seems more reasonable to conclude that for technics, the operative unit for selection must be humans plus artifacts—what I have been calling cognitive assemblages—along with all the social, economic, political, and psychological contexts in which humans operate. By importing technics into symbiotic discourse, Woods wants to emphasize the idea that the "tools" of symbiosis (that is, other species that form symbiotic connections) "have a biotic life of their own" (Woods 2022, 170), feeding back into the hosts to alter them as well as the symbionts themselves. The crucial analogy, then, is the idea that in symbiosis, as in Stiegler's view of technics, the symbionts coconstitute and coevolve with their hosts. By importing technics into symbiotic discourse, Woods creates the possibility for other analogies to be constructed as well. He already hints at this in suggesting that technics may evolve like symbiosis, through processes not entirely dependent on the individual wills (or even collective wills) of the entities who participate in it. "Prosthetic symbiosis" is thus a backformation that suggests two things at once: (1) that nonconscious mechanisms found in biological symbiosis may also be in play in technics, and (2) that the exponentially exploding trajectories of technics may also be entangled with symbiotics in the biological realm. In my terms, what his double entendre points toward are the second-order emergences of evo/techno relationalities.

I find this fascinating because his construct is the exact inverse of my argument for *technosymbiosis*. By this term, I mean to emphasize the coevolution and coconstruction of humans by technics. Obviously, the term works by importing symbiotic discourse into technics (the inverse of Woods's importing technics discourse into symbiosis). I too want to create a similar sense that the dynamics of biological symbiosis is now deeply entangled with the dynamics of technics, in the second-order emergences I designate through the micro/evo/techno composite. The technoevolution of humans has already put us in a position to intervene decisively in genetic transformations, including our own, as discussed in chapter 4. Further interventions in the dynamics of biological symbiosis are without doubt also possible, along with the corresponding feedback loops that will alter human life even as we alter the courses of more-than-human life. Where all this will lead is an open question: that very uncertainty should make us afraid—very afraid—for it has as much possibility for catastrophe as for beneficial outcomes, all the more reason to devote ourselves to understand its implications as fully, thoughtfully, and carefully as we can, which is the ultimate purpose of this book.

### Effects of Second-Order Emergences: Reversible Internalities

The chiastic relation observed above, between Woods's view of symbiosis as prostheses and mine of prostheses as symbiosis, is a specific case of a more general dynamic that, inspired by Jason Moore's "double internality" (2015, 105), I call reversible internality, which seems to me a more appropriate term.

As we will see, the figure of reversible internality has become a characteristic trope of our contemporary moment, popping up everywhere like dandelions after a spring rain. How and why did this formation become a talisman of early twenty-first-century thought, what are its implications, and specifically, how is it related to second-order emergences? To explore these questions, we will have to return to the implicit assumptions in analyzing emergences and work forward from there. With first-order emergences, the usual analytical stance is to presume that two or more entities interact in ways that lead to the emergence of a third, which is the perspective taken above when discussing first-order emergences in micro/evo/techno relationalities. First-order emergences position relationality as an active, generative process capable of opening pathways to something new.[21] Of course, as Simondon reminds us, this move assumes that there are independent entities that can interact in the first place. As his work testifies, this is an abstraction and simplification of the actual situations in complex adaptive systems, where (to echo the recent movie) "everything everywhere all at once" are interacting together. To call the effects "second-order emergences" presumes that this first interactive process has already happened and has subsequently created another tier of complexities by showing that emergent phenomena themselves interact to create something new. In this sense, second-order emergences build relations among relations. Note that implicit in this narrative is a linear temporality signified as "first" and "second."

Second-order emergences tend to subvert and complexify this linear story. The effects of the additional tier of complexity, building relations from relations, are to bring into question the ontological priority of the interacting terms presupposed to generate the first-order emergences. Thus the choices of "primary" entities interacting (measuring instruments and the flux for micro, unicellular organisms for evo, and technics and humans for techno) are revealed as analytical choices that could have been made otherwise. That is, the effect is to *reveal the relational frame used to analyze the emergence as a frame* and therefore as contingent on the chosen perspective. The transition from first-order cybernetics to second-order cybernetics shows this process in action. Whereas first-order cybernetics focused on central terms such as *information, feedback,* and *homeostasis,* second-order cybernetics, associated with von Foerster's *Observing Systems* (1984) showed that the description of any cybernetic system implied the presence of an observer who creates the frame, and that a change in perspective can bring the observer him- or herself into the picture (thus the pun of von Foerster's title, which can be read either as a system that is observed, in which case the observer is only implicit, or as the action of observation itself, which makes the observer explicit). Once the contingency of what counts as primary stands revealed, it becomes possible to make another choice. Reversible internality enacts precisely such a change of perspective.

Michael Snow's artist book *Cover to Cover* ([1975] 2020) illustrates this concept beautifully. As the title implies, the book begins with the front cover, showing a door; flipping the book over shows another door on the back cover, which readers will later realize is the other side of the same door shown on the front. The space between the two images—that is to say, the book's contents—are filled with pages meditating on how images are framed using different kinds of apparatuses and perspectives. For example, opening the cover reveals an image of the back of a man's head and torso standing beside the door. Succeeding pages show the man, seen from the back, entering the door as he incrementally advances through several scenes. But then the perspective shifts, the frame changes, and the man is seen from the front. Then the photographer who has taken the shots through which we have been paging stands revealed through several different kinds of frames and positions. Now another series of pages begins in which an image is shown on a page, while the facing page shows the apparatus that generated the image, including shots outside a window of the so-called natural world, now revealed as available to us as readers only through some form of mediation responsible for creating the image and framing what we see.

Yet another series begins with the man leaving the house and entering a truck and then driving around town, but now we readers have been so inoculated to expect a mediating apparatus to be revealed that we are forced to wonder about the apparatus creating the images, which has disappeared from view. Another series focuses on the man's head and hands, implying that mediation of some kind is inevitable, since the human mind and hands are the final mediators of all the mediations we experience as human beings. Between the front and back covers, then, lies a universe of reversible internalities.

Crucial to the concept of reversible internality are looming questions, such as what it means for something to be "internal" to something else. It means that System A, which is said to be "internal" to System B, is understood in terms dictated by the episteme of System B. Conversely, if System B is "internal" to System A, the inverse is true; that is, System B is now analyzed in terms subservient to the kind of universe that System A is assumed to generate. For example, technics may be seen as internal to humans (that is, humans make technological artifacts), in which case an artifact is analyzed in terms of human history, as contingent on human evolutionary developments such as bipedalism. Such a perspective might emphasize that when *Homo habilus* first walked upright, the hands were freed to make and use tools, which accelerated the evolution into *Homo sapiens*. But in the contemporary moment, this analytical choice would likely evoke its reverse, so that humans would be seen as internal to technics (that is, humans are created by the things they make). Now the context would no longer be the evolutionary history of the human but rather the evolutionary trajectories of technical objects, starting from flint flakes through inclined planes and all the rest of technics.

Putting both perspectives into play at once leads to a third, yet more complex realization, that both dynamics are happening simultaneously and coconstitute each other. This is the full effect of reversible internality: *to relationize the relations so that any presumption of ontological priority is undercut*. With this, the connection of reversible internality to second-order emergences becomes apparent, because it is when second-order effects are in active interplay with each other that the relationality of relations stands revealed as such. Hence the concept of reversible internality becomes not only a possible thought but a virtually unavoidable one, manifesting in many different discourses as a pervasive trope in early twenty-first-century formations.

A beautiful example of this pervasive trope is Jennifer Gabrys's recent work in "Smart Forests Atlas" (2022). Gabrys discusses how digital technologies are being used to measure and intervene in forest processes; in this perspective, the forest is internal to the digital sensing and computational devices that interpret its processes within the terms of digital computations. But as Gabrys and her collaborators point out, this perspective may be inverted so that the forest's processes are themselves seen as analogue computational media calculating, for example, the precise degree of moisture or nutrients in the soil. In this second perspective, computation is enfolded within the forest, which is constructed as a form of analogue vegetal computation ontologically prior to its sensing by digital devices (Gabrys 2023). Here the frame might foreground differences between analogue and digital computing, as well as the different ways in which analogue and digital data are stored, analyzed, and circulated. The same applies, of course, to the different cultural contexts for frames. In "Unsettling Participation by Foregrounding More-Than-Human Relations in Digital Forests," Gabrys and her collaborators discuss how different stakeholders have constructed frames with very different ethical and practical implications (Westerlaken et al. 2023).

As this example shows, the choice of analytical perspectives is never neutral. When the forest is internal (folded within) the realm of digital computation, it becomes subject to the rules of this episteme, with its processes understood through digital outputs and integrated into digital databases for preservation and analyses. Conversely, when digital computation is inside (enfolded within) the forest's own internal processes, the emphasis shifts to the evolutionary mechanisms that the forest has developed to sense and interact with its environment. But this is not really an either/or choice; rather it is both/and, because both perspectives are possible simultaneously and in fact coconstitute each other. Each entity can be seen as internal to the other, which leads to a higher-order realization that they are not merely dialectically joined but are two different constructions of the same reality. Nevertheless, the choice of frame is always consequential, so which frame is given priority becomes an ethical as well as an intellectual choice, as do the corresponding responsibilities for those choices.

Reversible internality, with its relationizing of relations, is a powerful tool with which to combat anthropocentrism, because it reveals with stark clarity the arrogance of presuming that humans are ontologically prior to, and thus privileged above and separate from, the phenomena they investigate. Anthropocentrism would eradicate reversible internality so that only one analytical choice is permitted: making everything else internal to the human-dominant externality. Technosymbiosis points in the opposite direction, highlighting interactions between different kinds of cognitive entities: humans, nonhuman lifeforms, and computational media. It emphasizes relationalities in general, and reversible internality specifically, relocating the human within the web of life while also recognizing that the web of life can be framed as internal to human interventions and perspectives.

### Technosymbiosis Defined and Situated: The Work It Can Do

By appropriating a biological term (symbiosis), I do not mean to imply that I regard computational media as alive.[22] Rather, I see the common trait justifying the splice between technological and biological categories as cognition. Together with symbiosis in the biological realm, which evolves life through life, *computational media and AI are prostheses evolving life by means other than life*. Now with the second-order emergences erupting, these two inversely related dynamics are merging to create explosive potentials to transform not just the meaning of life, but life itself.

In proposing technosymbiosis, my focus is on cognitive technologies, broadly understood as computational media. Computational media are much more than computers; they include chips of all kinds, routers to control and direct information traffic, networking hardware and software, and so on. Among all technologies, computational media are special because they are the controllers, interpreters, and disseminators of information. The more important information flows are to a society, the more computational media interpenetrate every other technology, and the more critical they become to the entire infrastructure's functioning. For example, a water supply facility uses many technologies, including the pipes that carry the water, the meters that measure water flow and pressure, the pumps that move the water from one location to another, and so forth. However, all these activities are regulated, coordinated, and controlled by computational media in the form of chips, software, and other computerized components. If a pipe breaks, the other pipes still work, and the water can be rerouted appropriately; if a meter malfunctions, all the other meters still measure accurately. But if the system controllers break down, the entire facility fails. The example illustrates that computational media have become dominant among technologies for the same reason that *Homo sapiens* have become the dominant mammals on the planet: their cognitive capacities.

My version of technosymbiosis takes a positive stance toward the homology between the cognitive capacities of *Homo sapiens* and computational media. Without being naive about the negative possibilities, technosymbiosis is dedicated to the proposition that the symbiotic relations in cognitive assemblages can be used to realize positive futures. The emphasis, therefore, is on ways to mitigate the worse effects and make better choices in the present that can lead to planetary flourishing. Donna Haraway has urged us to think about our "respons-ability" (Haraway 2012) as signifying our ability to respond to and empathize with other lifeforms (see also Haraway 2016, 28). To this excellent suggestion, I would add an insistence that it is not enough now simply to critique. Valuable and necessary as criticism is, the seriousness and urgency of the multiple crises confronting us require that we must also offer remedies and suggestions for corrections and improvements.

Technosymbiosis focuses on the ways in which cognitive assemblages operate in contemporary society. It refuses the assumption that humans are primary in these arrangements and instead investigates the dynamics of specific cognitive assemblages and how they operate. It interrogates the ways in which humans have appropriated the cognitive powers of other biological creatures to extend our reach, for example in gene editing, and urges humility in the face of such collaborations. It draws on technical discourses and critical approaches to computation to understand the implications and consequences of entering deeply into symbiotic arrangements with computational media. It turns to environmental discourses, especially the environmental humanities, for analyses of our contemporary crises and for suggestions about remedies. It foregrounds ways in which computational media and the environmental humanities intersect, arguing that some of the solutions to our worst problems require computational interventions. It explores the emergence of protosentience in sophisticated neural nets such as GPT-3 and GPT-4, asking how literary criticism should approach machine-generated literature, including novels, poems, essays, and other literary forms. It interrogates what kind of ethical frameworks are appropriate for cognitive assemblages, in which agency is distributed and not all the actors are human. Using literary texts, it explores through the trope of the conscious robot the possibility that computational media may someday achieve consciousness and speculates about the ethical questions that conscious robots will raise. Its primary mission is to think about environmental and computational issues together to achieve new insights and envision new solutions to our most pressing problems. In its most encompassing form, it leads to a philosophy of ecological relationalities, discussed above, which can be understood as a rethinking of liberal philosophy to avoid anthropocentrism and a reenvisioning of relations of *Homo sapiens* to other biological species and to computation media.

## Braiding the Chapters

Each chapter focuses on one of the micro/evo/techno threads, with added colorations from the others. In chapter 2, "Can Computers Create Meanings? A Technosymbiotic Perspective," the focus is on how cognitive lifeforms evolved from material processes. It traces their evolution using two complementary frameworks, Terrence W. Deacon's teleodynamics, emphasizing constraints and structural organizations (2011), and my own intermediation emphasizing lock-in effects created by moving from one level of organization to another (Hayles 2005). From there it engages the central question of how biological lifeforms use their cognitive capabilities to create and interpret signs. As indicated above, it draws on biosemiotics to argue that all living organisms, even unicellular organisms and plants, interpret signs in ways appropriate to their contexts. Although the biosemiotics framework advanced by such theorists as Jesper Hoffmeyer, Wendy Wheeler, and Terrence Deacon has considerable explanatory power, it has a significant blind spot in arguing that such signifying capabilities apply only to living organisms, not computers. However, many of the objections to computers creating meaning become moot if the relevant unit is considered to be human plus computer rather than either alone. Computers, embodied in ways profoundly different from ways that humans are, nevertheless have internal and external milieux within which they interpret information and create meanings relevant to their contexts and embodiments. In turn, these operations fundamentally transform the kinds of interpretations and meanings that humans create. The human species, this chapter argues, is in the midst of entering into a deep symbiosis with computational media. Still incomplete, this symbiosis has far-reaching implications, both positive and negative, for our human futures. By arguing for the meaning-making capabilities of computational media, this chapter performs a crucial step in redefining the relation of *Homo sapiens* to our computational symbionts, thus opening a path toward ecological relationalities.

In chapter 3, "The Emergence of Technosymbiosis and Gaia Theory," the focus turns to the biological. To better position and explain technosymbiosis, it returns to the origins of symbiosis in the works of James Lovelock (2016) and Lynn Margulis (1997), discussing how their collaboration furthered both of their goals. It interrogates the purposes that these pioneers envisioned for their work; for Margulis, this was revolutionizing contemporary biology by emphasizing the importance of symbiosis in evolutionary theory and the role that microorganisms, especially eukaryotic cells, played in the emergence of plants and animals; Lovelock's goals included emphasizing the ability of organisms to interact with their environments so as to make their survival possible. Like many theorists, Margulis and Lovelock looked to existing theories for support and development. Increasingly for Margulis, this role was played by the theory of autopoiesis developed by Humberto Maturana and Francisco Varela ([1972] 1980); the privileged predecessor theory on which Lovelock drew was cybernetics. Unacknowledged at the time were fractures between the desired goals Lovelock and Margulis wanted to achieve and the theories they turned to for support. For example, as we have seen, autopoietic theory is highly recursive and circular, creating closed systems that reflexively regenerate themselves, which fits uneasily with Margulis's view of the bacterial microcosm as constantly mutating and exchanging genes. Cybernetics, for its part, tends to downplay the differences between biological organisms and machines. Margulis consistently emphasized the distinctive features of the living, which is a large part of what made autopoietic theory attractive to her, whereas Lovelock tended to see the mechanical and biological as existing on a continuum. Through a careful analysis of these fractures and analyses of competing frameworks, this chapter discusses what components of symbiosis should be incorporated into a theory of technosymbiosis, and what kinds of inconsistencies and fractures should be left behind, thus clarifying what technosymbiosis includes and what it does not, as it both draws on and modifies these predecessor theories.

Chapter 4, "Cellular Cognition: Mimetic Bacteria and Xenobot Creativity," uses the concept of cognitive assemblages to discuss the extension of human reach through collaborations with bacteria. CRISPR-Cas9 gene-editing techniques, which use messenger RNA (mRNA) to target specific genes, work by exploiting bacterial responses to viral intrusions; bacteria cells remember these attacks by incorporating into their genome fragments of the viral DNA.

When the same virus attacks again, the cell deploys an enzyme that recognizes the virus by carrying within itself the copied fragment and then cuts the viral DNA at the corresponding point of its genome, effectively killing the virus. This series of events, I argue, qualifies as a kind of mimesis, which is typically understood as a mode of imitation that nevertheless is distinct from merely copying. By incorporating the viral genetic fragment into its own genome, the bacteria imitate the viral DNA but do not copy it. Rather, they recontextualize it by giving it a different context. In addition, they completely change its function, so that it used now not to replicate the virus but kill it. Thus the process constitutes what I call micromimesis.

Understanding it as a kind of mimesis highlights the fact that it is a cognitive act. CRISPR-Cas9 gene-editing techniques use this bacterial cognitive capacity to target, modify, or replace specific genes in a wide variety of animals and plants, giving humans the ability to direct evolution, not only for ourselves but for all the species on Earth. The ethical responsibility this implies can be reinforced by recognizing that this power results not from human cognition alone, but from a collaboration with bacteria to achieve results neither party could by itself. This implies an ethical orientation that is a crucial aspect of ecological relationality, for it emphasizes not human power alone but our collaborations with nonhuman others. The emphasis on collaboration provides a basis for a stronger sense of responsibility toward other species, to whom we are indebted for much of our powers and even our existence.

The theme of cellular cognition is further explored through the creation by biologist Michael Levin and various collaborators of "xenobots," explanted frog skin cells that then begin to exhibit new behaviors suppressed when they were integrated within the frog's cellular networks. Like bacterial cognition, the cognitive acts of these skin cells show that cognition goes "all the way down," as Levin puts it in an article with Daniel Dennett (Levin and Dennett 2020), down to the smallest unit capable of autopoietic behavior, the living cell. Consequently, Levin urges biologists to adopt a teleological approach to cellular behaviors; like all lifeforms, cells strive to continue their existence and will initiate new and creative behaviors to do so.

Chapter 5, "Rocks and Microbes: The Two Different Temporal Regimes of Biological and Mineral Evolution," begins with the article by Giuseppe Longo, Maël Montévil, and Stuart Kauffman (2012) mentioned earlier, arguing that physicochemical processes take place within a different temporal regime from biological evolution. Where material processes follow predictable trajectories that can be mapped in phase spaces, biological evolutions follow no such predictable paths, instead leaping from niche to niche by exploiting what Kauffman calls "adjacent possibles" (Kauffman 2000, 22). One way to understand these different temporal regimes is through the difference that cognitive capacities make in lifeforms compared to, say, in clouds and waterfalls. As stated earlier, living beings have stakes in what happens to them, and their desire to continue their existences leads them to exploit opportunities when they arise, even (or especially) on the microcosmic level. Clouds and waterfalls, by contrast, obey what is usually (and somewhat misleadingly) called the "laws of nature," that is, the underlying regularities that scientific experiments are designed to detect and formulate in abstract mathematical expressions.

A relatively new perspective within mineralogy embeds the changing frequencies and kinds of minerals within an *evolutionary* scenario. Traditionally, minerals have been classified taxonomically according to their chemical composition and crystalline structure. Mineral evolution adds a dramatic temporal dimension to this story, vivifying it and proposing new kinds of classifications based on when a development occurred within Earth's evolutionary history. Mineral evolution charts the changes through ten different periods from the solar system's beginnings through to the formation of Earth and its subsequent development. At Earth's beginnings, there were about three hundred different minerals present. Before life arose on Earth, physicochemical processes expanded that number to about six hundred, and then primarily through the action of water, expanded it again to about three thousand minerals during Earth's prebiotic eons.

As life emerged, microorganisms began interacting with minerals to create biominerals such as clays and other chemicals, again diversifying the evolutionary path of minerals to nearly six thousand different species—and counting. Thus mineral evolution constitutes a fascinating case study of what happens when the predictable temporality of physicochemical processes are interpenetrated by the unpredictable temporalities of the biotic evolution. The implications of these developments are still being explored within the field of mineralogy and lead to speculative comparisons of mineral, biological, and technological evolution. Now that three distinctly different kinds of evolution have been observed, we can ask what commonalities and differences they exhibit, and speculate about what these comparisons tell us about the probable futures of life and nonlife on Earth. These issues are explored in the chapter's conclusion.

Chapter 6, "Inside the Mind of an AI: Materiality and the Crisis of Representation," returns to technics to discuss the Transformer architectures central to large language neural net models such as OpenAI's GPT-3, asking crucial questions about how literary criticism (and other fields) should regard machine-generated literary works such as novels, poems, plays, essays, and other literary forms that are even now proliferating like rabbits on the internet. Among the approaches discussed is what I call the null strategy (by analogy with the null hypothesis in the sciences), which argues that a text is a text, regardless of who (or what) authored it. This chapter argues against the null strategy, claiming that it does indeed matter whether a text is created by a human or a machine.

The chapter further argues against the "stochastic parrot" view of these machine productions, a view that regards neural nets as so many parrots, producing texts that have no meaning in themselves, only the meanings that humans project onto them. The chapter proposes in contrast the view that a neural net can indeed detect and create meanings through the billions of indexical correlations that it performs on human-authored texts, thereby gaining some sense of human language as well as the human lifeworld. Nevertheless, these neural nets have a model only of language, not of the world, and hence their productions typically betray what I call a systemic fragility of reference. The chapter proposes four literary-critical strategies appropriate for machine-authored texts and illustrates them with selections from texts authored by GPT-3. The chapter concludes with the suggestion that although large language models such as GPT-3 are not sentient, they may be moving toward protosentience through the indexical correlations they create.

Chapter 7, "GPT-4: The Leap from Correlation to Causality and Its Implications," compares the performance of GPT-4 with GPT-3 and ChatGPT. Drawing from OpenAI's own technical report as well as an evaluation report by a team of Microsoft researchers, the chapter discusses GPT-4's performance on a number of standardized tests intended for humans—including a simulated bar exam, the SAT and GRE exams, and various other metrics—which shows significant improvements in mathematical reasoning, language comprehension, coding expertise, medical knowledge, and a range of other capabilities. Despite these improvements, the Microsoft report also discusses significant limitations, mainly due to the model's inability to "back up," that is, to engage in an inner dialogue about its guesses and assumptions, and its inability to revise its initial outputs. The proposed improvements in the Microsoft report would allow for longer-term memory, inner dialogue, the use of external tools such as search engines, and a hierarchical structure that combines the present bottom-up contexts with top-down contexts based on paragraphs and larger lexical units. Not coincidentally, these are precisely the qualities that also correspond to a stronger emergent sense of self. Drawing on an essay by Avery Slater (2020) about LLM creativity, this chapter speculates what it would imply if GPT-4 did develop a sense of self. While such a

AN INTEGRATED COGNITIVE FRAMEWORK                                      45

development would undoubtedly enhance its capacity to be creative, it would likely also increase its ability to do harm to humans.

This leads to a discussion of the contemporary proposals for regulating AIs in Europe and North America. Academic debates on the subject so far have not coalesced into a coherent, unified set of proposals, notwithstanding the recent passage of the Artificial Intelligence Act by the European Parliament (2022). Indeed, they do not even share a unified understanding of what artificial intelligence is. In the Unites States, the regulatory framework so far relies entirely on voluntary self-regulation, which numerous research articles find unsatisfactory. The chapter concludes by remarking that the absence of clear-cut regulatory frameworks makes works like Slater's speculative inquiry into the creative potentials of AIs—and fictions like those discussed in the following chapter—all the more necessary in our contemporary climate.

Chapter 8, "Subversion of the Human Aura: Three Fictions of Conscious Robots," turns to the near future, when our fictions are already imagining it will be possible to create robots with consciousness. Drawing a parallel between the aura of the artwork in Walter Benjamin's famous essay on technological reproducibility (2006) and the "human aura," I show how the latter can similarly be defined as a historical construction rooted in tradition and ritual. The human aura stems from the belief that each human is unique and irreplaceable, originally because of the belief that humans were created in the image of God and endowed with immortal souls, and more recently because of genetic combinations and cultural specificities that make each human unique in light of his or her inheritance and experiences. A crisis of representation lurks when the human aura is perceived to be diminished or undercut by AIs, which in some respects may have cognitive capacities superior to humans. Just as Benjamin hinted that the erosion of the artwork's aura may have salutary effects, this chapter holds out the possibility that the erosion of the human aura presents an opportunity to reconfigure it in more expansive terms, so that it is no longer so closely associated with human exceptionalism and dominance over other species.

The implications are explored through three contemporary novels, Annalee Newitz's *Autonomous* (2017), Kazuo Ishiguro's *Klara and the Sun* (2021), and Ian McEwan's *Machines Like Me* (2019). *Autonomous* features an apex predator, the military-issue robot Paladin, who in addition to being conscious also has a formidable array of weapons and sensory capabilities. The contradictions of being at once a conscious subject and a commodity to be bought and sold, designed and programmed, are explored primarily through narrative focalization on Paladin's consciousness, especially the growing erotic and romantic relationship with his (and then her) human partner Eliaz. Of central interest in this fiction is the author's attempt to imagine how a robot's consciousness may differ from typical human perceptions, and what differences these make in motivating the robot's actions in the fictional world.

A similar conundrum shapes Ishiguro's novel, which features a conscious robot narrator, Klara, whose special gift is her ability to observe humans closely and intuit their emotions. Bought as a companion to the ailing twelve-year-old Josie Arthur, Klara devotes herself to the child's welfare. Although the robot's thoughts do not differ in any significant way from a human's, she clearly occupies a subaltern position in the Arthur household, hovering in a gray zone somewhere between a person and a vacuum cleaner (as one of the characters puts it).

The effect of conscious robots on the human aura plays out at the individual level in Josie's mother's plan to make Klara "continue" Josie if she dies, thus replacing the original with a simulacrum. On the societal level, the introduction of AIs leads to massive job losses not only for service workers but also for middle- and upper-class white males, among them Paul Arthur, Josie's divorced father. In this way, robots and other AI systems disrupt the typical life patterns of people in developed countries, in which youth is occupied with preparing for a job or profession, midlife with establishing one's position in it, and older adulthood with transitioning to retirement. With few or no jobs to be had, youths are left rootless, older people unemployed or unemployable, and masses dealing with homelessness or worse. These are among the factors that put the human aura at risk in this fiction.

McEwan's *Machines Like Me* conveniently contrasts with both *Autonomous* and *Klara and the Sun* because it features a white midthirties Englishman, Charlie Friend, as narrator. Charlie avoids an office job by doing day trading, which earns him enough for a minimal lifestyle. Upon coming into an inheritance, Charlie on a whim buys an advanced-conscious robot, Adam. Charlie hopes that he can make the robot into a joint project shared by his neighbor Miranda, thus enabling him to advance from her friend to her lover. The plan is successful, but Adam thwarts their plans to buy a house and adopt a small child over abstract ethical concerns. Ethical issues come to crisis when Charlie kills Adam, raising serious questions about the rights that a conscious robot should have.

Taken together, these fictions make the case for the rights of a conscious robot to be considered a person. Inevitably such a move would profoundly affect how we understand humans as well. The recursive feedback loops between robot and human subjectivities fleshes out another crucial aspect of the ICF.

Chapter 9, "Collective Intelligences: Assessing the Roles of Humans and AIs," directly compares the kinds of cognitions, and hence creativity, that humans have with those of AIs. The main idea is that the technosymbiosis between human and artificial intelligence, if it is to be truly symbiotic in a positive sense, requires a deep understanding of how the very different embodiments and evolutionary histories of humans and artificial intelligences lead to complementary strengths and limitations.

The chapter takes issue with Yuval Noah Harari's *Homo Deus: A Brief History of Tomorrow* (2015), specifically his argument that machines can perform most tasks better and more efficiently than can humans. A key premise for this conclusion is his oft-repeated assertion that "current scientific thinking" tells us "organisms are algorithms" (372); therefore, since machines also operate on algorithms, there is no reason why they can't take over from us. This leads him to paint a dystopian future in which humans will inevitably be outthought, outperformed, and outclassed by computers. The chapter shows that the way Harari uses the idea of biological algorithms distorts and even falsifies what the scientific research and philosophical arguments actually say. The chapter investigates the implications of Harari's rhetorical formulation that "intelligence is decoupling from consciousness" (361) comparing it to the cognitive framework developed in this book, which makes an argument that appears very similar (consciousness is not necessary for cognition) yet has very different effects and leads to strikingly divergent conclusions.

To develop further the contrast between Harari's conclusions and the vision presented here, the chapter explores the arguments that Aden Evens (forthcoming) presents in *Discontents of the Digital*. Evens argues that computer ontology inevitably limits the creative powers of computational media, because computers understand everything solely in terms of binary numbers. Human creativity, by contrast, has evolved to deal with the messiness and unpredictability of real life, thus possessing a flexibility and creativity that computers can never achieve. A spectacular example of this creativity appears in the ability of humans to take advantages of accidents and unexpected events to create novel and unprecedented works.

A similar surface similarity and deep divergence between my work and Harari's conclusions occurs with Harari's prediction about the collapse of humanism. In *How We Became Posthuman* (Hayles 1999), published sixteen years before *Homo Deus*, I anticipated many of Harari's observations about how computational media (specifically virtual reality, robotics, and informatics) were undermining notions of free will, the autonomous individual, and the unified rational self. I argued that these developments were leading to a version of the human very different from the one we inherited from the

Enlightenment, which I called the posthuman. Without mentioning my work, Harari repeats each of these points in his own terms, culminating in a prediction that humanism will collapse and be replaced by a successor philosophy. The one he discusses at length is dataism, in which meaning is located in data flows rather than in people. Moving in the opposite direction, my purpose with the present book is to show how the liberal philosophy that undergirded the Enlightenment, rather than simply being jettisoned, can be reimagined to be more inclusive, less anthropocentric, and more conducive to positive futures. In this respect our intentions could not be more different.

At the very end of his book, Harari seems to want to position his arguments merely as provocations to imagine better and different futures. For that enterprise, however, he gives us no help whatever. What he does give us is a dystopian fiction that for all but the last ten of its 462 pages vivifies a future in which human life is meaningless, algorithms are all powerful, and ordinary people are powerless to change anything. This reminds me of a family anecdote in which my grandmother admonished her four-year-old son (my uncle), "Whatever you do, do not stuff those dried peas up your nose." The result was predictable, because she had just visualized for him a possibility that he might not have thought of by himself. So Harari depicts a world in which the worst tendencies of computational media have become dominant, and for that, he must accept responsibility for the effects of his own fiction—effects that he unconsciously acknowledges when he stresses the power of fictions to shape the world. This chapter provides a basis for turning away from dystopian futures toward the more positive outcomes of ecological relationalities.

Chapter 10, "Planetary Reversal: Ecological Relationality versus Political Liberalism," explores the complex and often conflicting possibilities for thinking through and with ecological relationalities. On the negative side, Lisa Lowe's *Intimacies of Four Continents* (2015) shows how deeply entrenched in democratic societies are the inequities of a discourse of rights. She reads archival administrative documents from roughly 1750 to 1850 to demonstrate that liberal political philosophy in practice always meant the enrichment of some humans at the expense of others, sometimes literally through slavery and indentured labor, and sometimes through the expropriation of resources and assets. The relation, she argues, is not merely correlative but causal; human rights, citizenship, and wage labor for some could be expanded precisely *because* others were disenfranchised and disempowered. However, I argue that our analyses cannot end there, because our responsibilities to ourselves and the planet require us to envision and implement positive futures, however difficult that may be.

Jason Moore's *Capitalism in the Web of Life: Ecology and the Accumulations of Capital* (2015) takes a small step in this direction by refusing apocalyptic scenarios. His main contribution, however, is the idea of double internality, the inspiration for my own theorizing about reversible internality. The double internality with which he is concerned is nature-in-capitalism and capitalism-in-nature. Whereas the former leads to a framework poised for exploitation of natural resources, the latter makes clear the accumulating costs of this perspective (hence the ironic tone of a phrase in his subtitle, "the accumulations of capital").

Important as Moore's work is, its main contribution is to analyze the problems rather than suggest solutions. For that project, Kim Stanley Robinson's *Ministry for the Future* (2020) is ideally suited. One way to think differently about the obstacles to progress, which can be summarized as an *us versus them* problem, is to focus on linking mechanisms that can serve to connect individual humans with larger global concerns, thus enfolding the local into the global and vice versa. *Ministry for the Future* does that in the form of a sprawling experimental novel that includes fictional narratives, factoids about our global situation, and an eclectic collection of theories and possibilities. It provides the resources to think deeply about how we got to where we are, the social, economic, and historical forces in play, and the kinds of choices and responsibilities that are possible. It forges a variety of linking mechanisms through its myriad of narrators, most human, some not, that span a huge range of scales from subatomic particles to cosmic entities such as the Sun. Without being in the least naive, it makes a case for ecological relationality by folding the future into the present, giving voice to those who have none, and testing out a variety of options, from geoengineering to radical political action and selective violence against those who would make ecological collapse a certainty. The chapter includes charts contrasting liberal political philosophy with ecological relationality, along with the various linking mechanisms that Robinson suggests can help to overcome *us versus them* attitudes.

This chapter works to connect and clarify the relation of cognitive assemblages to ecological relationalities, which has been prepared for and anticipated in the preceding chapters. One of the strongest aspects of the cognitive assemblage framework is its broad definition of cognition, formulated to include all biological lifeforms as well as computational media. This definition provides the basis for thinking about living creatures together with artificial intelligences, as exemplified in this introduction. In addition, by emphasizing interpretations and choices, this definition presents cognition as a spectrum of possibilities rather than a binary choice (avoiding such questions as, is this organism intelligent or not?). Moreover, it recognizes that there is continuous dynamic interplay between collectivities and individuals, exemplified in chapter 9 in the idea of GPT-3 as collective intelligence. This book turns to literary and cultural texts in chapter 8 to contest dysfunctional ideas and lay the foundation for ecological relationality, which chapter 10 develops as a philosophical and practical basis for action. Through these means, it opens the possibility for taking responsibility without at the same time reinscribing human dominance.

The "web of life" approach, seen in the work of Jason Moore (2015), Donna Haraway (2016), and others, is attractive but inadequate by itself to guide human futures, which will increasingly involve artificial intelligences and cognitive media. The framework this book forges overcomes that limitation, while still incorporating aspects of the "web of life" approach conducive to human and nonhuman long-term survival and flourishing. It looks seriously at the prospect that the human species is in the process of creating artificial species in intelligent robots and other forms of artificial intelligences that may become our evolutionary competitors as well as our symbiotic others. Technosymbiosis argues that our human futures are not predestined; they will evolve from the interpretations and choices we make, individually and collectively, human and nonhuman. Technosymbiosis, through its philosophy of ecological relationality, aspires to provide an encompassing, dynamic, and robust framework within which synergies can evolve and grow into positive futures for our human and more-than-human world.

# 2

# Can Computers Create Meanings?
# A Technosymbiotic Perspective

Prebiotic Earth roared with activity: volcanoes erupted, tectonic plates shifted, vast storms raged, chemical reactions seethed.[1] Yet none of this activity had any meaning at the time, because there were no living organisms to attribute meanings to it. With the emergence of life, meanings became possible. This chapter offers two frameworks within which the emergence of life and meaning-making practices can be understood: my own vision of emergence, which I call intermediation, and Terrence Deacon's constraint-based explanation. Viewed as complementary, the two frameworks account for more aspects of meaning-making practices than either can by itself.

The rise of computational media greatly extended the pervasiveness of meanings. Similar to the biosemiotic orientation that views meanings as behavioral responses to changing environments, a cybersemiotic approach will be used to argue that computers also engage in meaning-making practices specific to their internal and external milieux. For centuries technologies have blunted the edge of Darwinian natural selection, a process that became laser focused when humans, aided by computational media, teamed up with bacteria to intervene in genetic codes, human and nonhuman. The result is biotechno evolution, a hybrid process in which information, interpretations, and meanings circulate through the flexible human-computational collectivities that I call cognitive assemblages.

Our understanding of how cognitive assemblages work has been impeded by views of computational media that position them as mere calculators without the ability to create, disseminate, or participate in meaning-making activities. This view, widely shared among philosophers and even some computer scientists, is undergirded by questionable assumptions that have the effect of making humans the sole possessors of agency, value, and cognition. In addition to reinforcing a worldview in which humans have sole dominion over the Earth, this perspective badly distorts the actual effects of cognitive assemblages in contemporary developed societies and obscures the ways in which bio-techno evolution presents us with urgent issues about our collective futures. Articulating better ways to understand meaning-making practices in biological organisms, computational media, and their interactions in cognitive assemblages is the work of this chapter.

To make the case for computer cognition, it is necessary first to trace the dynamic transitions that enabled humans to become cognitive; this in turn requires a brief account of how life arose from nonlife. Although such moves may seem to risk infinite regress, they are useful because they locate the central question of computer cognition within the larger and older context of biological evolution. In broad scope, the first evolutionary leap bootstrapped complexity from simpler mechanistic reactions; the second emerged from the first, bootstrapping meaning-making practices from complexity; the third emerged from the second, bootstrapping artificial cognition from biologically derived signs and meanings.

The first explanatory strategy I call intermediation, introduced in my 2005 book *My Mother Was a Computer*. The basic idea is that emergence proceeds through a pattern that repeats at multiple scale levels. To emphasize continuities between biological and computational cognition, I call each level of organization "media," and "intermediation" refers to interactions between levels, both from higher to lower and from lower to high. For example, a low level of dynamical organization, say subatomic particles, forms quasi-stable patterns captured by the next level up through the formation of new entities, in this case atoms. Atoms in turn form quasi-stable patterns whose dynamics are captured by higher-level entities, here molecules; molecules in turn form patterns captured by proteins, and so forth. The change of media is crucial, because it "locks in" the lower-level dynamics and enables emergence to continue through successive levels of increasing complexity.

The second strategy draws on the work of Terrence Deacon (2011). Deacon has his own scheme for evolutionary emergence, and his explanation introduces an important new factor showing how meaning-making practices develop, first in very simple organisms such as bacteria, and then through progressively complex creatures up to and including humans. This way of thinking draws inspiration from the interdisciplinary field known as biosemiotics, "bio" denoting life, and "semiotics" referring to the science of signs. As mentioned earlier, biosemiotics developed from the semiotics of the philosopher C. S. Peirce, notable for using a triadic (rather than dyadic) view of signs. Writing from a biosemiotic perspective, Deacon enlarges biosemiotics' scope by arguing that signs in general become possible through the emergence of what he calls teleodynamics, end-directed processes that work by using something present to evoke or gesture toward something absent. Deacon excels at employing this kind of move, which we may call negative logic: using absence to determine and define presence.[2] The emergence of signs and their deep relationship to absence is crucial to the entire biosemiotics enterprise, for without signs, nonhuman organisms are limited to the world of the present. The past lingers and can be detected in the marks it leaves on bodies: rings on a tree, a dog's limp from an earlier leg break, wrinkles on a face. But without signs, the world of the future for nonhumans slips entirely from view: insofar as it has not yet happened, it can be anticipated and brought into the present only as a signifying absence.

The essential mechanisms enabling such gestures toward absent phenomena are constraints, and different forms of constraint function in his scheme similarly to different levels of intermediation in mine. Constraints necessarily refer to absences, because they limit the regions in which dynamic systems operate. Oil lubrication in an automobile engine, for example, creates the absence of states in which the engine would seize up because of friction between moving parts. Typically we think of constraints as defining the regions in which systems work as they should, analogous to focusing on the positive white image in a black-and-white drawing. Through the application of negative logic, however, such scenes can be psychologically flipped in their orientation so that they are seen to specify the *absence* of regions where systems break down. Analogously, it is as if one focused not on the white figure but on what had been taken as the black background as the defining structure, so it becomes the foreground figure and the white area becomes the background.[3]

Powerful though Deacon's strategy of negative logic is, it has a major blind spot noted earlier, shared by other theorists writing about biosemiotics such as Wendy Wheeler (2016a and 2006) and Jesper Hoffmeyer (2009 and 1997). Deacon argues that computers are not cognitive, engaging only in simple mechanical practices that are qualitatively different from cognition in biological organisms. What Deacon, Wheeler, and others do not take into account, however, is the possibility that computers evolve not through being autonomous (which clearly they are not) but precisely through their partnerships with humans. The assumption that beings must be autonomous to be cognitive is an example of what I call biologism, the faulty extrapolation of biological reasoning into computational media. Deacon, Hoffmeyer, and other biosemioticians are often deeply versed in biology but frequently manifest only a slight acquaintance with how computational media operate. Moreover, their focus on dynamics obscures the role that intermediation plays in the evolution of artificial cognition.

A principal contribution of this chapter is to show that arguments based in biologistic assumptions miss the mark and misunderstand cognitive processes as they are manifested in computational media. On the positive side, the chapter argues that a framework combining intermediation and constraints enables a deeper understanding of how computers generate meanings than either can by itself. It also illuminates how and why cognitive assemblages have become powerful actors in contemporary cognitive ecologies.

## Intermediation and Teleodynamics

Intermediation can be understood as an alternation between elements considered as individuals, which interact among themselves to create the quasi-stable patterns that are incorporated into the emergence of individuals at a higher level of complexity, creating a process that results in increasingly complex patterns. As Harold Morowitz (2002) has argued, this kind of emergence is how the universe actually evolved after the big bang. As the superheated plasma cooled, subatomic particles appeared, then chemical elements, on up to the formation of planets and the emergence of life on Earth.

Intermediation differs from Morowitz's account in referring to elements at each level as "media." Media," as John Guillory (2010) has reminded us, connotes both "materiality" (355) and "mediation" (326)"; "mediation" in turn connotes communication through distance in time and space (331). Technical media achieve this through different layers of code within a machine and networked functionalities between machines, but communication through distance can also be understood as the emergence of quasi-stable patterns as physicochemical entities "communicate" through their interacting dynamics. As noted earlier, intermediation denotes both bottom-up emergence, as new entitles are formed from lower-level patterns, and also top-down emergence as higher-level entitles interact with and change lower-level dynamics. Intermediation thus describes a recursive organization that I call a dynamic heterarchy (Hayles 2007b, 100), in which lower-level media produce the higher-level media, and the higher-level media simultaneously affect the lower-level media.

In informational terms, the specific levels formed by physicochemical media always contain more information within the level than they communicate upward or downward. Indeed, it is precisely this reserve of dynamic information that characterizes the levels as such. Subatomic particles, for example, manifest as probability distributions that get statistically "smoothed out" when incorporated into atoms, where they can be treated mathematically as point masses. However, the level-specific information is revealed through such phenomena as electron tunneling, explicable because there is a finite probability that an electron will appear on the other side of a barrier that it otherwise does not have enough energy to penetrate. Each level of physicochemical interactions has similarly specific dynamics in excess of its contributions to higher or lower levels.

A different strategy is at work in Deacon's scheme. His key terms involve dynamic patterns evolving through time (in this respect it overlaps with intermediation). In addition, he also introduces a novel component focusing on how constraints interact with structural organizations to produce emergent processes. He characterizes three different kinds of dynamics, which he calls homeodynamic, morphodynamic, and teleodynamic. Exploring this reasoning illuminates the contributions of his approach and also the complementary ways in which it overlaps with intermediation.

Homeodynamics characterizes mechanistic physicochemical systems that tend to become more random and disorganized over time. The second law of thermodynamics expresses this as a tendency for closed systems spontaneously to move in the direction of increasing entropy (or randomness). Cream poured into a cup of coffee spontaneously tends to dissipate; a sugar cube dropped into water tends to dissolve. As a homey example, consider a dresser drawer full of socks and underwear. Over time, as the owner roots among these items, they tend to mix evenly throughout the drawer, which the owner may perceive as having become "messy." Explanations of the second law frequently point out that what humans perceive as "ordered" represents only a small fraction of possible states; in this example, there are exponentially more ways for the items to mix together than there are ways for them to remain neatly separated.

A more rigorous way to state the second law, then, is to say that closed systems tend to move spontaneously toward more probable states, for example states that have lower potential energy than those that have higher potential energy. Because terms like "order" and "disorder" may have subjective implications (as in the dresser drawer example), Deacon usefully refines the terminology by calling processes that move in a direction consistent with what spontaneously happens as "orthograde," while processes that move against spontaneous tendencies are "contragrade."

To characterize further the three kinds of dynamic systems, Deacon introduces the idea of constraints. The orthograde tendency of homeodynamic systems is to dissipate constraints. Consider, for example, a reservoir in which a large body of water is being held back by a dam. The system's spontaneous tendency is to let the water be distributed evenly over the ground (there are many more possible states in which this could be achieved than the few states in which the water is contained, and the dam exerts a consistent pressure, or constraint, to prevent this from happening. If the dam is weakened by heavy rainfall or an earthquake, its constraint may not be sufficient to restrain the water, and in that case the system's orthograde tendencies take over, and the water bursts through, cascading onto the ground.

So far these explanations do not go beyond a high school physics textbook, but their explanatory power emerges when Deacon (2011, 27) points out that constraints function as presences that point to what is absent, namely what is excluded by virtue of the constraints. For example, a numerical system may be constructed using the constraint that the only integers allowed are odd numbers. This constraint becomes visible only through absence of even integers, which are the phenomena gestured toward by the odd-number constraint. Such exclusionary constraints Deacon calls "absential" (2011, 323), phenomena that function by gesturing toward what is not present.

This view of constraints enables Deacon to distinguish further between homeodynamic systems and the next kind of system up the evolutionary ladder, which he calls morphodynamic. Morphodynamic systems are self-organizing phenomena such as the red eye of Jupiter, where storms circle continuously in consistent patterns that fluctuate within constrained dimensions of time and space. As Ilya Prigogine and Isabelle Stengers explain (1984), dissipative systems are far-from-equilibrium phenomena in which the entropy production is so large that pockets of order can form without violating the second law. Self-organizing systems are characterized, in Deacon's terminology, by the production of constraints, the spontaneous appearance of attractors that define the limits within which the system's phase space trajectories move. A famous example is the Lorenz attractor characteristic of weather patterns (described in Gleick 2008, ix), with its butterfly-wing shaped pattern traced by trajectories unpredictable as individual instances but nevertheless constrained by the attractor's parameters in phase space.

Deacon argues life could never have arisen directly from homeodynamic systems, because life depends on the continuing existence of many different kinds of constraints, from cell membranes to energy and temperature requirements. Morphodynamic systems provide the necessary stepping-stone that allows constraints to be produced and preserved. Rephrasing Prigogine and Stengers's insight, we can say that morphodynamic systems are encapsulated within homeodynamic systems, whose strong entropy production serves to protect and buffer them against the constraint dissipation that is orthograde within the larger homeostatic system but contragrade within the morphodynamic system.

The next evolutionary step is the emergence of teleodynamic systems, systems directed toward some goal or end point. For living organisms, the foundational goal is survival and reproduction. Again, organizational dynamics and associated constraints are essential in understanding how such systems can emerge from self-organizing phenomena. When morphodynamic systems strongly couple together, as in chemical networks that produce the catalysts that spark further reactions extending and strengthening the reactions, recursive patterns emerge in which the system's output becomes its input, leading to further catalysis as the input cycles through to produce more output.

Stuart Kauffman (1996, 47–48) calls such systems autocatalytic networks and argues they are the immediate precursors to life. In Deacon's terms, such systems connect and extend constraints in complementary fashion, coevolving and cocreating each other through their mutual interactions. They recursively self-constitute one another and mutually reproduce a system of correlated constraints, leading to such complex organizations as a living cell. The cell as a whole is constituted through the constraints and organization of its parts, but the parts in turn are constituted through the constraints and organization of the whole.[4] Deacon explains, "Life is characterized by the use of energy flowing in and out of an organism to generate the constraints that maintain its structural-functional integrity. Organisms additionally need to constantly impede certain forms of dissipation. Organisms take advantage of the flow of energy through them to do work to generate constraints that block some dissipative pathways as compared to others. Organisms don't just block constraints, they generate new forms with new constraints" (Deacon 2011, 263).

Before proceeding to the next evolutionary development, in which constraints in living systems enable the emergence of signs, let us pause to consider the relationship between the two explanatory frameworks of intermediation on the one hand, and dynamical organization and constraints on the other. Both frameworks emphasize the existence of levels and the relations and transitions between them. Both identify mechanisms that "lock in" previous patterns and allow further reorganizations. Both emphasize recursive patterns of organization in which the whole produces the part and the parts simultaneously produce the whole.

There are also significant differences. Whereas the levels in intermediation are understood in terms of the patterns and entities forming them, Deacon's dynamical framework emphasizes the structural organizations and constraints that characterize the three different levels culminating in teleodynamic systems. Thus the two frameworks offer different explanations for the "lock-in" effect. In intermediation it is the change in media, whereas Deacon identifies it as a "polarity reversal" in how constraints operate: "the orthograde signature of thermodynamic change is constraint dissipation, the orthograde signature of morphodynamic change is constraint amplification, and the orthograde signature of teleodynamic change is constraint preservation . . . and correlation [among constraints]" (2011, 323). In a sense, intermediation focuses on what is present, whereas the constraint framework emphasizes what is absent.

A related divergence appears when we consider what the two frameworks take to be the ultimate phenomena to be explained. For Deacon, it is not so much the emergence of life as the emergence of sign relations in living organisms. The framework's great strength, especially in relation to intermediation, is its explanation of how signs emerge from the absential phenomena already implicit in the formation of constraints. It thus builds in signs and meaning production from the beginning, whereas intermediation simply assumes that signs can operate through dynamics similar to the intermediating processes that brought the cosmos into existence.

This strength is also a limitation, however, because it ties emergence so strongly to dynamical organizations that it has little purchase when dynamics ceases to be the important evolutionary driver in computational media, where intermediations between different levels of signs take center stage. Constraints do not need to evolve in computational media, because they are programmed in at a foundational level through logic gates. Similarly, sign elations are also programmed through bit (binary digit) patterns. However, such programming is possible only because constraints and signs had previously evolved in humans; computational media thus presuppose and depend on biological emergence. In this sense, intermediation picks up the story of bio-techno evolution where the dynamical framework leaves off. Viewed as complementary rather than antagonistic, the two frameworks, with their different emphases, strengths, and limitations, provide an encompassing way to understand our present situation. Moreover, keeping both frameworks in view allows important distinctions between computers and humans to appear without needing to subsume one under the other. To explore further the emergence of meaning-making practices, let us return to consider how signs evolved biologically.

## The Emergence of Signs from Constraints

Once living cells emerge, natural selection operates to connect constraints and couple them with environmental signals of change. This gives the interrelating dynamics a forward motion in the form of anticipation. For example, as the temperature drops in the fall, this environmental signal is interpreted by dynamic processes within a deciduous tree that set in motion a series of interrelated changes. Tree hormones begin a process known as abscission, in which specialized cells cut off the flow of nutrients, which are withdrawn from the leaves and stored in the roots. We might say the tree "anticipates" winter, although obviously no such abstract conception is at work, only correlated changes that in the evolutionary history of the species have proven reliable indicators of future events. This exemplifies what Deacon calls "absential" phenomena (2011, 473), where something that is not present (winter) causes something that is present (leaves on the tree) to undergo changes that otherwise could not be explained.

Wendy Wheeler elaborates on the idea of absential phenomena when she relates it to meaning. "The meaning of something is to be discovered in what it does in the world, in how it allows things to be and also to change. Thus we say that a biological meaning is a function. But this process, while certainly always carried by, and often (although not exclusively) about, material objects in fact involves something other than substances alone" (Wheeler 2016b, 7). This "something other" is a relation between what is present and what is absent. "What happens is that a relation is established, and that relation has a value of one kind or another for the organisms that experience it" (8). In biosemiotic terms derived from Peircean triadic sign relations, the value for the organism is understood as emerging between a phenomenon in the world (a "representamen," in the above example, an autumn temperature drop), a function or process that responds to this event (the "intrepretant," here the abscissional process), and the "object," absent but anticipated (the approach of winter) (see Peirce 1998, 5–7, for a summary of these terms). "Expectations," Wheeler writes, "are relations to no-things which have real causal and shaping powers" (2016b, 13).

This approach has proven fruitful in explaining how meaning-making practices emerge in the nonhuman as well as the human world, even for organisms with no central nervous system such as trees and for unicellular organisms such as bacteria. As humans initiate changes in our global ecology that have plunged the world into the sixth mass extinction and created epic levels of pollution and global warming, the calls to rethink anthropocentric assumptions about how the world is organized and how it operates have become increasingly urgent. Accepting that meaning making is not an exclusively human prerogative is a crucial step in the right direction. Biosemiotics should be celebrated for its central contributions to this effort.

That these contributions are claimed to apply only to biological organisms is an unnecessary limitation based on flawed reasoning. The crucial point to be interrogated is the assumption that to be capable of meaning-making practices, an entity must be autonomous, self-organizing, self-maintaining, and self-encapsulated. While these assumptions are warranted in the biological realm—indeed, they collectively constitute the requirements for life to emerge—they do not necessarily apply to artificial cognition in networked and programmable machines. Cognitive media's evolutionary requirements include their necessary interactions with humans, already existent through biological evolution. These interactions do not preclude artificial media from being cognitive; they only indicate that bio-techno evolution now continues through a cognitive machine-human dynamic rather than through biological evolution alone.

## Limitations of Biologistic Reasoning

As noted above, the major objection within biosemiotics to the proposition that computers can participate in meaning making is their lack of autonomy. Wheeler makes this explicit in a footnote: "Of course, computers can be programmed to 'notice' relations. But 'programmed' is the point. A computer without a programme which originated from a human being is just an inert collection of metal, plastic, and silicon" (2016a, 25n2). Deacon observes, "Today's computers are conduits through which people (programmers) express themselves. Ultimately, then, software functions are human intentions to configure a machine to accomplish some specified task" (2011, 100). He uses this and similar statements to conclude that computation is only a "descriptive gloss" and that "computation only transfers extrinsically imposed constraints from substrate to substrate, while cognition (semiosis) generates intrinsic constraints that have a capacity to propagate and self-organize" (2011, 497).

It is of course true that computers are not autonomous in the ways that biological organisms are. But missing from this picture is the possibility that evolution, having produced life, may now proceed by different mechanisms that rely on human-computer interactions to form semiotic relationships that exceed the limits of biological cognition alone. If one uses only the criteria that enabled biological cognition to emerge—the interaction of constraints with dissipative dynamics—computers will of course come up lacking because they do not rely on attractors and dissipative systems to achieve cognition. But the requirement that this must be the only way through which cognition emerges is an arbitrary limitation that flies in the face of what computer-human interactions actually achieve.

The extraordinary blindness to which this kind of biologistic reasoning leads is illustrated in this passage from Deacon: "There is nothing additionally produced when a computation is completed, other than the physical rearrangement of matter and energy in the device that is performing this interpretation" (2011, 525). He is emboldened to make this claim because he has previously emphasized work as the output created when biological constraints operate on energy flows—work that moves things in the world and causes events to happen that would not otherwise be possible. His reasoning, however, assumes that the "work" must be done by an autonomous entity rather than a human-technical dyad. The fallacy here can be exposed if we ask instead what would be required for humans to do the cognitive work performed by computers and use that measure to determine how much has changed in the world. Just to pose the question reveals how absurd Deacon's claim is, for the cognitive tasks now performed by millions of networked computers would arguably require full-time labor by all 7.4 billion human inhabitants of the planet to produce equivalent results. Moreover, many tasks performed by computational media could not be done by humans no matter how long and hard they worked, for the scope of the data analyses, pattern recognitions, and correlations that networked and programmable machines routinely provide for their human partners are simply too vast, too fast, and too complex for human cognition alone to encompass.

Many of Deacon's arguments about the limitations of computation are aimed at critiquing the computational theory of mind, which argues that biological brains operate like computers, so that human cognition "is understood in terms of a rule-governed mapping of specific extrinsic properties or meanings to correspondingly specific machine states" (Deacon 2011, 490). I agree wholeheartedly with this critique; a wide range of arguments by neuroscientists, cognitive scientists, and others has shown the importance of embodied and enworlded situatedness for human cognition (Varela, Thompson, and Rosch 1991; Nuñez and Freeman 2000; Damasio 2000; Edelman and Tononi 2001). However, an important asymmetry emerges here, for the fact that biological brains use input from their environments, sensory systems, and bodily functions to achieve cognition does not mean that computers must also be enworlded and embodied in the same way as humans to engage in meaning-making practices.

Wheeler uses a similar line of biologistic reasoning when she argues that "no human-produced machine has any choice about its inputs; it has no body responding to its environment both nonconsciously and consciously. . . . In distinction, organisms, even the simplest, *do* make choices about their inputs and outputs, *do* have feeling bodies, *do* reproduce themselves and grow their themselves and environments" (2016a, 139). But again we see arbitrary limitations imposed here, in the stipulation that choice must pertain to "inputs and outputs" to serve a cognitive function, and the assumption that feelings, reproduction, and growth are necessary for an entity to be cognitive. Computers do have bodies, just differently organized than are biological bodies; networked programmable machines equipped with sensors and actuators do make choices about their inputs and outputs. Nevertheless, the thrust of Wheeler's comment is correct insofar as it points to the myriad differences between humans and computers; what is lacking is a willingness to explore what those differences might mean for computer cognition.

Both Deacon and Wheeler have moments when they come close to realizing that biologistic reasoning may not be appropriate for artificial cognition. Wheeler, for example, follows the above passage by remarking, "We should either be suspicious of the machine metaphors [used to describe biological brains] or radically rethink what we understand machines to be" (2016a, 139). But the proffered alternative—"radically rethink[ing] what we understand machines to be"—is not developed and remains a dead end in her text. Similarly, Deacon explains that "one fundamental difference" between biological cognition and machines is "obvious. The principles governing physical-chemical processes are different from those governing computation" (2011, 104); in another passage, he argues "we need to think about the origin of organic mechanisms quite differently than designed mechanisms" (2011, 426). Like Wheeler, however, he fails to see how this might serve as a critique for his biologistic assumptions and rather uses it to further his argument against computational theories of mind.

Since the emphasis in both Deacon and Wheeler is on biological organisms and the emergence of meaning-making practices in biota, succeeding brilliantly in this regard, why does it matter that they fall short when talking about computers? It matters because the frameworks they develop for interpretation, purpose, and meaning in the biological realm can become, with some necessary modifications, a powerful store of ideas to understand how computers achieve meanings and how computer-human interactions have dramatically transformed our potential to affect our world, for better and for worse. Showing how computational media create, communicate, and disseminate meanings is the next section's focus.

## Intermediation and Constraints in Computational Media

The useful ideas emerging from biosemiotics that can be modified for computational media include the notion that interpretation (in Peirce's semiotics, the "interpretant") consists of an action or behavior that has been consistently linked to an interior or external sign relation established through a historical connection, such as a young bird hearing the song characteristic of its species from its parents. The parental song, Hoffmeyer writes, serves as a semiotic scaffolding that allows the young bird to accomplish what it otherwise could not, and by learning the song, the young bird interprets an acoustic signal to initiate a new behavior, which in turn provides the basis for further interpretations (Hoffmeyer 1997).

In von Neumann computational media, interpretation is ultimately based on the logic gates, which generate the constraints that in Deacon's scheme were accomplished through orthograde and contragrade dynamics. This signals one of the fundamental differences between biological and artificial cognition. Whereas with organisms the crucial point to explain is how they are able to generate and respond to sign relations, with computational media, designed to implement symbolic processes through their logic gates, the crucial point is how these deterministic operations can create a basis for flexibility, adaptability, and evolvability, the hallmark achievements of biological cognition. To explore this question, let us consider how logic gates operate (see Petzold 2000, 86–131, for a useful introduction to Boolean logic, logic gates, and switches).

Logic gates typically have either one or two inputs from transistors, which if switched on (accomplished through a voltage signal of five volts) register as a 1 in binary digits (a bit), or if switched off (accomplished through a voltage signal of zero), register as a zero. From the input combinations and the kinds of gates through which the inputs pass, various logical operators are generated. For example, the AND gate requires that both inputs be switched on to generate an output of 1; otherwise, the output is zero. The OR gate requires that either of the inputs be switched on to generate an output of 1; only if both are off does it output a 0. The NOT gate, or inverter, has only one input and one output. It reverses whatever the input is and generates the opposite, so if the input is switched on, for example, it outputs a 0. Other kinds of gates include the XOR (exclusive OR), NAND (AND with outputs reversed), and so forth. These logical operators can be combined in many different ways, resulting in increasingly complex commands through the layers of code that sit on top of them. These include the machine and hardwired code that calls the machine's instructional set architecture, the system software that runs the specific hardware (for example, the OS10 operating software that runs the MacBook on which I am typing), the assembly language that encodes the operating system, up to high-level languages such as Python, Java, or C++ that encode the operating system software that runs the specific hardware processor and chipsets, on up to executable programs such as Microsoft Word, which is the level at which most users interact with the machine. Ultimately, however, all commands must be translated downward to the logic gates to be processed (either line by line as needed in interpreted languages, or in prepared batches in compiled languages), and then translated upward again to the high-level languages to communicate their results.

The distinctions between these different levels of code function as the different media in an intermediating dynamic heterarchy, with information communicated up from the logic gates to the higher code levels, and down from the executable programs through the different code levels to the logic gates, in consistent patterns of interpretation that respond either to internal cues (endosemiosis) when the levels communicate with one another or to external cues (exosemiosis) when a user inputs a command.

Biosemioticians writing about interpretation in a biological context often emphasize that for an interpretation to have the capacity to be right, it must also be possible for the interpretation to be wrong; otherwise there is no choice involved, only factual statements. The predator chasing a bird that appears to have a broken wing may discover, too late, that he was wrong when the bird flies away after drawing the predator away from the nest; the bacterium moving toward a presumed food source may discover it is a toxin instead. Such a requirement makes sense, for interpretation requires at least two choices to be pertinent; otherwise the situation can be parsed as a simple causal chain with no interpretation required.

In what contexts does interpretation operate within a computer, and how can it be wrong? Although the logic gates are deterministic, they can and do make mistakes. For example, the varying voltages that turn the transistor switches on and off typically manifest trail-off errors in which the signal decays over time, so at the end of the interval, the voltage may actually be, say, 4.7 volts rather than 5.0. Since 4.7 is much closer to 5.0 than it is to 0, electronics are used to rectify the signal and interpret it as "on" rather than "off" and reinstate the 5.0. More significant voltage variations may cause an error in this rectification process.

Another source of uncertainty occurs when cosmic rays flip bits within a computer, analogous to when cosmic rays cause a mutation in a gene. Consequently, computer crashes are more frequent at thirty thousand feet than at sea level, because the cosmic rays are stronger there. A study of IBM in the 1990s suggested that such errors typically occur once every 256 megabytes per month, but as chips decrease in size, they need less voltage and less charge to set a bit and so are more at risk from cosmic rays (StackOverflow.com, n.d.). An Intel patent application made this prediction: "Cosmic ray induced computer crashes have occurred and are expected to increase with frequency as devices (for example, transistors) decrease in size in chips. This problem is projected to become a major limiter of computer reliability in the next decade" (Hannah and Intel 2007). Sometimes bit flipping by cosmic rays has consequences in real life (see Ziegler 1996; Ziegler and Lanford 1979). The video *The Universe Is Hostile to Computers* instances several of these, including the time a plane plummeted from the sky because of a flipped bit, and an election miscount in Brussels on May 18, 2003, when an obscure candidate, Marie Vindevogel, was given exactly 4,096 more votes than a recount indicated she should have had (*Universe* 2021). The number 4,096 is a significant one because it is a power of 2 (2 to the twelfth power). Forensics revealed that the thirteenth bit had been flipped from a 0 to a 1 because of a cosmic ray, resulting in the miscount. Routines such as parity checking and other programs able to check themselves may detect a flipped bit and correct it, rightly interpreting it as an error. Conversely, the flipped bit may cause certain commands to be misinterpreted, crashing the computer. Like mutations, it is possible (although extremely unlikely) that a flipped bit would actually improve the computer's performance, but if so, it would likely remain undetected.

These errors, however, are externally caused and do not undercut the deterministic quality of the logic gates. To arrive at interpretation understood as choices among options, it is necessary to move up to the level of executable programs, where complex interactions between logical operators combine with unpredictable inputs to create possibilities for interpretations. For biological organisms immersed in unpredictable environments, contingency is a fact of their existence. In computer programs, by contrast, contingencies have to be deliberately introduced to increase variability and flexibility. There are two basic strategies to accomplish this, recursivity and pseudorandomness.

Recursive dynamics are known to be powerful drivers of evolutionary processes in biological organisms, resulting in increased cognitive complexity.[5] In computational media, recursivity introduces contingencies because outputs are used as inputs; and in self-learning systems, the system itself determines how these inputs will be weighted for the next cycle, creating a kind of artificial evolution capable of proceeding in unpredictable directions. Randomness (more accurately, pseudorandomness) is produced when values are interjected into algorithms through contingent processes. For instance, one strategy takes a value from the computer's clock at a specified point, deterministic in the sense that the algorithm and clock always operate in the same ways, but the result of their contingent conjunction becomes unpredictable in a specific instance because the relation between the clock value and the algorithm is uniquely established each time the algorithm runs.

In evolutionary programming, randomness is sometimes achieved by creating variations in algorithmic structures and then testing the variants against fitness criteria to discover which performs best. Since it is not known in advance which one will succeed, this introduces unpredictability in the sense that the only way to determine the result is to run the program through hundreds or thousands of successive trials and see what emerges. An artistic implementation of this idea occurs in an evolutionary program designed by Swedish artists Johannes Heldén and Håkan Jonson to "evolve" Heldén's poetry through algorithmic variations. As with the clock example, the program consults various data sets to introduce pseudorandomness into the algorithms. In an amusingly eclectic mix, these include the "mass of exoplanetary systems detected by imaging," "GISS [Geographic Information Science and Systems] surface temperature" for various locations and dates, and "cups of coffee per episode of Twin Peaks" (Heldén and Jonson 2014, n.p.). Each time the program is run, it produces different results because of the injection of these pseudorandom values (Heldén and Jonson 2014; see Hayles 2018 for an analysis of the work).

Anticipations are prominent in biosemiotics because they establish a relation between something present and something absent, which nevertheless has real causal powers. Deacon and Wheeler point to Peirce's definition of a sign as something "which is in a relation to its object on the one hand and to an interpretant on the other, in such a way as to bring the interpretant into a relation to the object, corresponding to its own relation to the object" (Peirce 1958, 8:322). As a relational operator, a sign creates possibilities for meaning making, because it brings a behavior (the interpretant) into relation with the object, the absential phenomenon for which the sign stands. It is this relation that invests the interpretant with "aboutness," its meaningfulness in the context in which interpretation occurs. Each organism has its own semiotic niche, which in turn is coordinated with all the other niches operating in its environment. The sum total of all the semiotic niches constitutes what Hoffmeyer (1997) calls the semiosphere, the dynamic interactions of sign relations enveloping the planet.

Anticipations also operate in computer codes. A basic structure in algorithms is the "if/then" construction: if A, do B; if C, do D. Even the simplest programs, such as the "Hello World" algorithm used to introduce beginners to programming, has an implicit if/then relation with the monitor, for the message will appear only if the algorithm completes its calculation and instructs the monitor to display the specified string. More complicated programs often require a result from a subroutine in order to proceed. In some complex programs, a supervening program needs a result from a subroutine,

CAN COMPUTERS CREATE MEANINGS?                                                                67

but to save time, it anticipates the result and goes on to the next step. If the result is other than what it anticipated, it will back up and redo the calculation.

No one doubts that signs operate within computers, but the cogent objections have to do with "aboutness," the computer's ability to recognize the absential phenomena so important for biosemiotics. In brief, a computer can manipulate signs, but does it know anything about what those signs signify? The classic challenge here is John Searle's "Chinese Room" thought experiment, paraphrased as follows (originally in Searle 1984; concisely restated in Searle 1999). Imagine that a man who does not read, write, or speak Chinese sits in a room. There is slot in the door through which strings of Chinese letters are passed. The man uses a rule book and a basket of Chinese letters to compose a reply and slips it back through the door. His interlocutors are convinced that he knows Chinese, even though all his replies are incomprehensible to him. The man, of course, represents the computer, and Searle's point is that the computer achieves its results simply by matching patterns and understands nothing about what it does. There have been hundreds of replies to this challenge, including ones pointing out that the computer is not simply the man but the rule book and characters as well as the room itself; other replies imagine that an embodied robot takes the man's place, capable of connecting input and output through embodied interactions. As rebuttals have proliferated, Searle has elaborated the challenge in ways that he claims effectively answer these (see Cole 2015 for a summary of the various responses and counterresponses by Searle).

Suppose that we pose the question differently, adapting from Jakob von Uexküll (2010) the idea that every organism has an umwelt (roughly translated, a world horizon) through which it makes sense of world. Then we may ask what the computer has knowledge about in its internal milieu—in effect, what the computer's umwelt is. This shifts the ground from whether the computer "understands" as a human would to the internal states of the computer and its capacities for sensing and responding to these states. Here a brief reflection on "knowledge" in the context of biological organisms may be helpful in setting the scene. Many organisms have information about ("know") how to interpret environmental cues, but this information is not necessarily present in every part of the organism. For example, like many people, I have the ability to set my internal alarm clock to wake me at a specified time, say 5 a.m. Invariably I will wake within a minute or two of the set time, but how does my body "know" what the time is? It would not be difficult to figure this out: simply count how many times my heart beats per minute, multiply by the minutes that need to pass, count that many beats, then send a wake-up signal.

But who is counting? Obviously not my consciousness, since I am asleep. Nevertheless, something in my body accomplishes this task, no doubt with nonconscious cognition. "I" might not know, but somewhere in my body as an integrated system, this information exists.

Similarly, a computer has many interacting parts, so a better way to ask what it "knows" is to ask what information exists somewhere within the computer's integrated system, although not necessarily in every part. A computer has information about time through its internal clock, and it has information about how many clock cycles have passed for any given operation. It has information about the different layers of its codes and how they interrelate. It has information about how to interpret the algorithms it uses to solve problems, and it is able to anticipate the next steps in those algorithms and construct the proper sequences that a given problem requires. It may also have a program able to discern when it is infected with a virus, and it has information about how to cure itself by deleting the malicious software. All this implies that instead of the biological imperative to survive and reproduce, the computer is designed for certain purposes (or self-designed, for computers that have evolved on their own beyond their initial design parameters), and its "umwelt" consists of the functions, architectures, and procedures that enable these purposes to be achieved.

Returning to Searle's thought experiment, I want to make explicit the anthropocentric assumption embedded in his claim that the man "understands" nothing, an assumption underscored by figuring the interlocutor in the room as a human. Perhaps the single most important contribution of biosemiotics is its challenge to anthropocentrism and its reconceptualization of "meaning" as a response to an environmental cue that benefits the organism in some way. Wheeler writes, "meaning is always a kind of doing. The *meaning* of a sign is to be found in the *changes* . . . which it brings about" (2016a, 121). With design and purpose displacing biological imperatives, a computer achieves meaning in this behavioral sense when it processes an algorithm, reads a data set, performs the calculations indicated, and produces results that it "understands" through its anticipations, interpretations, and information flows.

Of course, this is not all that computational media do, because humans have designed them to perform useful work in the context of hybrid human-computer assemblages. Rather than a stand-alone computer, used above to explore the ground-state capacities of logic gates, contemporary computational media are networked with each other and connected to a wide variety of sensors and actuators, permeating contemporary infrastructures in ways often invisible to the humans who rely on them. As soon as a computational system includes sensors, it is exposed to the kinds of contingencies that organisms evolved to cope with; similarly, the system's algorithms must also be able to cope with uncertain or ambiguous data. The results of these computational systems are unquestionably teleodynamic—end directed toward purposes that their designs enable and that (presumably) benefit the humans interacting with it. Granted that these computational media do not understand in the same sense as do humans; nevertheless, they are capable of meaning-making practices within their umwelten, performing actions in response to internal and external cues, making interpretations, and constructing relations between what is present and what is absent through their anticipations and operating constraints. As von Uexküll (2010) emphasized, umwelten of different species may overlap, but they are never completely coincident with one another. Similarly, computational systems have information about their internal and external milieux, which overlap with but do not coincide with what the humans know. Moreover, in self-designing programs such as evolutionary computations, genetic algorithms, and neural net architectures, it is extremely difficult, if not impossible, for humans to reconstruct after the fact everything that has happened within the computational processes, so in this sense the computer "knows" more than humans.

Including more complex forms of artificial cognition into the conversation returns us to the issue of bio-techno evolution. As we have seen, the next evolutionary leap after the emergence of human cognition has been achieved through the development of artificial cognizers, communicating with each other and with humans through planet-wide networks of information flows, interpretations, and meaning-making practices. This strategy has resulted in a corresponding explosion in sign exchanges, extending the scope from the semiosphere, used to denote sign relations in the biological realm, to the cognisphere, which includes all these as well as all the sign relations created and communicated through computational media (see Hayles 2006 for a more extended discussion). To explore this development, I turn now to cognitive assemblages and their implications for our human and nonhuman futures.

## Cognitive Assemblages: Technosymbiotic Perspectives

As I have argued (Hayles 2017), cognitive assemblages are everywhere in our contemporary world, from satellite imaging software to railroad switching controllers to electrical grid components and operators. Perhaps the most obvious and visible is the internet and the web it hosts, facilitating worldwide traffic of enormous reach and complexity. Assuming that only the human participants in these assemblages are capable of meaning-making practices is as erroneous and anthropocentric as believing that the only species in the biosphere capable of making meanings is *Homo sapiens*, a viewpoint that has become not only untenable but dangerously skewed in its implicit acceptance of human domination. Urgently needed are alternate perspectives that recognize the contributions of other species to our planetary semiosphere, as well as theoretical frameworks that underscore the importance of cognitive media in creating the meanings that guide hybrid human-technical action, perception, and decision making in the contemporary world horizon of the cognisphere.[6]

The evidence supporting this claim for the meaning-making capabilities of cognitive media is pervasive. Choosing one example out of thousands, I instance this proposal for a cognitive vision system for robotic docking with a free-flying space satellite. Here is the authors' description:

> Our system functions as follows: (Step 1) captured images are processed to estimate the current position and orientation of the satellite. (Step 2) behavior-based perception and memory units use contextual information to construct a symbolic description of the scene. (Step 3) the cognitive module uses knowledge about scene dynamics encoded using *situation calculus* to construct a scene interpretation, and finally (Step 4) the cognitive module formulates a plan to achieve the current goal. The scene interpretation constructed in Step 3 provides a mechanism to verify the findings of the vision system. The ability to plan allows the system to handle unforeseen situations. (Qureshi et al. 2005, 1)

Interpretation, cognition, knowledge, planning, and flexibility characterize the design parameters for this system, created, implemented, and overseen by humans but with the autonomy necessary to operate on its own in a space environment. The researchers make clear that this system does have an umwelt, a cognitive horizon encompassing the perceptions, actions, and anticipations created by its sensors, actuators, and cognizing modules. Within its umwelt, it acts as a teleodynamical system with the end-directed goal of successful docking with a free-flying satellite. Within the larger human-technical assemblage, it serves the purpose of protecting astronauts from hazardous forays outside spacecraft required by manual docking, or conversely from the risks of automated docking that follow detailed control scripts, which are error prone and inflexible, without the ability to cope with unexpected developments.

As Bruno Latour (2002) and Peter-Paul Verbeek (2011) have argued, such cognitive systems are much more than neutral tools, for they alter the horizon of what can be known and thereby have far-reaching consequences for social, cultural, and economic practices. Louise Amoore (2016) highlights these implications in her work on cloud computing, especially ICITE (pronounced "eyesight"), the Amazon Web Services $600 million contract for data security and intelligence infrastructure. She writes, "the cloud promises to transform not only what kinds of data can be stored, where, and by whom, but most significantly what can be discovered and analyzed of the world. The cloud's capacity to extend 'big data' to 'infinite data' opens new spaces of what I have elsewhere called the politics of possibility, where security practices act upon possible horizons" (Amoore 2016, 7). Among ICITE's platforms is Digital Reasoning software, a machine-learning program for "analyzing and deriving meaning from information." She quotes from the software's description, which touts the program's ability to extract "value from complex, often opaque data . . . [empowering] the analyst with advanced situational awareness, enhancing cognitive clarity for decision-making" (Amoore 2016, 13). As she demonstrates, such programs "make possible the distributed analysis of big data across data forms," for example, human language in emails and social media sites. The upshot, as one analyst she cites observed, is that "it allows us to say that correlation is enough" (Amoore 2016, 14)—enough, that is, for actionable intelligence in a variety of surveillance contexts, ranging from detaining "persons of interest" suspected of terrorism to spotting insider trading in the financial industry.

This example, and many others, raise troubling questions about the ethics of employing such hybrid human-technical systems when they threaten long-standing traditional values, such as the right to be free from surveillance and to opt out of having one's data analyzed without one's permission or knowledge. In her book *Cloud Ethics: Algorithms and the Attributes of Ourselves and Others* (2021; the article cited above appears here as a chapter), Louise Amoore explores the ways in which algorithms function to supply the flexible norms that govern such activities as who can pass a border crossing and who cannot. Wendy Hui Kyong Chun in *Discriminating Data: Correlation, Neighborhoods, and the New Politics of Recognition* (2021) explores the relation between data collection and storage and practices of discrimination; Antoinette Rouvroy (2012) broadens the scope to critique what she calls algorithmic governmentality in "The End(s) of Critique: Data-Behaviourism vs. Due Process"; Erich Hörl in *General Ecology: The New Ecological Paradigm* (2017) links these concerns with environmentalism. Following a similar line of thought to that of Amoore, he writes that if "being is relation" (Hörl 2017, 122), then relation encompasses all the incredible variety and nuances of human interactions as well as our interactions with environments, nonhuman lifeforms, and cognitive media. The idea that this complexity could be adequately captured as a series of identifiable attributes, each with its own correlative probability, is a dream (or better, an illusion) of control that any simple walk in a park could disprove. To tackle these complex issues, we must recognize that computational media do act as ethical agents in our contemporary world, although the locus for ethical responsibility remains with the humans who design, implement, and maintain the systems. What we can no longer afford is the illusion that computational media simply perform calculations, devoid of interpretations, anticipations, and meaning making.[7]

It seems obvious to me, and to many who write about our human futures, that (short of environmental collapse or catastrophic global war) our journey into symbiosis with computational media, which has already begun, is likely to accelerate in the coming decades. Stuart Kauffman has written about the "adjacent possibles" (2000, 22) within biological evolution. When there is a very large possibility space, he argues, the relevant issue is not the space's theoretical size but the strong likelihood that pathways adjacent to the existing one will be followed, giving a teleological impetus to evolutionary developments. Through this reasoning, he demonstrates that it was not a statistical freak that life evolved on Earth but rather a probable outcome produced by following a series of adjacent possibles. Given where we are now, the adjacent possibles include increasing use of intelligence augmentation, in which the cognitive capacities of humans are enlarged and enhanced by computational media as in the examples above, which will sit side by side with the development of more powerful, flexible, and pervasive forms of artificial intelligence.

Among these are neural net architectures and deep learning algorithms, already mentioned above as programs that make extensive use of recursive dynamics. The potential of neural net architecture is exemplified in the development of AlphaGo by DeepMind (recently acquired by Google; DeepMind 2017). I previously have discussed this instance:

> Go is considered a more "intuitive" game than chess because the possible combinations of moves are exponentially greater. Employing neural net architecture, AlphaGo used cascading levels of input/output feedback, training on human-played games. It became progressively better until it beat the human Go champions, Lee Sedol in 2016 and Ke Jie in 2017. Now DeepMind has developed a new version that "learns from scratch," AlphaGoZero, [which uses no human input about previously played games, limiting the input to only] the basic rules of the game. Combining neural net architecture with a powerful search algorithm, AlphaGoZero played against itself and learned strategies through trial and error. At three hours, AlphaGoZero was at the level of a beginning player, focusing on immediate advances rather than long-term strategies; at 19 hours, it had advanced to an intermediate level, able to evolve

and pursue long-term goals; and at 70 hours, it was playing at a superhuman level, able to beat AlphaGo 100 games to 0, and arguably becoming the best Go player on the planet. (Hayles 2018)

There are utopian as well as dystopian possibilities in developments like these. Our best hope for navigating these tricky waters is to start with a clear-eyed view of how human umwelten overlap with and differ from those of the nonhuman others, including biological organisms and computational media. Understanding computational media as capable of meaning-making practices contributes to a philosophy of ecological relationality because it breaks the illusion that only humans are able to create, disseminate, and understand meaning (perpetuating this illusion was precisely the point of Searle's "Chinese Room"). Moreover, it emphasizes that computational media are involved in intense feedback loops with humans, affecting everything from how human cognitions perform in the world to the (infra)structures of human sociality, economics, politics, and much more.

The cognisphere in which we all interact grows ever more dense and complex, a sure sign that micro-evo-techno evolution continues to accelerate toward an unknown future. It is up to us to create, recognize, and interpret the sign exchanges that will enable all of us cognizers, human and nonhuman, to thrive and flourish. Adopting a philosophy of ecological relationality is a step toward realizing those kinds of futures.

# 3

# The Emergence of Technosymbiosis and Gaia Theory

Technosymbiosis, like most other theoretical constructs, builds on previous theories even as it suggests innovations and changes. As the name implies, it specifically builds on Margulis's idea of symbiosis as a primary evolutionary driver for life on Earth. As we saw in chapter 1, there are a number of unresolved tensions between symbiosis, autopoiesis, and systems theory, including where (or if) to make the cut between system and environment, whether to characterize organisms as individuals or as symbionts, whether organisms have operational closure or not, and whether objects created by technics can be considered to evolve or not. This chapter returns to these questions to place them in fuller context and to show which characteristics technosymbiosis carries over predecessor theories and which it rejects, and why.

Among the researchers explored in this chapter are James Lovelock, inventor of the Gaia hypothesis; Humberto Maturana and coauthor Francisco Varela, who articulated the concept of autopoiesis; Lynn Margulis and her frequent coauthor Dorion Sagan, who focused on symbiosis, expanding the concept to include symbiogenesis; Bruce Clarke and his articulation of neo-cybernetic systems theory (NTS); the reference frame theory (RFT) of Chris Fields and Michael Levin (2020), and of course my own integrated cognitive framework, ICF. Although biosemiotics is not generally evoked in the discussions, it appears as a player near the chapter's end. The chief revisions are a broader view of cognition that links it with meaning and also with computational media. This view emphasizes how organisms transduce information rather than asserting that no information flows between an organism and its environment, which is a more dynamic and evolutionary-friendly view of organisms interacting with their environment, facilitating a return to seeing Earth systems as dynamically interacting rather than assimilating them into a single entity named Gaia. The aim is to create a stronger, more coherent framework that makes computational media and AI not a late add-on sitting uneasily on top of a biological theory but an integral part of the emerging worldview of integrated cognition.

### Lynn Margulis, Brilliant Rebel

Arguably no one has done more to argue for, extend, and develop the concept of biological symbiosis than Lynn Margulis. Starting from her seminal 1967 paper "On the Origin of Mitosing Cells," Margulis argued that cell organelles such as mitochondria and chloroplasts were the remnants of previously independently living organisms that had been absorbed by the cell, were only partially digested, and eventually merged so that they became part of the cell proper. Her persistence in the face of stiff opposition is remarkable. The "Origin" paper, for example, was rejected some seventeen times before it was accepted, and one of her grant applications received the comment, "Your research is crap. Do not bother to apply again" (see "Lynn Margulis" 2021). Nevertheless, her work on microbial symbiosis and evolution was eventually recognized as largely correct. Her ideas about the origin of mitosing cells received strong confirmation when genetic testing determined that mitochondria have different DNA from that of the cell's nucleus and reproduce on a different time schedule. Since mitochondria are crucial factors in the cell's respiration, producing ATP (adenosine triphosphate), which converts glucose into energy for the cell's metabolism, this endosymbiosis was extraordinarily important in the evolution of life on Earth, for it was responsible for the emergence of eukaryotic cells (cells with a nucleus) from the predecessor prokaryotic cells. "Without mitochondria," she wrote with coauthor Dorion Sagan, "the nucleated cell, and hence the plant or animal, cannot utilize oxygen and thus cannot live" (Margulis and Sagan 1997, 31).

Moreover, the presence of atmospheric oxygen (dubbed the Great Oxidation Event) was itself the result of microbial activity, as was photosynthesis, which Margulis and Sagan call "the most important single metabolic innovation in the history of life on the planet" (Margulis and Sagan 1997, 79).

They explain, "the first photosynthetic organisms were bacteria that used hydrogen gas or hydrogen sulfide for the process and never produced oxygen" (Margulis and Sagan 1997, 79). Originally these bacteria took hydrogen directly from the atmosphere, but as it was depleted, "more photosynthetic bacteria used hydrogen sulfide produced as waste by fermenting and sulfide-breathing microbes. Photosynthetic bacteria—then as now—used light energy to cleave off the hydrogen molecules. They excreted unused yellow pellets of congealed sulfur" (79). But as the purple and green photosynthetic microbes grew "frantic for hydrogen," they "discovered the ultimate resource, water, and its use led to the ultimate toxic waste, oxygen" (99). Margulis and Sagan observe that "this single metabolic change in tiny bacteria had major implications for the future history of all life on Earth" (101) because it led to the emergence of macro lifeforms capable of using oxygen such as plants and animals. In *Acquiring Genomes*, they reiterate the point. "The creative force of symbiosis produced eukaryotic cells from bacterial. Hence all larger organisms . . . originated symbiogenetically" (Margulis and Sagan 2002, 56). They assure readers that the "creation of novelty by symbiosis did not end with the evolution of the earliest nucleated cells. Symbiosis still is everywhere" (56). As these passages indicate, Margulis went beyond her initial observations to argue that, rather than random mutation, symbiosis, in its species-producing form as symbiogenesis, was the primary driver of evolutionary changes.[1]

Brilliant as this work was and powerful as it proved to be in changing perspectives so that the microcosm was accepted as driving evolutionary changes in the early history of life on Earth, there remained in Margulis's theories a gap between what she could demonstrate in the microbial world and what she claimed about the evolution of visible (macro) life. "No evidence in the vast literature of heredity change shows unambiguous evidence that random mutation itself, even with geographical isolation of populations, leads to speciation," she and coauthor Sagan claimed (Margulis and Sagan 2002, 29). They acknowledge that "the major difference between our view and the standard neodarwinist doctrine today concerns the importance of random mutation in evolution" (11). Despite the prevailing consensus to the contrary, they doubled down on their assertions, claiming that "mutation accumulation does not lead to new species or even to new organs or new tissues," and asserting that "99.9 percent of the mutations [as shown in experiments with fruit flies] are deleterious" (11–12). Of course, this argument does not take into account that over evolutionary time scales, even .1 percent can result in significant changes.

With customary bravado, Margulis invited Ernst Mayr (whose work she cites and also criticizes) to write the foreword. Mayr evidently found her assertions about macro lifeforms too much to take. Pursuing the courteous route of attributing misunderstanding to the reader rather than to the authors, he writes, "Some of their statements might lead an uninformed reader to the erroneous conclusion that speciation is always due to symbiogenesis. That is not the case. Speciation—the multiplication of species—and symbiogenesis are two independent, superimposed processes. There is no indication that any of the 10,000 species of birds or the 4,500 species of mammals originated by symbiogenesis" (Mayr 2002, xii–xiii). Wanting to be as generous as possible to the authors (as befitting his role as writer of the foreword), he concludes that "given the authors' dedication to their special field, it is not surprising that they sometimes arrive at interpretations others of us find arguable. Let the readers ignore those that are clearly in conflict with the findings of modern biology. Let him concentrate instead on the authors' brilliant new interpretations and be thankful that they have called our attention to worlds of life that, despite their importance in the household of nature, are consistently neglected by most biologists" (xiii–xiv). Good advice, which I intend to take.

### James Lovelock, Cybernetics, and Homeostasis

When Margulis discovered the work of James Lovelock, after some initial hesitations she saw that their collaboration could bring significant benefits to both partners, as did Lovelock as well. While Lovelock's analysis was focused mostly on interactions between organisms and the environment, Margulis offered a much richer and temporally deeper account of how microorganisms evolved over time, through interactions both with each other and with their environments. Lovelock, for his part, provided a larger sense of the environment, including the oceans and atmosphere, and thus provided a means by which Margulis's arguments about symbiosis could be extended beyond microorganisms to the planet as a whole. Increasingly, Margulis began to write about coupled systems, which culminated in the Gaia hypothesis. For example, in "Big Trouble in Biology" (Margulis 1997), she remarks that "of all the organisms on Earth today, only prokaryotes (bacteria) are individuals. All other living beings ('organisms'—such as animals, plants and fungi) are metabolically complex communities of a multitude of tightly organized beings" (273).

The tensions between their different approaches did not entirely dissipate, but after about 1980, Margulis consistently referenced Lovelock's work, and Lovelock returned the favor. In *Gaian Systems: Lynn Margulis, Neocybernetics, and the End of the Anthropocene* (2020), Bruce Clarke has extensively traced the development of their collaboration and has suggested how it can be integrated with neocybernetic systems theory (NST). He has also edited the scientific correspondence between Lovelock and Margulis in *Writing Gaia: The Scientific Correspondence of James Lovelock and Lynn Margulis* (Clarke 2022), an importance resource for understanding the nuances of their relationship. My own approach is to explore the tensions in the theory with a view to showing how technosymbiosis can revise, strengthen, and add to the existing theory. I propose taking Gaia theory in a different direction by interrogating the points where fractures appear as concepts were developed and amalgamated.

Independently of Margulis, Lovelock had developed the idea of Earth as a single self-regulating system in which organisms coupled with their environments through innumerable interactions, changing their environments even as they were changed by them. Following a suggestion by his neighbor, novelist William Golding, Lovelock named the Earth system Gaia, the Greek name for the Mother Goddess Earth. In two books *Gaia: A New Look at Life on Earth* (2016) and *The Ages of Gaia* (1988), Lovelock proposed three major pieces of evidence for his Gaia hypothesis: (1) the oxygen level in the atmosphere has remained relatively constant, despite being far from thermodynamic equilibrium (because oxygen is a reactive gas); (2) the salinity of the ocean has remained relatively constant, whereas evaporation and other factors should cause the salinity continuously to rise; and (3) the temperature of the Earth has remained within a range consistent with life on Earth, despite the Sun having grown larger and hotter over the three billion years of life's existence on Earth. These anomalies could exist, he argued, only because organisms were interacting with their environments so as to stabilize conditions such that they remained consistently within the parameters where life was possible.

The choice of Gaia as a name for the Earth systems was, and remains, controversial. Margulis herself had misgivings about the appropriation, although she grew to accept it. The naming of Gaia introduced complications for both Lovelock and Margulis. For example, calling the integrated Earth systems Gaia obligated Lovelock to clarify that he did not intend thereby to imply intentionality. Although his original formulations sometimes implied a teleological impetus of the Earth systems to preserve life, he eventually disowned this implication as well, clarifying that Gaia consisted of innumerable subsystems, each following its own directive to continue existing.[2] In contrast to intentionality and teleology, an *intended* effect was to imply that the critical zones of Earth inhabitable by lifeforms could be regarded as a single entity. This obviously had the advantage of emphasizing interactions between the Earth systems, but it had the important disadvantage of encouraging unwarranted inferences, for example being seduced by the siren song of misplaced concreteness. Bruno Latour, generally very sympathetic to Margulis and Lovelock's Gaia project, warns against just this kind of move, specifically against conceptualizing Gaia as a superorganism. "No matter how tempting it is to lump all life forms into one huge, unified, and continuous biosphere of some sort, or to invoke a superorganism, any idea of a giant composite planetary body should be resisted as much as the myth of the machine. Those who project onto Gaia the image of a global body, or even worse that of a female body, simplify too much Lovelock and Margulis's common project" (Latour and Lenton 2019, 7).[3] A case in point is the way that Bruce Clarke handles Gaia in his excellent book *Gaian Systems*. He knows perfectly well that "placing the emphasis on 'the oneness of Gaia' has tended to undervalue the complexity of Gaia's planetary aggregation and to blur the manifold of elemental cycles and ecological subsystems needed to buffer the operations of the 'whole system'" (Clarke 2020, 17). Nevertheless, he finds it irresistible to follow Margulis when she talks about planetary "sentience," which Clarke interprets as "planetary cognition."

Planetary cognition would require a planetary cognizer, but there is no such thing, only innumerable individual cognizers, each following its own dictates to continue to exist. Moreover, these cognizers do not operate to a single effect, nor do they cooperate to achieve unitary goals. As a collective, they also do not have a single enclosing membrane, the requirement for all lifeforms to exist, from the lipid surface of a cell's membrane to an elephant's wrinkled skin (unless the planet's atmosphere is considered such, which stretches the meaning of membrane beyond recognition). When Clarke attempts to rescue the idea of "planetary cognition" by referring to Niklas Luhmann's observation that "the world itself, as [the] co-occurring other side of all meaning forms, remains unobservable" (Luhmann quoted in Clarke 2020, 43), he asserts that "Gaia per se is not a *meaning* system; its planetary cognition does not make 'sense.' Rather, planetary cognitions maintain the conditions through which Gaia constructs its own continuation, and thus, the possible continuations of its systemic elements" (Clarke 2020, 43). The slippage here between the plural and singular ("planetary cognitions" versus "Gaia") registers the fracture between Gaia conceived as a single entity and Gaia as the name of an dynamic collectivity of competing and cooperating organisms interacting with each other and their environments.

Although the thrust of Lovelock's argument was to position biological organisms and environments as a single interacting system, with organisms modifying their environments so as to favor their own continuation, he nevertheless recognized that the catalyzing force of biota, not physicochemical processes in themselves, was what brought about the changes. In fact, this was precisely his point in comparing the lifeless planet of Mars with the life-filled Earth. The atmosphere of Mars was at thermodynamic equilibrium, whereas the atmosphere of Earth was not, a contrast that could be explained, he argued, only by the active interventions of lifeforms in stabilizing their own environmental conditions.

In my lexicon, this difference authorizes a distinction between actors and agents. While material processes undoubtedly possess agency and can bring about enormous and even violent changes (e.g., volcanic eruptions), their trajectories are explainable within the realms of physics and chemistry, through what Longo, Montévil, and Kauffman call "entailing laws" (2012, 1). Biological systems, by contrast, swerve from predictable trajectories through actions such as symbiogenesis that cannot be predicted in the same way. Because of this unpredictability, I designate biological lifeforms as actors, which implies selections, unpredictable evolutions, and capacities to sense and respond to environmental conditions. Margulis agrees that lifeforms must be able to sense their environments, which is a precondition for performing actions: "All living beings, not just animals but plants and microorganisms, perceive. To survive, an organic being must perceive—it must seek, or at least recognize, food and avoid environmental danger" (Margulis and Sagan 1995, 32). I will return to this point later when explaining the reference frames used by Fields and Levin (2020) to analyze how organisms create meaning. For the moment, I note that while introducing the actor/agent distinction does not actually contradict Lovelock's Gaia hypothesis, it does change the emphasis by assigning different roles to biological organisms as actors in contrast to material processes as agents.

In *Gaia: A New Look at Life on Earth*, Lovelock indulges in a science fiction scenario that vividly portrays the powerful effects of biological actors. He imagines that a Dr. Eeger develops a "motile micro-organism with a capacity to gather phosphate from soil far more efficiently" than other organisms (Lovelock 2016, 39). Although Eeger has the best intentions of increasing crop yields, the organism eludes his control, and "within six months more than half of the oceans and most of the land surfaces were covered with a thick green slime which fed voraciously on the dead trees and animal life decaying beneath it" (40). Reminding us that the story is a fiction, Lovelock observes that its plausibility depends on the organism being able to "exert its aggression without check or hindrance" (41). His point is to suggest that "our continuing orderly existence over so long a period can be attributed to yet another Gaian regulatory process, which makes sure that cheats can never become dominant" (43).[4] Clarke interprets this tendency as "planetary immunity" and links it with the theories of Roberto Esposito (Clarke 2020, 213–42). In Lovelock's later books on the climate crisis, *The Revenge of Gaia: Earth's Climate in Crisis and the Fate of Humanity* (2006) and *The Vanishing Face of Gaia: A Final Warning* (2009), he warns that Earth's immunitarian tendencies may be insufficient to keep humans from destroying themselves, although he, like Margulis, remains confident that life itself will continue, if only in the microcosm.

Lovelock's recognition of the severity of the climate crisis demonstrates how far he has come from his early emphasis on homeostasis and first-order cybernetics in *Gaia*. Very much in the vein of control engineering, in *Gaia* he argues that "one of the main characteristic properties of all living organisms . . . is their capacity to develop, operate, and maintain systems which set a goal and then strive to achieve it through the cybernetic process of trial and error" (Lovelock 2016, 45). In this construction the influence of first-order cybernetics is clear, with flows of information, feedback loops both negative and positive, and regulative mechanisms looming large. He proclaims that information is "an inherent and essential part of control systems [and] in another sense, that of memory. They must have the capacity to store, recall, and compare information at any time, so that they may correct errors and never lose sight of their goal" (57). He never altogether gives up this emphasis, and its primacy in his thought marks a growing difference between his perspective and that of Margulis. After about 1980, she increasingly turned to the autopoietic theories of Maturana and Varela as defining the distinctive qualities of the living. Lovelock, for his part, continued to believe that the "only difference between non-living and living systems is in the scale of their intricacy, a distinction which fades all the time as the complexity and capacity of automated systems continue to evolve" (Lovelock 2016, 58). This attitude explains why, in his last book (published when he was ninety-nine!), *Novacene: The Coming Age of Hyperintelligence* (2020), he predicted that AIs would be the likely successors to humans.

### Fractures in Autopoiesis: Amalgamating Circularity with the Living

Following a different trajectory, Margulis made autopoiesis a central part of her theory after about 1980, acknowledging her indebtedness to Maturana and Varela's seminal work *Autopoiesis and Cognition: The Realization of the Living* ([1972] 1980). A central premise of Maturana's development of autopoietic theory is that all organisms, from cells up to *Homo sapiens*, have the ability to self-organize their operations and self-reproduce their organizations.[5] The theory's circularity marks its affinity with the recursivity characteristic of second-order cybernetics: the organization produces the parts, and the parts produce the organization. The operational closure implicit in this view is emphasized by Maturana and Varela's assertion that no information passes from the environment to an organism. As we saw in chapter 1, in their view an environmental stimulus merely acts as a "trigger" for an organism's actions (Maturana and Varela [1972] 1980, 95). The nature of those actions, and indeed even the selection of what kinds of information have the ability to serve as triggers, are determined only and always by an organism's own organization. For example, the apparent causality of a rabbit fleeing because it sees a fox exists only in the mind of an observer, who puts the two actions together to form a causal link. For Maturana and Varela, the focus is entirely on biological processes—what actually happens in the organism as distinct from inferences that an observer might draw from his or her observations.

No doubt a major influence in Maturana's development of this position was a 1959 paper on which he was the second coauthor, "What the Frog's Eye Tells the Frog's Brain." This highly influential paper (which has the distinction of being among the most frequently cited ever, according to *Science Citation Index*) had as additional coauthors J. Y. (Jerry) Lettvin from MIT as first author, Warren McCulloch as the "senior member, IRE" (Institute of Radio Engineers), and the brilliant Walter Pitts as the final contributor.[6] Discussing the specificity of the frog's visual processing, the authors note that "the frog does not seem to see, or at any rate is not concerned with the detail of stationary parts of the world around him. He will starve to death surrounded by food if it is not moving" (Lettvin et al. 1959, 1940). Focusing on the four different kinds of nerve fibers in a frog's eye, they determined that the different fibers register correspondingly diverse aspects of the environment such as convexity detectors and moving edge detectors. The researchers summarized their conclusion as follows. "What are the consequences of this work? Fundamentally, it shows that the eye speaks to the brain in a language already highly organized and interpreted, instead of transmitting some more or less accurate copy of the distribution of light on the receptors" (Lettvin et al. 1959, 1950).

There are tantalizing hints in the article of discussions that the research team must have had. For example, in discussing their conclusions, they remark that "the operations thus have much more the flavor of perception than of sensation if that distinction has any meaning now. That is to say that the language in which they are best described is the language of complex abstractions from the visual image. We have been tempted, for example, to call the convexity detectors 'bug perceivers.' Such a fiber . . . responds best when a dark object, smaller than a receptive field, enters the field, stops, and moves about intermittently thereafter" (Lettvin et al. 1959, 1951). This is precisely the kind of causal connection that Maturana, when he returned to his home in Santiago and the University of Chile, would disavow. We can speculate that the published phrasing, "we have been tempted," was inserted because Maturana even then objected to this inference, so it remains a "temptation" articulated in a provisional rather than a conclusive tone. Similarly, at the article's beginning the authors declare that "this work has been done on the frog, and our interpretation applies only to the frog" (Lettvin et al. 1959, 1940), a caution that Maturana would throw to the winds when he generalized the findings to ground his theory of autopoiesis.

The circularity of autopoietic theory is at once a strong point in accounting for the distinctiveness of living organisms and a weak point in terms of drawing connections with evolutionary processes, including Lovelock's argument that organisms and environments constitute a single system. By insisting that no information passes from the environment to the organism, the theory downplays, and indeed comes close to dismissing, the active role that the environment exercises through its interactions with an organism. The idea that no causal connection exists between the actions of organisms and environmental conditions makes it difficult to see how they could operate as a single interactive system, since strictly speaking, according to autopoietic theory there *are* no interactions between them.

In saying that the environment acts as a "trigger" for an organism's actions, the theory employs a metaphor whose literal meaning contradicts the implication they want to draw, since firing the trigger of a gun unmistakably initiates a causal chain leading to what happens as a result. In my view, it would make more sense to say that an organism *transduces* the information that it receives from the environment. This formulation acknowledges the crucial point made by "What the Frog's Eye Tells" (Lettvin et al. 1959), that sensory organs do not merely passively convey information but actively transform it according to evolutionary processes that emerged through the organism's adaptation to specific contexts and environmental conditions. It is likely that the transduction possibility occurred to them, but Maturana was unequivocal in continuing to insist that there is an epistemological barrier insulating the organism from its environment, and indeed from any goal-directed activity. "A living system is not a goal-directed system; it is, like the nervous system, a stable, state-determined and strictly deterministic system closed on itself and modulated by interactions not specified by its conduct. These modulations, however, are apparent as modulations only for the observer who beholds the organism or nervous system externally, from his own conceptual (*descriptive*) perspective, as lying in an environment and as elements in his domain of interactions" (Maturana and Varela [1972] 1980, 50). Thus the idea that an environment could impinge on an organism is phrased as a "modulation," which sounds like it could have some kind of effect, but that supposition is then withdrawn by asserting that it can be perceived as such only by an external observer.

However, it is far from clear that activity in organisms with nervous systems and brains constitute a "stable, state-determined . . . system" (Maturana and Varela [1972] 1980, 50). Walter J. Freedman (1992), for example, has shown that nervous activity in the olfactory bulbs of rabbits has regions of chaotic dynamics, which are not "state-determined" because their trajectories within phase space are unpredictable. Moreover, Varela in *The Embodied Mind:*

*Cognitive Science and Human Experience* (Varela et al. 1991), along with coauthors Evan Thompson and Eleanor Rosch, develops the concept of "enaction," which sees the active engagement of an organism with its environment as open-ended and transformative, and indeed crucial for the organism's development. Ironically, it would be more appropriate to call a computer with von Neumann architecture a "stable state-determined . . . system" (Maturana and Varela [1972] 1980, 50) rather than so designating the nervous activity of a rabbit—or a human.

In contrast to Maturana's unwavering commitment to the idea that no information passes from the environment to the organism, Varela was later willing to concede that information, coding, and messages might be "valid explanatory terms" through which to construct a complementary account of an organism's actions (Varela 1981, 39). Even so he continued to give top priority to autopoiesis, objecting that information and information processing should not be in "the same category as matter and energy." Transporting these ideas from engineering fields to descriptions of living systems, he maintained, was an epistemological "blunder" (45). "To assume in these fields [of natural systems] that information is some *thing* that is transmitted, that symbols are *things* that can be taken at face value, or that purposes and goals are made clear by the systems themselves is all, it seems to me, nonsense. . . . Information, *sensu strictu*, does not exist. Nor do, by the way, the laws of nature" (45). He may be right about goals, purposes, and the laws of nature (as Bruno Latour has famously argued in *We Have Never Been Modern* [1993, 45]), but from a contemporary viewpoint, he is not correct in saying "information . . . does not exist." Indeed, some physicists declare that information is a more fundamental category in explaining how the universe works than is either matter or energy (see, e.g., Zuse 1969; Fredkin 2003; Wolfram 2002). Even Lovelock, in his last book, *Novacene*, suggests that "information may indeed be the basis of the cosmos," speculating that in the future his view may be validated, "that the bit is the fundamental particle from which the universe was formed" (Lovelock 2020, 88–89).

We can assess Varela's belief about information from a quick survey of the categories at issue. Seen from the bird's-eye view, matter must have been quantified very early in human history, certainly by the time of agriculture's advent, when it became necessary to measure grain, for example, so that one could determine crop yields and create commensurate values with other commodities. In the Newtonian era, the distinction between mass and weight emerged. Energy, for its part, was the focus for the great thermodynamicists of the nineteenth century; they quantified it; codified the differences between free energy, potential energy, and energy dissipation; and formulated the three laws of thermodynamics. As matter and energy were quantified and their properties explored, they became progressively to be accepted as things that were "real" and even obvious, a perception strengthened when Einstein showed that they were interconvertible. It was only in the twentieth century, however, that information was quantified and subsequently provided the basis for the development of electronics and computational media. Thus it may be an accident of historical contingency that matter and energy seem more "real" than information, rather than an accurate account of their relative abstractness.

The fundamental nature of information has emerged in new contexts with speculations about the nature of dark matter and dark energy. In 2019, the physicist Melvin Vopson of the University of Portsmouth proposed that information is equivalent to mass and therefore to energy, existing as a separate state of matter, a conjecture known as the mass-energy-information equivalence principle (Vopson 2019). This would mean that every bit of information has a finite and quantifiable mass and energy equivalent. More recently, Michael Paul Gough has proposed that information energy is crucial to understanding the role that dark matter and dark energy play in the expansion of the cosmos, leading to the prediction that the universe will eventually stop expanding (Gough 2022). My point here is not to endorse these conjectures, which range far beyond my expertise, but rather to indicate that it is by no means accepted within the physics community that information "does not exist."[7] We may also note the commonsense observation that the entire information infrastructure of contemporary society indicates that information does indeed exist, and moreover that it has become a concept more powerful and far-reaching than autopoiesis ever was. All this is a further indication that it would be better to say that information from the environment is transduced by an organism's autopoietic organization rather than asserting that no information flow exists between the environment and organism.

### Autopoiesis and Evolution: The Fractures Deepen

Another fracture in autopoietic theory appears with the phenomena of evolution. If an organism's organization continually reproduces itself, how does change happen? It is not a coincidence that there is little to no discussion of evolution in *Autopoiesis and Cognition*. A primary motive for its publication was to counter the narratives of Neo-Darwinism with a different account focused on the distinctive qualities of living organisms located in actual cellular processes rather than abstractions such as the "selfish gene" and "fitness criteria." "Reproduction and evolution are not essential for living organisms,"

Maturana and Varela assert in *Autopoiesis and Cognition* ([1972] 1980, 11). In a retrospective account, Varela made clear that he and Maturana were consciously aware of wanting to provide an account that would not depend in any of its essentials on the idea of a genetic code (Varela 1981, 36).

Although Varela later had second thoughts about this approach, Maturana remained committed to his position. In the 1980 article "Autopoiesis: Reproduction, Heredity, and Evolution," Maturana wrote, "I claim that *nucleic acids do not determine hereditary and genetic phenomena in living systems*, and that they are involved in them, like all other cellular components, according to the particular manner in which they integrate the structure of the living cell and participate in the realization of its autopoiesis" (Maturana 1980, 61). Granted that Maturana here is responding to the exaggerated importance assigned to DNA by such Neo-Darwinists as Richard Dawkins, Maturana's move of relegating DNA to being merely one of many cellular processes with no determinate role in heredity is arguably just as distorting in the opposite direction.

In *The Tree of Knowledge*, written as an introduction to autopoietic theory for a general audience, Maturana and Varela (1987) attempt to suture autopoiesis to evolution by insisting that at every point as organisms evolve, autopoiesis is conserved. Rejecting the customary term "genetic drift" (denoting variation in the relative frequency of genotypes in a population), they substitute natural drift, which apparently also refers to a change in genotype frequency without mentioning genetics. Later the term drifts (so to speak) to structural drift. However, if an organism's structure drifts and thereby changes, how is autopoiesis conserved? Here they return to the difference between structure and organization that they employed in *Autopoiesis and Cognition*: "*Organization* denotes those relations that must exist among the components of a system for it to be a member of a specific class. *Structure* denotes the components and relations that actually constitute a particular unity and makes its organization real" (Maturana and Varela [1972] 1980, 47).

If organization is conserved at every point (implicit in the idea that autopoiesis is conserved), then how does any organism ever change to become something else? In critiquing these ideas in *How We Became Posthuman*, I offered this analogy. "Consider the case of an amoeba and a human. Either an amoeba and a human have the same organization, which would make them members of the same class, in which case evolutionary lineages disappear because all living systems have the same organization, or else an amoeba and a human have different organizations, in which case organization—and hence autopoiesis—must not have been conserved somewhere (or in many places) along the line. The dilemma reveals the tension between the conservative circularity of autopoiesis and the linear thrust of evolution. Either organization is conserved and evolutionary change is effaced, or organization changes and autopoiesis is effaced" (Hayles 1999, 152). If change happens by random mutation combined with selective fitness pressures, one could argue that a newborn mutant might have an organization different from that of its parents, but nevertheless is able to survive because its autopoiesis is conserved. In this case, one could say that autopoiesis is conserved *within a given organism*, but not necessarily between generations of the same species. Maturana and Varela, however, offer no such compromise position. In my view, autopoietic theory serves the valuable function of identifying what living systems must possess to be alive, but by itself it is inadequate to account both for the dynamic vitality of interactions between organisms and their environments, and for the changes that emerge in organisms through their interactions with their environments. This makes the fact that Margulis, an evolutionary biologist, chose to align her ideas with autopoietic theory more than a little curious. The next section explores how this tension is managed in Margulis's writing and speculates about the nature of her attraction to autopoietic theory.

## Margulis and Autopoietic Theory

No doubt one of the attractions of autopoietic theory for Margulis was its focus on cellular processes and its emphasis on autopoiesis as the defining quality of the living. As a microbiologist, Margulis felt a far deeper commitment to the differences between living and nonliving systems than did her collaborator James Lovelock. Another attraction was the inclusion of the observer, which called into question the notion of scientific objectivity as the view from nowhere. Not coincidentally, including the observer also opened scientific practices to cultural critique, for it implied that the observers could be, and often were, influenced by factors such as funding priorities and peer pressure to conform to disciplinary norms. In "Big Trouble in Biology," she made the contrast explicit: "In the autopoietic framework, everything is observed by an embedded observer; in the mechanical world, the observer is objective and stands apart from the observed" (Margulis 1997, 277).

Her description shows her awareness of the importance of framing moves, and her contrasting formulation of a mechanical versus an autopoietic worldview precisely matches the "reversible internality" concept analyzed in chapter 1. Margulis's entire body of work can be understood as *reversing the internality of microcosm-within-the-human world*, where microbial activity is largely interpreted as contributing (or not) to the health and well-being of multicellular organisms, and inverting the perspective so that *humans and other multicellular organisms are positioned as internal to the microbial world,* which in this view far exceeds multicellularity in biomass, ecological importance, and evolutionary significance. Here multicellularity is interpreted as originating from, and entirely beholden to, the unicellular organisms that were their evolutionary progenitors and that remain crucially important for the maintenance of life.

### Fractures between Margulis's Views and Autopoietic Theory

Notwithstanding Margulis's attraction to autopoietic theory, there are fractures between it and her work. The circularity that the theory emphasizes in the self-regulation and self-reproduction of cellular organization sits uneasily with her focus on symbiogenesis, with its implication of constant mutations. Moreover, the dynamics of a cellular organization calmly reproducing itself gives a picture of the microcosm far different from that articulated by Margulis and Sagan in *Microcosm* (1997) and *Acquiring Genomes* (2002), and by Margulis in *Symbiotic Planet* (1998), and in their other texts—these are worlds full of hungry, starving organisms desperate to eat or invade their neighbors and constantly bombarded by horizontal gene transfers from every direction. "Genes and only genes may pass into the recipient cell from anywhere: the water, a virus, or a donor dead or alive," Margulis writes in *Symbiotic Planet* (1998, 88). Despite these discrepancies, after about 1980 Margulis adopts the phrase "autopoietic Earth" and uses it often. What might be the likely motivation for this (conceptually uneasy) alliance?

I think that autopoiesis is attractive for her in part because it reflects her belief that the microcosm is enormously resilient. She even declares that if there were a nuclear war, the bacterial world would scarcely notice (Sagan and Margulis 1997, 239). In keeping with her internality reversal that places multicellularity within the microcosm, her writings make clear that her deepest sympathies lie not with *Homo sapiens* but with the microcosm. "Our own role in evolution is transient and expendable in the context of the rich layers of interliving beings forming the planet's surface. We may pollute the air and waters for our grandchildren and hasten our own demise, but this will exert no effect on the continuation of the microcosm" (Margulis and Sagan 1997, 67). It would be going too far to say that she really does not care if humans face extinction—as a human herself, not to mention as a mother and grandmother, self-interest alone would seem to dictate that she must care. Nevertheless, her writings are remarkably free from anthropocentric assumptions and leanings.

Moreover, she must have felt considerable sympathy for the antigenetic position staked out by Maturana and Varela. In "Big Trouble in Biology," she makes her anti-Neodarwinist attitude very clear. Contrasting "physiological autopoiesis" with "mechanistic Neo-Darwinism" (as the subtitle proclaims), she accuses mainstream biologists of engaging in a Fleckian "thought-collective" anxious to police its ideological boundaries and reject any heretic that disagrees with its tenets (Margulis 1997, 271). Noting the abstractness of such characteristics as genetic drift, random mutations, and selective fitness, she writes that "evolutionary biology" has become populated by "computer jocks (former physicists, mathematicians, electrical engineers and so forth), with no experience in field biology," with the sad result that a mechanistic worldview has replaced an attitude of deep respect for the vitality and resilience of the natural world. "Science practitioners widely believe and teach—explicitly and by inference—that life is a mechanical system fully describable by physics and chemistry. Biology, in this reductionist view, is a subfield of chemistry and physics" (266). Rejecting mainstream biologists "smitten by physicomathematics envy," she urges a return to a "life centered alternative worldview called autopoiesis, which rejects the concept of a mechanistic universe knowable by an objective observer" (266). Focusing on cellular metabolism as "the detectable manifestation of autopoiesis," she writes that "all the bacteria on Earth form a worldwide living system—a huge autopoietic entity" (269).

She contrasts this life-affirming worldview with the "neo-Darwinists [who] proffer formal mathematical explanations for the ways in which organisms evolve. Neo-Darwinism has produced a large body of professional literature that is the sacred text of most evolutionary biologists. Self-identifying neo-Darwinists control what little funding for evolutionary research exists in this Christian country. Since the seventies, learning heavily on computer simulations, the neo-Darwinist religious movement has generated subfields called population genetics, behavioral ecology, sociobiology, and population biology" (Margulis 1997, 270). Speculating on why Neo-Darwinists have "become famous darlings of the life scientists today," she writes that she attributes "their popularity in part to the soothing effects of their assertions of mathematical certainty" (270). She further points out that the "neo-Darwinist mechanistic, nonautopoietic worldview is entirely consistent with the major myths of our dominant civilization," citing William Irwin Thompson's observation that our "materialistic civilization . . . is concerned almost exclusively with technology, power, and wealth" (Thompson quoted in Margulis 1997, 270).

In light of these arguments, it is perhaps easy to see why Margulis was not too concerned with the fractures between her evolutionary emphasis on symbiogenesis and the operationally closed worldview of autopoietic theory. She may have felt that the relatively weak ties that autopoietic theory forged with evolutionary processes was more than compensated for by her own strong theories of symbiosis, including her argument that symbiogenesis was the primary driver for evolutionary change. Moreover, she may have felt, as "Big Trouble" testifies, that the real enemy at the gates was a Neo-Darwinist mechanist view, and she may have concluded that the enemy of her enemy was her friend.

Whatever the reasons for why Margulis adopted autopoietic theory and amalgamated it together with symbiogenesis, the joint was far from seamless, although the fractures do not appear to have concerned Margulis enough to prevent her from adopting an autopoietic vocabulary. The fractures widen, however, when Margulis, along with her son and frequent coauthor Dorion Sagan, speculate on the possibility that machines can also evolve. Then the restrictions of autopoietic theory become a major problem in trying to join autopoiesis with the idea of machine evolution. The widening fissures run directly through how autopoietic theory conceptualizes cognition.

## Cognition and "the Evolution of Machines"

One consequential achievement of autopoietic theory is its conception of cognition as an essential characteristic of living organisms. Defying centuries of tradition in which cognition was understood as virtually synonymous with human thought, Maturana and Varela defined cognition in much broader terms that emphasized its essential quality as a process intrinsic to, as their subtitle proclaimed, the "realization of the living": "A cognitive system is a system whose organization defines a domain of interactions in which it can act with relevance to the maintenance of itself, and the process of cognition is the actual (inductive) acting or behaving in this domain. Living systems are cognitive systems, and living as a process is a process of cognition. This statement is valid for all organisms with and without a nervous system" (Maturana and Varela [1972] 1980, 13). This definition makes very clear that all living organisms, including those without nervous systems or brains such plants and microorganisms, have cognitive capacities that are realized through the process of living, which also is the process of autopoietically producing and reproducing both the organism's distinctive organization and the components that constitute the organization. Whereas "the living" could be interpreted as a gerund, with its implication of a noun-like quality, cognition is unmistakably constructed as an activity, a "process" constantly in dynamic motion. Articulated more than forty years ago, this view of cognition is as bold as it is forward looking. It anticipates developments such as biosemiotics, the emerging field of plant cognition, and the cognitive capacities of bacteria essential to the field of genetic engineering. It remains a remarkable achievement of significant, indeed momentous, proportions.

These achievements notwithstanding, this view of cognition has the disadvantage of limiting it to processes within the living organism, rather than something that could be shared between organisms. Such a move was virtually unavoidable, given the autopoietic premise that no information flow connected the organism to its environment. It also has the disadvantage of making it difficult to extend cognitive capacities to artificial media, a problem that has become increasingly urgent as computational media are developed that clearly have cognitive capacities.

There are tantalizing hints of how these difficulties may be overcome in a brief (seven-page) article, "Gaia and the Evolution of Machines," published in *Whole Earth Review* (successor to the *Whole Earth Catalog*). There Sagan and Margulis (1987) speculate on how machines could be incorporated into an autopoietic worldview. Sagan's name appears as the first author, in contrast to the scientifically oriented texts, which always list Margulis first. It is tempting to speculate that perhaps Margulis felt less invested in the article's speculations (or conversely, Sagan felt more invested). Whatever the reason, they suggest that humans are "technologically dependent biological organisms" (Sagan and Margulis 1987, 17) and that a human is an "obligate technobe" with a "weak body entirely dependent" on machines (18). After reiterating the main ideas of autopoiesis, including its focus on "self-assembly" and "self-maintaining," they suggest that although machines do not possess these qualities, they are nevertheless "*parts* of autopoietic systems" (19). Observing that "the prowess of machines and their interdependence with human beings" is far greater today than in the nineteenth century, they argue that "from a biospheric view, machines are one of DNA's latest strategies for autopoiesis and expansion. The classification of machines as non-autopoietic does not negate the fact that they produce, and reproduce with mutation, as avidly as viruses. Like beehives, termite mounds, coral reefs, and other products of the activity of life, machines—if indirectly through DNA and RNA—make more of themselves. Through us they make other machines" (19). Moreover, when machines mutate, they also evolve, which they can do much more quickly than humans, since they are not limited by the generational timelines that govern human reproduction. Sagan and Margulis continue: "that machines apparently depend on us for their construction and maintenance does not seem to be a serious argument against their capacity to evolve . . . using autopoietic us as an integral part of their evolutionary mechanisms" (20). Continuing in a speculative vein, they assert that "civilization as we know it can no longer survive without machines. It seems likely that humans will survive to govern the transition from an organic to a technobic biosphere. . . . Human beings in association with machines have a great selective advantage over those alienated from machines" (20). "We may compare the future evolution of machines to the evolution on dry land 400 million years ago. Life may continue to expand by technobic autopoiesis" (21).

Although Sagan and Margulis write about "machines" without qualifying them further, one can argue that in the present era, the most important machines are computational media, for increasingly they communicate with and control myriad other machines: diesel locomotives, waterfront cranes, stoplights, self-driving cars, and many more. As we saw in chapter 1, computational media are not just one technology among thousands of others: they are distinguished from other electromechanical devices precisely because they have the capacity to process, communicate, store, transform, and disseminate information. The collectives that Sagan and Margulis propose in saying that machines mutate and evolve through "autopoietic us" are what I have called "cognitive assemblages."

Their speculative leap into the evolution of machines sits uneasily with other ideas they espouse, including autopoiesis. To develop further the ideas articulated in "Gaia and the Evolution of Machines," it would be necessary to define cognition in ways that are applicable both to biological organisms and to computational media; to accept that information flows between and among humans, nonhumans, and computational devices; and to make clear that cognition applies to computational media as well as to biological organisms. As we have seen, these requirements are either forbidden by autopoietic theory or entirely absent from its vocabulary. To the extent that Margulis aligns her work with autopoiesis, this would imply making significant revisions in her approach, revisions to autopoietic theory, or both. None of this is attempted in "Gaia and the Evolution of Machines" (Sagan and Margulis 1987), which for this reason seems more of a one-off, highly speculative piece than a sustained effort to develop a theory of evolution that would include machines.[8] This positionality is further indicated by the fact that the relatively brief article appears in a popular magazine rather than a peer-reviewed scientific journal.

Given the article's relatively thin argument, it is somewhat strange to find it emphasized in Bruce Clarke's *Gaian Systems*, where he refers to it several times and devotes an entire section to discussing its implications (Clarke

2020, 165–69). His approach is to encompass Margulis's and Lovelock's work and enfold them into neocybernetics systems theory (NST). Moreover, he wants to extend the notion of autopoiesis into social and psychic systems, via Niklas Luhmann's social systems theory, and absorb the technosphere as well. For this project, "Gaia and the Evolution of Machines" (Sagan and Margulis 1987) serves a very useful purpose, for it forecasts how an autopoietic emphasis could be coupled together with the technosphere. Carrying this project further, Clarke positions neocybernetic systems theory (NST) as encompassing biological, technical, social and psychic systems, for which he coins the phrase "metabiotic." "In the metabiotic formula for Gaian operation, Gaia's own autopoiesis couples with open material-energetic processes of abiotic systems [that is, the environment] and the metabiotic extrusions of technological structures [that is, the technosphere] to the self-bounded operational processes of biotic systems with the physiochemical medium of matter" (Clarke 2020, 43).

In this amalgamation, the role of cognition—and also of meaning—is underspecified, as when Clarke refers to the "planetary cognition" of Gaia (2020, 43), without indicating how such cognition would come about, or through whom. In the passage about "planetary cognition" cited earlier, Clarke (2020, 17) proclaims that the "cognition" of Gaia has no meaning, that it is not a meaning-making system. Surely, however, when organisms strive to survive, when they attempt to eat their neighbors, when they struggle to continue their existence even as they become the prey of other organisms, these activities have meaning for them. In my view, the most potent meanings for all organisms are those bound up with their attempts to maintain their existence.

Clarke is surely aware of the problems here. In critiquing Donna Haraway's "cyborg" Gaia, for example, he remarks on "a factitious unity of systems operations" that "calls forth its own deconstruction into distinct structural couplings maintaining heterogeneity of functions." Gaian systems theory, by contrast, does not point toward its own deconstruction, he argues, because it observes "material and operational distinctions among 'the geological, the organic, and the technological' in order to bring out finer orders of attention in the construction of their systematic couplings and composition" (Clarke 2020, 52). To be fair to Clarke, he often acknowledges the tensions and fractures created when he attempts to bring everything together under the big tent of neocybernetic systems theory, including, for example, Maturana's own misgivings about whether the idea of autopoiesis could be validly extended to social and psychic systems. When he comes up with the inscrutable phrase "planetary cognition," he writes with disarming honesty that it is "an ur-medium of corporeal meaning awaiting adequate remediation" (Clarke 2020, 43).

In the spirit of following up on this invitation to remediate "planetary cognition," I reiterate the definition of cognition offered earlier in chapter 1: cognition is a process of interpreting information in contexts that connect it with meaning. Note that this definition converges with Maturana and Varela's insistence that cognition is a process, a dynamic and ongoing event emerging from an organism's embedding in an environment. Unlike Maturana and Varela's definition, however, this one asserts that information does flow between and among organisms and their environments. Arguably even more important is the assertion that through these flows of information, interpreted contextually through an organism's specificities and those of the environment, meanings are created, modulated, transmitted, and transformed. Finally, the definition is crafted so that it applies as well to computational media, which also have contexts and interpret information accordingly.

I have already discussed how biosemiotics contributes to this project by framing, through Peircean semiotics, the idea that an organism's umwelt emerges from the specific ways in which it is coupled with the environment, and that an organism's access to meaning comes through its performances in relation to the environment, that is, its behavior. Some have objected to the biosemiotics view on the grounds that it interprets behaviors through a semiotic lens, whereas nonconscious organisms themselves surely have no ideas about signs, so it is imposing a human-oriented view on nonhuman organisms. Chris Fields and Michael Levin (2020) have responded to this critique by developing an alternative description of meaning that does not rely on signs at all but only on the organism's own frames of reference. Citing the work of Maturana and Varela, Fields and Levin focus on the "embodied-embedded-enactive-extended cognition" of contemporary cognitive science (the 4E approach), comparing it with biosemiotics to conclude that "meanings characterize structural and functional, but in most cases explicitly nonrepresentational, capacities of an embodied system" (Fields and Levin 2020, 2). By proclaiming that meanings are explicitly nonrepresentational, they rule signs (and thus semiotics and biosemiotics) out of court. Moreover, they make clear that they intend their analysis to apply to the entire span of biological organisms, from cells to humans. They continue, "We can approach questions about the evolution, development, and differentiation of meanings as questions about the evolution, development, and differentiation of RFs [reference frames]," and they subsequently describe how internal and external reference frames evolve. Although they do not employ the vocabulary of biosemiotics, they nevertheless come to similar conclusions. For example, they point out that information is "meaningful to an organism in a context that requires or affords an action, consistent with sensory-motor meaning being the most fundamental component of language as broadly construed. It is in this fundamental sense that meaning is 'enactive'" (Fields and Levin 2020, 3). Their innovation, then, is to develop and deploy a vocabulary other than biosemiotics, showing that similar conclusions emerge from reference frame theory without using any allusions to signs.

In developing their theory, Fields and Levin cite an idea central to Luhmann's systems theory (although they do not mention him by name), namely that one defines a domain of interaction by making a "cut" separating a system from its environment.

> The "cuts" that separate the observed world of any system into "objects" are purely epistemic and hence relative to the system making the observations. Understanding what "objects" S [a given organism] "sees" as components of its E [environment] thus requires examining the internal dynamics of S. These internal dynamics, together with the system-environment interactions, completely determine what environmental "objects" S is capable of segregating from "background" of E and identifying as potentially meaningful. Whether it is *useful* to S to segregate "objects" from "background" in this way is determined not by the internal dynamics of S, but by those of E. Meaning is thus a game with two players, not just one. It is in this sense that it [S] is fundamentally "embedded." . . . In the language of evolutionary theory, it is always E that selects the meanings, or the actions they enable that have utility in fact for S, and culls those that do not. (Fields and Levin 2020, 4)

It is easy to connect these remarks with the results articulated in "What the Frog's Eye Tells the Frog's Brain" (Lettvin et al. 1959). Here S might be rendered as F (for Frog), where the "internal dynamics" of F are investigated through the four different nerve fibers in its eye. The environment E obviously plays a role in the evolution of F, for if there were no bugs in F's E, then it would not be *useful* for F to develop "bug detectors," since it would have no payoff in food provision. For example, those aspects of E that F constructs as objects and discriminates from the background must be capable of moving; to rephrase, "if a frog is surrounded by food that does not move, it will starve to death." It is in this sense that evolution enters the picture. Through environmental specificities as they emerge through interactions with the organism, the environment is constituted as one of the (two) players in constructing meaning.[9]

To illustrate how reference frames emerge, Fields and Levin take the case of an *E. coli* bacterium engaging in chemotaxis, moving toward or away from the concentration gradient of a substance. "Bacterial chemotaxis has long served as a canonical example of approach/avoidance behavior and hence of the assignment of valence to environmental stimuli" (2020, 8). They point out that to sense the gradient accurately, the bacterium must have an internal reference frame that distinguishes between "good" gradients (for example, food) and "bad" ones (for example, toxins). Although the sensing mechanisms that develop the reference frame are complex (as determined by previous research on *E. coli*), the point is relatively simple: the bacterium must have an internal reference frame in order to makes the distinctions activated in chemotaxis. Moreover, "implementing internal RFs requires energetic input from the environment. This energetic input is necessarily larger than the energy required to change the pointer state associated with the RF. Any RF is, therefore, a dissipative system that consumes environmental free energy and exhausts waste heat back to the environment. Every RF an organism implements requires dedicated metabolic resources" (Fields and Levin 2020, 8). Since RFs are energetically expensive, the authors conclude that "only meaningful differences are detectable," since "organisms do not waste energy acquiring information that is not actionable" (6). They conclude that "at every level, RFs specify actionability and therefore meaning" (7), suggesting that exploring the emergence of internal RFs and their linkages to external RFs may provide answers to the "fundamental question of an evolutionary theory of cognition" (6), for instance, my earlier question about how an amoeba can evolve into a human through myriad intermediate stages.

Note that in this theory, E does not serve as just a "trigger" for the organism but is an active force intrinsic to selecting the features of E that will be seen and acted on by S (or F). Thus there is some overlap with Maturana and Varela's concept of autopoiesis, but also some significant changes, chief among them the introduction of "objects" versus environmental background, the emergence of meanings through perceptions and behaviors, and the connection between organism and environment that determine how the organism will evolve. My definition of cognition, and therefore ICF, sits easily both with biosemiotics and with reference frame theory, because it introduces the critical concepts of context, information, interpretation, and meaning.

### Summarizing Conflicting Claims

At this point it will be useful to summarize the preceding distinctions and arguments through a table of "conflicting claims of theories and theorists" (table 3.1). No table can capture all the nuances, but I hope that this table will give the flavor of the major differences and overlaps between the different players.

THE EMERGENCE OF TECHNOSYMBIOSIS AND GAIA THEORY 97

TABLE 3.1. Conflicting Claims of Theories and Theorists

| | |
|---|---|
| 1. All the lifeforms of Earth and their environments constitute a single entity called Gaia.<br>Lov, Mar+S, NST | The lifeforms and environments of Earth constitute innumerable interacting dynamic systems.<br>RFT, ICF, BioS |
| 2. No information passes from the environment to an organism.<br>Mat+V1, NST | Organisms sense their environments, transducing the information through the specificities of their systems.<br>Mar+S, V2, RFT, ICF, BioS |
| 3. Cognition is synonymous with the process of living and occurs only in the living.<br>Mat+V1, BioS, Mar+S (?) | Cognition is a process of interpreting information in contexts that connect it with meaning and is done by biological lifeforms and computational media.<br>ICF, RFT |
| 4. Autopoiesis is preserved at every point as organisms evolve through structural drift.<br>Mat+V1, NST | When organisms mutate and evolve, their organization changes.<br>Mar+S, ICF, RFT |

Key:
BioS: BioSemiotics
ICF: Integrated cognitive framework
Lov: Lovelock
Mar+S: Margulils and Sagan
Mat+V1: Maturana and the early Varela
NST: Neocybernetic systems theory
RFT: Reference frame theory
V2: The later Varela

## The Emergence of Technosymbiosis

The researchers discussed above each had specific goals in developing their ideas. Maturana and Varela ([1972] 1980) wanted to provide an account of "the living" that emphasized self-reproduction and minimized Neo-Darwinism; James Lovelock wanted to insist that organisms change their environments even as they are changed by it; Lynn Margulis wanted to reposition narratives about evolution so that the microcosm was recognized as the foundational layer of all higher lifeforms and to develop in rich detail the strategies that make symbiogenesis the primary driver of evolution. My goal in discussing these ideas is to use them to develop an account of what technosymbiosis is—and what it isn't. From Maturana and Varela I keep the idea that life is special, and specifically different from machines, because it is autopoietic. From Lovelock I draw the implication that life is the catalyzing force that keeps the Earth from arriving at thermodynamic equilibrium. I draw on Margulis and

Sagan for their accounts of the strategies employed in symbiotic interactions, which range from horizontal gene transfer, to opportunistic exchanges between different species that are not (yet) permanently joined, and finally, to the complete absorption of one species by another to create a new organism. From RFT I adopt the position that cognition in cellular organisms does not need to be construed solely through the semiotic filter of language. Finally, I endorse some speculations from Lovelock, as well as Sagan and Margulis, projecting the kinds of futures that await humans, now that we are on the cusp of developing computational media with the cognitive resources to converse with humans.

*Not included* in my account is Maturana and Varela's claim that no information passes between an organism and the environment; their implication that cognition occurs only in the living; Margulis's claim that the *only* way in which speciation develops is through symbiogenesis; and Bruce Clarke's implication that planet-wide cognition has now emerged and has no necessary connection to meaning.

What then is technosymbiosis? Starting from a definition of cognition that enables all biological lifeforms and some computational media to be seen as cognitive entities, technosymbiosis proceeds to the realization that in developed countries, humans are already in symbiotic relationships with intelligent machines. How far have we already gone into a symbiosis from which there is no return? Recall that chapter 1 posed the question of what would happen if all computational media were to crash tonight. Developed societies would be rendered helpless: trains could not safely run; food delivery networks would be disrupted; banking systems would crash; basic services such as electrical grids, water lines, and sanitation operations would fail; and communication networks of all kinds would suddenly become inaccessible. These disastrous consequences suggest that we have already gone far enough so that any retreat would cause enormous loss of human life and apocalyptic-level social disruptions.

One advantage of thinking about our relation to cognitive media as symbiosis rather than as the emergence of the cyborg is the flexibility and range of different options that symbiosis offers. Whereas the cyborg typically implies some kind of physical fusion, symbiosis can designate our present state of living in close proximity with computational media through interdependent relations. Symbiosis can also authorize future scenarios that envision human-computational hybrids, humans absorbed into computational media surviving as "remnants," much as mitochondria survive in cells (Sagan and Margulis 1997, 246), or humans as the domesticated familiars of independently living robots, among other possibilities. Sagan and Margulis write in "Gaia and the Evolution of Machines" that "civilization as we know it can no longer survive without machines. It seems likely that humans will survive to govern the transition from an organic to a technoboic biosphere" (Sagan and Margulis 1987, 20). Here the implication is that humans will give way to machines as our evolutionary successors, which they elsewhere call our "descendants" (21). "The future of our machines, provided they can remain part of the autopoietic biosphere, is less bleak than that of ourselves," Sagan and Margulis write (1987, 21). Lovelock for his part foresees a future epoch, the Novacene, populated by artificial knowers (which he calls, somewhat atavistically, "cyborgs") (Lovelock 2020, 94). They "will have designed and built themselves from the intelligence systems we have already constructed" (29). Moreover, he predicts that "they will be entirely free of human commands because they will have evolved from code written by themselves" (94). Perhaps the *least* likely scenario is Lovelock's wistful prediction that future knowers "will be obliged to join us in the project to keep the planet cool . . . [and] will be eager to maintain our species as collaborators" (106). This assumes humans would be able to join together in the great project Lovelock imagines of keeping Earth cool. Given the anthropogenic contributions to global warming that we see in the present, it seems far likelier that this project will fail to materialize and that future knowers would on the contrary want do away with humans as soon as possible.

Whatever futures may unfold, technosymbiosis implies that, short of environmental catastrophe or nuclear war, our present symbiosis with computational media will intensify, expand, and undergo transformations perhaps now unimaginable for us. "The outcome of information exchange between computer, robotic, and biological technology is not foreseeable," Margulis and Sagan write, adding "Perhaps only the most outlandish predictions have a chance of becoming true" (Margulis and Sagan 1997, 252). Noting their wise caution, I nevertheless think that some consequences entailed by technosymbiosis are already evident and are likely to remain valid for the foreseeable future. These include distributed cognition, which always implies distributed agency, disrupting the belief that agency is invested principally in human actions. Through the collectivities formed by cognitive assemblages, information, interpretations, and meanings will continue to circulate between humans, nonhumans, and computational media. Since the processes of interpretation are key, technosymbiosis emphasizes the specificity of an entity's context, which includes sensors and actuators (for computational systems), nervous systems and sensory organs (for biological organisms), and the development of internal and external reference frames (for both lifeforms and computational media). It also emphasizes that all observers are embedded, a condition that makes knowledge possible (for embeddedness is what makes it possible to sense anything at all) and that also limits what kinds of knowledges can be produced, excluding, for example, the supposed objective or God's-eye view critiqued by Donna Haraway (1988). The emergence of cognitive assemblages implies flexible, temporary arrangements that direct information flows through an enormous variety of interfaces, from self-driving cars to the web to aircraft control centers, and much else. It recognizes that, like the evolution of biological lifeforms, the evolution of computational media cannot be predicted and does not fit within prestated parameters, as is the case for chemicophysical material processes. As discussed in chapter 5, the temporality of evolution, whether biological or technological, is simply of an order different from that of the temporality of material processes. Unpredictable in its evolution and given to spontaneous swerves that may lead to enormous changes very quickly through technological mutations and evolutions, technosymbiosis is capable of altering the future of humanity, and the future of life on this planet, in ways unimaginable to us in the present.

In this and many other ways, technosymbiosis implies a thorough-going repudiation of traditional ideas such as human dominance of the planet, humans as the most evolved species, and humans as the sole practitioners of meaning-making practices. Recognizing the potential for great changes that may be uncontrollable in whole or in part and that have consequences beyond what we can imagine, it points toward a humbler, more accurate perception of humans as partners with our nonhuman symbionts in the project to make a livable future for us all.

# 4

# Cellular Cognition:
# Mimetic Bacteria and Xenobot Creativity

One of the central premises of the integrated cognitive framework is that all lifeforms have cognitive abilities.[1] This chapter expands on that idea by engaging with recent research in evolutionary biology, along with the emergence of CRISPR-Cas9 gene editing, that has revealed the power of individual cells and cellular networks to perform actions indicating they are capable of cognitive achievements, specifically mimetically representing viral threats and creatively repurposing cellular components. Combined, these results suggest that evolutionary theory should be revised so that cognition is understood to begin, not with brains and nervous systems, but much earlier and lower down in scale, at the cellular level. Not only is this perspective a game changer in terms of evolutionary theory; it also has significant implications for the future of human collaborations with cognitive cellular actors.

This chapter focuses on two areas where cellular cognition has been investigated and its implications for human futures explored. The first is CRISPR-Cas9 gene editing, pioneered by Emmanuelle Charpentier and Jennifer Doudna and their collaborators (Jinek et al. 2012). Their work initiated the technological capture of the nonconscious cognitive processes of bacteria defending themselves against viral attacks. The bacterial responses show how the mimetic processes of the nonhuman realm can be modified and manipulated to serve human purposes, including producing posthuman bodies. Mimesis, my analysis concludes, is not only about art but also about fundamental strategies of survival for humans and nonhumans alike.

The second part of my analysis focuses on work by biologist Michael Levin of Tufts University, evolutionary computation scientist Josh Bongard at the University of Vermont, and their collaborators, toward the creation of synthetic entities they call "xenobots." Xenobots are constructed organisms, synthetically created and sometimes artificially configured, that are able to perform actions requiring cellular cognition, such as reproducing themselves, navigating their environment, and joining with other cells to create larger entities. Levin and collaborator Rafael Yuste (2022), director of the NeuroTechnology Center at Columbia University, in "How Evolution 'Hacked' Its Way to Intelligence from the Bottom Up," explore the implications of xenobots for evolutionary theory and for possible future applications for humans, including such science fiction scenarios as the regeneration of organs and spontaneous repair of birth defects. Levin has also teamed up with Daniel Dennett in "Cognition All the Way Down" (Levin and Dennet 2020) to argue that the next great task for biologists is to understand how cells, tissues, and organisms can be considered "agents with agendas," a provocation with special relevance to cognitive assemblages between humans and other organisms.

The point of this chapter is to show not only that biological cognition is present at the cellular level but also that it possesses specific characteristics. Biological cognition is organized through modules nested inside one another, starting at the cellular level and progressing upward to cellular networks, tissues and organs, and finally whole organisms. In this sense, it is analogous to the dynamic heterarchies discussed in chapter 2 in the context of computational cognition, with each level organized so as to bootstrap the process to the next level up. Secondly, cellular cognition has been shown to have the capacity to create and store memories, including recognizing viral threats and, with xenobots, remembering patterns that enable cells to complete bodily regeneration when only a small part of the pattern is present. Moreover, Levin and Dennett (2020) argue that nested levels of cognizers, from individual cells to networks, tissues, organs, and organisms, can be understood through what they call the "cognitive horizon" of the entity. Presumably brains, and human brains especially, have a much wider cognitive horizon than, say, the human liver. Thinking in these terms provides a useful way to talk about what happens in cognitive assemblages. When computational media partner with humans in such assemblages, the cognitive horizon of both entities expands. Further, the nested approach fits well with technosymbiosis. When the cognitive capacities of bacteria are captured through technological interventions by humans—that is, through technosymbiosis—the human cognitive horizon is increased. Such strategies, and the thinking that underlies them, promise not only to revolutionize our understanding of how evolution works but also to open human futures to new and unpredictable horizons that will change the very nature of what being human means.

## Micromimesis: Collaborations with Bacterial Actors

Throughout the millennial-long history of the mimesis concept—notwithstanding its variations in meanings, applications, and examples—one generalization holds true: *mimesis* has almost always been applied to human art forms.[2] However, if nonhuman organisms can engage in meaning-making practices, as biosemioticians have convincingly declared, then it may be useful to consider what mimesis may signify in the nonhuman realm. As explained in chapter 1, my definition of cognition is "*a process that interprets information within contexts that connect it with meaning*" (Hayles 2017, 22). This definition opens up a wide-ranging reconsideration of nonhuman mimetic practices (26).

Mimesis has long been understood not as mere copying but as selective representation of aspects of nature. Selection and recontextualization incorporate into the concept the scent of the artist, the whiff of creativity that makes a mimetic re-creation different from the original. Moreover, since Aristotle, mimesis has also been associated with both distance and empathy, the former catalyzing and empowering the latter as viewers, partially insulated from personal threats by the differences, are enabled to recognize similarities between themselves and re-presented others, thus facilitating identification and catharsis. How might the dynamics characteristic of mimesis change when the organisms are nonhuman? How are the borders between copying and re-presenting negotiated when the media are not art forms but organismic responses? What purposes would mimetic processes serve, and what functions would they enable if the organisms engaging them were not conscious? What if the stakes in mimetic re-creation are not only an individual's fate but that of the collective and even the species?

Biologists have known for some time that when viruses attack a bacteria colony, some bacteria are likely to survive and develop immunity. Only within the last few decades, however, have the mechanisms involved been revealed. The discovery began with the recognition that the DNA of such microorganisms as *Haloferax mediterranei*, which live in salt pools where the salinity is ten times that of the ocean, consisted of repeating small fragments that were the same, interspersed with spacers that had different configurations. Francisco Mojica, a microbiologist at the University of Alicante in Spain, explains the acronym CRISPR (**c**lustered **r**egularly **i**nter**s**paced **p**alindromic **r**epeats) to describe the phenomenon. This discovery was coupled with the realization that while the CRISPR clusters were identical, consisting of the same fragments of DNA, the spacer portions were highly variable, as well as different from the repeating clusters. Mojica and other researchers then realized that the CRISPR sections were copied from the DNA of attacking viruses. Serving as the cell's memory of the event, the information of the repeats was transferred via RNA to an enzyme, Cas9, that had the ability to cleave nucleotide sequences, that is, to act as a nuclease. When it found a match to the information it carried, it severed the corresponding viral DNA, thus effectively killing the virus. "So clever," Mojica remarked of the strategy in "Human Nature," an episode of the PBS documentary series *Nova*, quickly adding that of course it came about through "evolution," not through any conscious insight (Bolt 2019).

This strategy exemplifies what I call micromimesis. It fulfills the traditional requisites for a mimetic act. It re-presents something found in nature—the virus DNA—but does not merely copy it. The re-created DNA fragment differs from the original because it has been recontextualized into a CRISPR and incorporated into the bacterium's DNA. Moreover, there it performs a function completely different from its function in the original. Now it doesn't operate to replicate the virus as it hijacks the cell's machinery to turn out copies of itself, an action that kills the cell, but rather is used to counterattack and kill the virus instead. Seen from the perspective of human actions, the high stakes of this drama are worthy of a Greek tragedy, being nothing less than life or death. For the bacteria, of course, the action is guided not by conscious thought but by its evolutionary history in which a species has developed this adaptive mechanism through mutation and natural selection. The affective responses of humans to this microbial battle should rightly be understood as entirely a projection without basis in what the bacteria themselves experience.

The micromimetic drama intersected with human acts and intentions, however, when researchers realized its significance for human-initiated gene editing. Several seminal papers, coauthored by Emmanuelle Charpentier and Jennifer Doudna and their collaborators, made the connection between Cas9 cutting of the DNA of viruses and its potential for gene editing, work subsequently recognized when they were awarded the 2020 Nobel Prize in Chemistry. Calling Cas9 a "programmable" nuclease, they created an all-purpose gene-editing tool by fusing the RNA fragment that recognized the viral DNA with the RNA that matches it to a gene (transactivating CRISPR RNA or tracer RNA), thus creating guide RNAs that direct Cas9 to a specific DNA sequence to cut (Jinek et al. 2012). They further pointed out that the mechanism could be used for any gene, not simply to cut viral DNA. The concluding sentence of their "immortal" paper made the point clear.[3] "We propose an alternate methodology [to existing methods] based on RNA-programmed

Cas9 that could offer considerable potential for gene-targeting and genome-editing applications."

Additional research showed that the cutting could be accomplished by another process that inserted a new DNA sequence at the site of the cut. When a cell encounters a disruption in a DNA sequence (a process that happens with considerable frequency during normal cell division), the cell will try to splice the two ends together. Doudna and Samuel Sternberg (2017) point out the haphazard nature of the process, because the nucleotide sequences may lose a few bases to create the correct pair joining (A joins with T, G with C). By contrast, if the guide RNA offers the cell a template that exactly matches the two cut ends, the cell will preferentially use that instead. This means not only that specific sites in DNA can be targeted for cuts but that repairs can also be made to correct disease-causing mutations. Other procedures have expanded CRISPR's functionality so that it now enables a suite of gene-editing processes, including deletion, addition, insertion, rearrangement, knock-out (which disables a gene's ability to code for a protein), and knock-in.

The huge advantages of CRISPR-Cas9 gene editing are its precision, efficiency, and relatively cheap cost (in part because CRISPR relies on DNA-RNA pairings, rather than other approaches that synthetically engineer proteins that bind to specific DNA sequences). For example, modifying a single gene with ZFN (zinc finger nuclease, an older gene-editing technology) cost $25,000; CRISPR can replace several genes at once for a few hundred dollars or less.[4] It is now possible to submit online the DNA of the gene to be edited, and a laboratory will create CRISPR RNAs targeted to that gene and ship them out in a few days. The result has been to open gene editing to a large variety of organisms, including dogs, pigs, mice, butterflies, and mosquitoes, among other animals. Plants have also been treated with a large variety of gene edits designed to increase yields and to make them more resistant to diseases and more tolerant of climate changes. CRISPR gene editing has also been used with humans to cure sickle-cell anemia, beta-thalassemia (a blood disorder that reduces the oxygen level in blood cells), inherited blindness, and even certain types of cancer.

Although my focus here is on gene editing, it is worth noting that the recent development of the Moderna and Pfizer vaccines against Covid-19 viruses also use techniques that enlist the cognitive powers of cells. Different in details, both vaccines are similar in using messenger RNA (mRNA) encased in lipid delivery systems. (The low temperatures are required because the nucleotide chains of RNA sequences are fragile, lacking the robustness of DNA's double helix.) The vaccines are injected into muscle cells in the arm, where the messenger RNA sequences travel into the cell interiors. There they are read as instructions to create a spike protein on the cell wall and are afterward disassembled by the cell. The spike protein, a copy of those on the surface of the Covid virus, is easily recognized by the immune system as not properly belonging there. Here again, the interpretive capacities of cells are crucial in creating the immune effects.

### The Capture of Nonconscious Cognition by Consciousness

There are several areas of controversy with gene-editing techniques. First is the concern that gene editing on wild-type animals and plants may get out of control and result in unintended consequences or, alternatively, be used to eradicate some species altogether. These concerns have been catalyzed by the development of gene drives, techniques designed to spread mutations rapidly in a population. One such technique uses CRISPR to inject an organism with the CRISPR sequence itself, thus giving the organism a recursively doubled mutation ability. Ethan Bier and his student Valentino Gantz at the University of California, San Diego, used the technique in 2015 with fruit flies, editing a gene that determines color, so that the mutated flies were yellow instead of the usual brown (Gantz and Brier 2015). The technique was so successful that they estimate if a single mutated fruit fly had escaped the laboratory, it would have spread the yellow color to between 20 and 50 percent of all fruit flies worldwide.

The technique has also been used with mosquitoes to give them resistance to the parasite *Plasmodium falciparum*, responsible for malaria (see Gantz et al. 2015). Other researchers have used gene drives to spread genes that cause female sterility in mosquitoes (see Hammond et al. 2016). Since the affected gene is recessive, the mutation would spread without apparent consequences until enough females carried two copies of the recessive gene, whereupon the population would suddenly crash. While making malaria-causing mosquitoes extinct might be cause for celebration for the millions of humans afflicted with the disease, the concern is that the technique could be extended to other species, including humans.

Another looming problem is the availability of CRISPR amateur gene-editing kits to anyone who has a hundred bucks or so. Although the kits are designed for gene editing in bacteria and yeast (making them attractive for craft-beer aficionados), the potential for biohacking them to apply to other organisms is unknown. Given the ingenuity shown by hackers with computer code, what mischief they could do with genetic code is cause for real concern.

In light of these considerations, it is not difficult to see why, in the 2016 *Worldwide Threat Assessment* report by US intelligence agencies to the Senate

Armed Services Committee, they identified under the "Biological Warfare" section the "biological materials and technologies, almost always dual use, [that] move easily in the globalized economy, as do personnel with the scientific expertise to design and use them" (National Counterproliferation Center 2016). The tip-off that they had CRISPR gene editing in mind is the phrase "dual use," which denotes the potential for benefit as well as devastation. Conventional biological weapons, for example nerve agents or poison gas chemicals, have no such beneficial aspects.

Perhaps the most intense controversies concern editing the human genome, especially germ-line editing in which changes will be inherited by all subsequent individuals coming from that genetic line. While editing an individual's genome in somatic cells to correct single-gene inherited diseases such as sickle-cell anemia has already been done and is largely considered to be ethically acceptable, many researchers have argued for a moratorium on germ-line editing. The risks are clearly laid out in "Don't Edit the Human Germ Line" (Lanphier et al. 2015). Addressing its use in a human embryo, the authors explain that "it would be difficult to control exactly how many cells are modified. Increasing the dose of nuclease used would increase the likelihood that the mutated gene will be corrected, but also raise the risk of cuts being made elsewhere in the genome" (Lanphier et al. 2015). Moreover, there are also possibilities that only one copy of the double-helix target gene would be modified or that the cell could start dividing before the corrections are complete, resulting in a "genetic mosaic." Consequently, the "precise effects of genetic modification to an embryo may be impossible to know until after birth" or even for several years afterward.

A case in point is research published in *Protein and Cell* by Junjiu Huang's lab at Sun Yat-sen University in China (Liang et al. 2015). They used CRISPR with eighty-six human embryos to edit the gene that produces beta-globin, intending to demonstrate that it was possible by this means to cure the blood disease beta-thalassemia, caused by a defect in this gene. Of the eighty-six embryos, only four were successfully mutated. Other embryos had their gene sequences edited at off-target sites, resulting in haphazard mutations other than the one intended. In addition, several embryos were converted into genetic mosaics (just as Lanphier and his colleagues feared) because the CRISPR intervention took place after the cells had already started dividing. In this research program, all the embryos were nonviable and would never mature into human persons, but the results clearly showed the potential for disastrous outcomes if the CRISPR editing is not done with scrupulous care.

Not everyone agrees that germ-line editing is a mistake. On the other side of the issue, the bioethicist Julian Savulescu and colleagues in 2015 published an article, also in *Protein and Cell*, arguing that there is a moral imperative to *continue* research into human germ-line editing. They pointed out that about 6 percent of all children born worldwide have serious birth defects arising from genetic or partly genetic causes; making gene editing available for them would benefit about eight million children per year (Savulescu et al. 2015). Indeed, close reading of Lanphier et al. (2015) reveals a rather large gap between the abstract concepts they espouse and real human suffering. It is one thing to argue, as they do, that germ-line editing is unnecessary because editing somatic cells can accomplish the same goal without the risks. Imagine, however, making that argument to a man who has suffered life-destroying levels of pain because he has sickle-cell anemia. He and his wife want to have a child. They want to have an embryo edited in vitro to correct the gene that causes the disease, thus freeing the baby and all the child's subsequent progeny from the disease. Telling them to wait until the baby is born and then, when the child is old enough to have an operation, do gene editing on the somatic cells would mean to them condemning their child to the same racking pains the father has endured. Moreover, the father would be acutely aware of the cellular damage that the disease causes. Why wait until the disease has already started and has caused intense suffering, when germ-line editing could eliminate it relatively easily, when only a few cells (instead of millions) would need to be modified?

More conundrums arise if (or rather when) CRISPR techniques are used for human enhancement, that is, for editing genes controlling traits such as height, eye and hair color, endurance, musculature, or need for sleep, or for even more complex multigene inheritances such as intelligence, creativity, or flexibility. Further complicating the issue is the fact that the line between correction and enhancement is far from clear-cut. Already in use are techniques in which a couple uses their eggs to create a number of embryos and then employs genetic testing to choose among them, a technique called preimplantation genetic diagnosis (PGD). What if, in addition to screening for heritable diseases, they are also choosing for sex and other attributes?

These kinds of concerns, already present with older gene-editing techniques, have been greatly exacerbated by the accessibility, ease of use, and economics of CRISPR. They go far beyond the boundaries of scientific research into questions about ethics, social justice, economic inequalities, and the futures of humans and nonhumans. Micromimesis thus intersects with macro concerns that have traditionally been the province of the humanities. To address these issues, we can start by distinguishing between a tool (as gene-editing mechanisms are often called) and the cognitive acts of the bacteria. Traditionally, a tool is considered to be without its own agency; a hammer, for example, needs a person to activate it and decide what object to hit. By contrast, the bacteria and Cas9 are interpreting information from their environments and proceeding according to their own agency, attributes that in my definition make them cognitive. Within their contexts, their actions are meaning-making practices, with *meaning* understood, as in biosemiotics, not as an abstract concept but as an adaptive response to an environmental signal.

As we saw in Chapter 2, biosemioticians employ a Peircean triadic semiotics in which an interpretant intervenes between the sign vehicle and the object to connect the two. For nonhuman species without consciousness, the interpretant is understood as an embodied action contextualized by the interpretive capacities and sensoria of an organism. As Wendy Wheeler puts it, "meanings are the result of a process of discovery and interpretation. Life is process, and all organisms must be capable of change in response to changing conditions" (Wheeler 2016b). In this view, meaning is created when an organism interprets an environmental signal and subsequently performs actions that have consequences for the organism. Seen in this light, CRISPR gene-editing mechanisms represent a technosymbiosis between the cognitions of conscious humans and the nonconscious cognitions of bacteria to accomplish something that neither party could by itself.[5] Both create and rely on semiosis: humans through verbal and symbolic languages such as English and mathematics, and bacteria through actions that function as interpretants joining the sign vehicle (mimetic recognition of DNA) with the object (cutting the DNA at the appropriate site).

What are the implications of considering CRISPR gene editing not as a clever tool invented by humans but as collaboration between human conscious cognitions and bacterial nonconscious cognitions, with both partners relying on the generation and interpretation of signs? Such a shift opens onto much broader and deeper issues about the human place in the ecosphere of life on Earth. As argued throughout this book, combating anthropocentrism and the belief that humans have the right to dominate all other species is an urgent task. Recognizing the cognitive capacities of bacteria and other nonhuman lifeforms is a step in the right direction. So is the realization that nonhuman lifeforms create, transmit, and interpret signs, always in embodied contexts that connect their interpretations to meaning *as it exists for them*. This view powerfully counteracts the mistaken belief that only humans (and perhaps a few other mammals) are capable of meaning making, opening the entire biosphere, nonconscious as well as conscious, to intersecting, overlapping, reinforcing, and contesting realms of meaning.

### Functions of Micromimesis

One of the traditional functions of mimesis has been to facilitate social cohesion by re-presenting others different from oneself in ways that invite identification and understanding. Micromimesis serves to protect the individual bacterium and community from predatory outsiders that would kill the colony. In this sense, it represents an action very familiar in human dramas. Perhaps no other theme, stretching across millennia from *The Iliad* to *Independence Day* (Emmerich 1996), is so pervasive. That micromimesis can now be technically captured and made into a mechanism to alter the genomes of humans and nonhuman organisms bestows an awesome power godlike in scope and significance, for it can literally change the relation of humans to our species and to nonhuman organisms. If we focus only on the "dual use" of CRISP technology—its potential for good as well as devastation—we may mistakenly decide that the technology itself is neutral. Such a view misses the larger point that the technology is absolutely not neutral in its effects, for it radically transforms the dynamics of evolution to make it a human-directed process. Unprecedented in human history, it bestows the "unthinkable power to control evolution" (as the subtitle of Doudna and Sternberg's book [2017] proclaims). What wisdom can micromimesis offer to help us navigate these uncharted waters, both for our own species and for those with whom we share the planet?

I think there are two large lessons that micromimesis can provide. One has already been touched on—the need for caution and slow development, accompanied by broad and deep conversations among scientists, bioethicists, legislatures, and citizens on how to proceed. Here we may emphasize the evolutionary history of micromimesis, which developed over millions of generations and was tested continually in the contexts of changing and fluctuating environments. Only the most robust responses compatible with the environment and the complex requirements for survival could pass such tests. Gene editing, by contrast, can result in significant changes that take place suddenly and without extensive environmental testing. When applied to an individual organism, the consequences can possibly be anticipated adequately, for example when gene editing is used to cure someone with sickle-cell anemia. Even here, however, gene editing involves unknown risks, for interactions between genes are not well understood, and changing one gene may have cascading and unpredictable effects.

These risks apply also to gene editing of nonhuman organisms, where the results may affect ecological interactions and result in disastrous unintended consequences. The Covid-19 pandemic may be the most powerful

CELLULAR COGNITION                                                                 111

object lesson yet on that issue, since the US Intelligence Agencies found there was a "plausible" possibility that it emerged from a chimeric virus created in the Wuhan Virology lab and escaped through various "lab leak" mechanisms (National Intelligence Council 2021). Even if this proves not to be the case, the warning signs are clear, and the potential risks to human life are enormous, for Covid-19 is closing in on seven million people as of this writing. When the human germ line is at issue, the long-term risks to the *Homo sapiens* species are correspondingly greater and mitigate against introducing new modifications not already present in normal human genomes. This may mean proceeding with curative germ-line editing but drawing the line, however difficult to determine, between curing inherited diseases and introducing human enhancements.

The second large lesson has to do with the age-old function of mimesis to facilitate identification and empathy. As practiced in Greek drama and in micromimesis, the range of the mimetic identification is relatively narrow, never reaching beyond the limits of one's species. But the collaborations between humans and bacteria that resulted in CRISPR gene editing have radically changed that. The awesome power these collaborations create must be matched by a correspondingly intense responsibility not only to members of our own species, but to every species with whom we seek to intervene. Moreover, the goal of CRISPR interventions must be not only to increase profits for capitalist enterprises but to increase the health, vitality, and robustness for the species involved, nonhuman as well as human. After all, the bacteria never had the opportunity (or the wherewithal) to give informed consent for our use of them. We have simply appropriated their powers without adequately acknowledging their contributions as cognitive meaning-making lifeforms. We can begin to repay them by understanding micromimesis as an opportunity, and obligation, to recognize our responsibilities to them and other nonhuman species.

The technocapture of micromimetic actions should thus enlarge the scope of mimetic meaning far beyond its original limits of increasing empathy for other humans and protecting the community. Now micromimesis invites—no, compels—us to empathize with and consider the value of all species. Such empathic bonds further a philosophy of ecological relationality because they spring from the recognition that our human powers are not ours alone but emerge from our collaborations with nonhuman organisms.

Reversible internality, already begun in the perspective of Lynn Margulis on the microcosm as noted in chapter 3, applies to gene editing as well. One perspective would see humans as the primary agents of change, locating bacteria within the human world of gene editing, in which they are seen as human-invented tools of enormous power. Reversing the internality would locate humans within the microcosm of unicellular organisms, within which human multicellularity is recognized as derived from, and still absolutely dependent on, unicellular organisms that carry on their business in their own terms, oblivious to human desires and needs.

It is clear that we humans are the ones who must take responsibility for the results of our partnerships with bacteria, because only we have the power to do so.[6] The final lesson that micromimesis offers is reflected in the necessity to take responsibility, acknowledge kinship, and feel humility in the face of the enormous complexity of the biosphere, which makes us startlingly aware of the limits on the human ability to control consequences.

## Xenobots and Cellular Cognition

Whereas gene editing necessarily focuses on genes and gene sequences, the research of Michael Levin and various collaborators explores the creative innovations initiated by cellular cognition other than through DNA and genetic processes. To demonstrate the remarkable plasticity of organisms in coping with unique environmental challenges, one team transplanted a tadpole's eye from its head onto its tail (Blackiston, Vien, and Levin 2017). Despite its novel location, the eye found a way, when stimulated by serotonin, to restore organismic vision by connecting to the spinal cord rather than directly to the brain. Through this indirect route, the eye's signals were interpreted by the brain in ways functional in its environment. Moreover, when the tadpole underwent metamorphosis, the result was a frog with eyes in the correct head position. This suggested that somehow the cells stored a pattern for growth that could be reestablished, even when significant disruption had taken place.

A parallel research track explored the potential of cells to repurpose components in ways functional within an altered environment. This time Levin and his collaborators (Blackiston, Lederer, et al. 2021) excised skin cells from a frog embryo (*Xenopus laevis*) and placed them in a petri dish with a liquid medium, thus constructing them as xenobots. The name is appropriate, not only because of the frog's genus designation as *Xenopus*, but also because the Latin prefix "xeno" translates as "strange." Xenobots are constructed autonomous organisms demonstrating behaviors different from those of the original organism from which they came. In this sense, they are "strange" because they demonstrate the capacity for cellular innovation and creativity.

The skin cells repurposed the cilia on their surface—the small, slender, hair-like structures on the surface of most animal cells. In the frog, the cilia are used to distribute the mucus that protects the animal's skin. However, "liberated" (my term) from the cellular networks that interact with skin cells, the excised cells began to use the cilia for motility—a function often seen in wild-type cells that achieve motion by having the cilia beat in the same direction (Blackiston, Lederer, et al. 2021). The researchers hypothesized that when the cells were within the cellular network, this primitive mode of motility was suppressed by the network and directed to other ends. Freed from these constraints, the cell's programming reverted to what may have been an older stage of its evolution.

Even more surprising were the actions of the now-mobile skin cells. They began to form clumps, acting as unified collectives capable of moving through their environment. Moreover, the collectives were able to self-heal themselves when damaged through external manipulation. Calling the xenobots "synthetic living machines" (Blackiston, Lederer, et al. 2021, 5), the researchers showed that the collectives were able to navigate through their environment, for example stay in the middle of a liquid-filled tube when rounding a corner. Although the typical xenobots lived only for nine to ten days (a good safety feature), surviving by metabolizing the maternal yolk platelets present in all early embryonic *Xenopus* frog tissues, the researchers found they could extend their lifetimes to more than ninety days if they raised them in a *Xenopus*-specific culture media.

The xenobots also demonstrated the capacity for self-repair. Working with microsurgical tools, the researchers inflicted wounds on the spherical xenobots, in some cases cutting them almost in half, with the two halves joined by a single "hinge." "In all cases," the researchers report, "individuals were capable of resolving the wound, closing the injury site, and re-establishing a spherical shape" (Blackiston, Lederer, et al. 2021, 10). "In no case was mortality observed; all individuals (n = 15) persisted for the remainder of the experiment." Not only do these data indicate a "robust ability to self-repair"; they also "repair to their characteristic xenobot morphology, not to a frog embryo–specific shape." In this respect, they show behaviors very different from the modified tadpole that reverted to the usual frog morphology after metamorphosis. In addition to the movements of individual xenobots, the researchers also observed cohorts of xenobots demonstrating emergent group behaviors. They released multiple xenobots into arenas "covered with 5 micrometer silicone-coated iron oxide spheres," then videoed time-lapse sequences of their movements. "Over the course of 12 hours, regions of the arena were swept clean of the particles, creating piles of debris across the field" (11). Although it is unclear what utility this emergent behavior would have, it does hint that the xenobots' capacities for spontaneous action could conceivably be harnessed for useful work.

This hypothesis was tested in another experiment in which frog embryo skin cells were placed in a petri dish with dissociated skin cells (Kriegman et al. 2021). Their combined movement aggregated some of the dissociated cells into piles. "Piled cells adhere, compact, and over 5 [days], develop into more ciliated spheroids capable of self-propelled movement" (Kriegman et al. 2021, 1). The researchers call this action "kinematic self-replication," claiming that it is "previously unseen in any organism." Normally the skin cell aggregates can produce only one generation of ciliated spheroids, but when the researchers used computer algorithms to design the assemblies for optimal shapes, the aggregates were able to generate replication through two generations. Moreover, using a simulation, the researchers showed that the assemblies could produce useful work by closing an electrical circuit through random action. They conclude, "This suggests other unique and useful phenotypes can be rapidly reached through wild-type organisms without selection or genetic engineering, thereby broadening our understanding of the conditions under which replication arises, phenotypic plasticity, and how useful replicative machines may be realized."

Because the complexity of the frog embryo skin cell behaviors indicated cognitive activity, the researchers tested them for the presence of neurons using a panneuronal marker from immunohistochemistry They determined that "no neurons were present at the surface or internally in any of the individuals imaged, indicating that their behavior is driven entirely by the signaling and functional dynamics of non-neural cells." The researchers concluded that the "xenobots are thus a synthetic model for understanding preneural life forms and their capacities" (Blackiston, Lederer, et al. 2021, 7).

## Implications of Modular Cognition

At least since Darwin, biologists have been trying to avoid teleological assumptions. As Gillian Beer showed in her superb study *Darwin's Plots: Evolutionary Narrative in Darwin, George Eliot and Nineteenth-Century Fiction* (2009), Darwin struggled, sometimes on a sentence-by-sentence basis, to avoid the teleology inherent in the then-pervasive idea of a Divine Creator who has already plotted out the trajectory of human history. Evolution, Darwin argued, does not have a predestined end, as teleological reasoning assumes, but rather works through differential survival and reproduction within specified environments, which may or may not be considered progress (who makes this determination is only one reason this view is problematic).

It is therefore startling to read Levin and his collaborator Rafael Yuste advocate for what they call "the intentional stance," arguing that asking "why" an evolutionary development occurs is sometimes even more important than asking "how" (Levin and Yuste 2022, 17). In short, they want to legitimate talking about goals in evolution, because they theorize that evolution "doesn't come up with answers so much as generate flexible problem-solving agents that can rise to new challenges and figure things out on their own" (2). This notion is central to their argument that cells can improvise new ways to arrive at traditional solutions, for example when a tadpole's transplanted eye achieves vision by connecting to the spinal cord. In lectures on YouTube, Levin and Bongard invoke William James's definition of intelligence: "Intelligence is a fixed goal with variable means of achieving it" (quoted in Cross Labs AI 2022). This definition authorizes Levin and collaborators to claim that plasticity at the cellular level is evidence of cellular "intelligence" (Levin and Yuste 2022, 1). Moreover, by focusing on goals at multiple levels, from individual cells to cellular networks to organs and tissues—the "why" rather than the "how"—they build on their discoveries showing that cells and cellular networks can improvise novel solutions to environmental challenges.

This leads directly to their first design principle for evolution: that cognition (or "intelligence") is modular in nature, each level building on its capacities to enable the larger goals of the next level up to emerge. "Single cells can process local information about their environments and their own states to pursue tiny, cell-level goals. . . . But networks of cells can integrate signals across distances, store memories of patterns, and compute the outcomes of large-scale questions (such as 'Is this finger the right length?' or 'Does this face look right?')" (Levin and Yuste 2022, 4). With feedback going upward to the next level and downward from that level to the one below, the goals within each level are nested inside one another for maximum efficiency. "Feedback loops within feedback loops, and a nested hierarchy of incentivized modules that can be reshuffled by evolution, offer immense problem-solving power" (7). Artificial life programs and computer modularized programming have independently discovered similar solutions, using recursive and nested architectures to exploit the same "immense problem-solving power."[7]

In addition, modularity in biological organisms facilitates change because it reduces the costs of innovation. "Modularity means that the stakes for testing out mutations are reasonably low. . . . Modularity is about self-maintaining units that can cooperate or compete to achieve local outcomes, but end up collectively working toward some larger goal" (Levin and Yuste 2022, 5). The cost is low because "each level doesn't need to know how the lower levels do their job, but can simply motivate them (with things like reward molecules and stress pathways) to get it done" (5). Modularity, Levin and Yuste conclude, "is part of what enables the emergence of intelligence in biology" (6).

Like turtles "all the way down" in the fable about the world resting on a turtle's back, intelligence in their view similarly goes all the way down. "When unicellular organisms joined up to make multicellular bodies, each module didn't lose its individual competency. Rather, cells used specific proteins to merge into every more complex networks that could implement large objectives, possess longer memories and look further into the future" (Levin and Yuste 2022, 7). A case in point of cells not losing competency is xenobot skin cells recovering the capacity to use cilia for motility.

The second design principle for evolutionary processes that Levin and Yuste identify is what they call pattern completion. Their analysis of this capacity focuses on how the modules are controlled. "Encoding information in networks requires the ability to catalyse complex outcomes with simple signals. This is known as *pattern completion*: the capacity of one particular element in the module to activate the entire module" (Levin and Yuste 2022, 8). Referring to the well-known "master regulator" genes that trigger coordinated expression of other genes, such as the *Hox* genes that specify the body plan for bilaterally symmetrical animals, they relabel such entities as pattern completion genes. "The key is that modules, by continuing to work until certain conditions are met, can fill in a complex pattern when given only a small part of the pattern" (8). This greatly facilitates evolutionary change, because "evolution does not have to rediscover how to specify all the cell types and get them arranged in the correct orientation—all it has to do is activate a simple trigger, and the modular organization of development (where cells build to a specific pattern) does the rest. Pattern completion enables the emergence of developmental complexity and intelligence: simple triggers of complex cascades that make it possible for random changes to DNA to generate coherent, functional (and occasionally advantageous) bodies" (9).

Pattern completion is also evident in the ability of some organisms to regenerate their bodies when amputation or damage has occurred; for example, a flatworm uses a bioelectric network of cells to store a pattern memory that controls individual cells to restore the entire body from a small segment. Pattern memory works because it "enables connections between modules at the same and different levels of the hierarchy, knitting them together as a single system. A key neuron in a lower-level module can be activated by an upper-level one, and vice versa." In conclusion, they write that "life has ratcheted towards intelligent designs by exploiting the ability of modules to get things done, in their own way" (Levin and Yuste 2022, 15).

The researchers point out that the promise of this kind of research is immense. "If we truly grasp how to control the setpoints of bodies, we might be able to repair birth defects, induce regeneration of organisms, and perhaps even defeat ageing (some cnidarians and planarian flatworms are essentially immortal, demonstrating that complex organisms without a lifespan limit are possible, using the same types of cells of which we are made)" (Levin and Yuste 2022, 17). Added to these visionary prospects are "new approaches to machine learning and AI" by exploiting architectures "based on ancient and diverse problem-solving ensembles beyond brains, such as those of bacteria and metazoan" (17).

With this, we go beyond the creative potential of symbiosis in the biological realm championed by Lynn Margulis to the wider field of technosymbiosis, which envisions the technological capture of microcognitions through gene editing and xenobot synthesis. Notwithstanding their differences, the underlying premises of symbiosis and technosymbiosis remain the same: that microorganisms have cognitive capacities that bestow remarkable abilities for innovations, rapid changes, and creative adaptations to challenging environments. Levin and Yuste conclude their essay by calling on biologists to embrace "the intentional stance," which would envision "cells and cellular processes as competent problem-solving agents with agendas, and the capacity to detect and store information" (Levin and Yuste 2022, 18). Such a perspective, they argue, is "no longer a metaphor, but a serious hypothesis made plausible by the emergence of intelligent behaviour in phylogenetic history." Although obviously still at a much more nascent stage than gene editing, the creation of xenobots in their view also has major implications for our understanding of our relations to other organisms, especially to the microcosm. "If we can come to recognize intelligence in its most unfamiliar guises, it might just revolutionise our understanding of the natural world and our very nature as cognitive beings."

I began this section with a quotation from William James about intelligence, so it is fitting to conclude with another of his aphorisms: "If you can change your mind, you can change your life" (quoted in Cross Labs AI 2022). In an article that Levin coauthored with Daniel Dennett, they provide additional reasons for changing our minds about cellular cognition. "Agents," they write, "need not be conscious, need not understand, need not have minds, but they do need to be structured to exploit physical regularities that enable them to use information . . . to perform tasks" (Levin and Dennet 2020, 5). They acknowledge that a single cell's "knowledge is myopic in the extreme. Its *Umwelt* is microscopic . . . but thanks to communication—signalling—with its neighbors, it can contribute its own local competence to a distributed system that does have long-range information-guided abilities, long-range in both space and time" (7). This is an excellent description of one kind of cognitive assemblage (down to even using "*Umwelt*"). They continue, "In large cell collectives, these are scaled massive in both space (to a tissue- or organ-scale) and time (larger memory and anticipation capabilities, because the combined network of many cells has hugely more computational capacity than the sum of individual cells' abilities)" (8). Moreover, they suggest that it was just this kind of system "that later became coopted to an even larger and faster goal-directed activity of brains, as ancient electrical synapses . . . evolved into modern chemical synapses, and cells . . . became speed-optimised lanky neurons" (8). "The key dynamic that evolution discovered" they write, "is a special kind of communication allowing privileged access of agents to the same information pool" (9). Indeed, in cognitive assemblages in which humans and computational media participate, the "special kind of communication" is achieved through coding architectures and network protocols that allow computers to collaborate with each other and with us. They conclude that "the point is not to anthropomorphise morphogenesis—the point is to naturalise cognition" (10).

By focusing on cognition, they achieve the kind of integration that ICF is intended to facilitate. They suggest that "there is one continuum along which all living systems (and many nonliving ones) can be placed, with respect to how much thinking they can do," which they call a system's "cognitive horizon" (Levin and Dennett 2020, 11). "The central point about cognitive systems, no matter their material implementation (including animals, cells, synthetic life forms, AI, and possible alien life) is what they know how to detect, represent as memories, anticipate, decide among and—crucially—attempt to affect." Consequently, the scale on which different kinds of cognitive systems can be measured is "the scale of its goals." This idea is enormously useful to the present project, because it allows the direct comparison of different kinds of cognitive assemblages by providing a uniform scale along which cognitive entities can be measured, from bacterial colonies to the most sophisticated human-computer systems. With advanced AIs and the technocapture of bacterial cognition, the scale of possible interventions has just widened enormously, both in language abilities and in the species transformations possible with gene-editing technologies.

Changing our collective minds about cognition as a modular capacity that appears at every level in biological and synthetic entities, not only in humans, opens onto breathtaking possibilities for human futures with our nonhuman symbionts (for an exploration of the wide implications, see Blackiston, Kriegman, et al. 2023). Fraught with dangers as well as replete with positive developments, our futures urgently require us to engage in ecological relationalities in which we see ourselves not as independent autonomous actors exercising free will, but as collaborators and partners with nonhuman entities that can contribute, in life-affirming ways, to our mutual flourishing far into the future.

# 5

# Rocks and Microbes:
# The Two Different Temporal Regimes
# of Biological and Mineral Evolution

Swept away by the successes of Newtonian celestial mechanics, the French scientist Pierre Simon Laplace (1749–1827) is credited with the ultimate reductionist boast.[1]

> We may regard the present state of the universe as the effect of its past and the cause of its future. An intellect which at a certain moment would know all forces that set nature in motion, and all positions of all items of which nature is composed, if this intellect were also vast enough to submit these data to analysis, it would embrace in a single formula the movements of the greatest bodies in the universe and those of the tiniest atom, for such an intellect nothing would be uncertain and the future just like the past would be present before its eyes. (Laplace [1814] 1951, 4)

Sometimes called "Laplace's Demon" (although Laplace called it only "une intelligence"), the entity he imagined can be seen as the extrapolated end point of what can be accomplished today by supercomputers modeling complex simulations that predict the mysterious movements of plate tectonics, the swirling complexities of ocean currents, and the weather patterns of tremendous storms. Yet, according to Giuseppe Longo, Maël Montévil, and Stuart Kauffman (hereafter LMK), there is one large (and for humans, particularly significant) domain where reductionist enterprises are doomed to fail: biological evolution (Longo et al. 2012). They argue that the methods central to physics and chemistry, which depend on analyzing trajectories in phase spaces, are inapplicable for evolutionary change, which cannot be predicted because the variables that would hypothetically define a phase space keep transforming. Hence their argument posits two distinct temporal regimes: one epitomized by celestial mechanics, still used to calculate trajectories for rocket flights to Mars using Newton's laws of motion, and another for biological evolution, where events remain forever unpredictable because there are no entailing laws, only what they call "enablement" (Longo et al. 2012, 1).

The rapidly emerging field of mineral evolution offers an ideal case for evaluating this claim. Pioneered by Robert Hazen and his collaborators (Hazen et al. 2008), the case for mineral evolution is built on identifying ten epochs in which the original three hundred or so minerals that emerged as the cosmos cooled and expanded after the Big Bang have now proliferated into the nearly six thousand minerals observed on Earth and in meteoritic samples. This approach has revolutionized how mineralogists think about mineral diversity by giving it a temporal dimension that the field previously lacked (or at least underappreciated). In a word, it has given mineralogy a *story*. The characters in this story include not only the minerals themselves, but also the agents and actors that interact with them to create a dynamic that moves minerals toward increasing complexity and diversity. The most important are water, plate tectonics, and lifeforms. Two of these fall under the auspices of physics and chemistry, while the third, of course, emerges from biological evolution. Thus mineral evolution can be read as a case study in the kinds of forces and possibilities that produce novelty. Given this, can a case be made that biological evolution is incommensurable in temporal terms with water and earth as agents of change?

I accept the broad distinction for which LMK argue, but the work of Hazen and his colleagues persuade me that temporality is better seen as a spectrum, with one end anchored in celestial mechanics, where interactions between entities are damped down and simplified by the vast distances between them, and the other end rooted in the hot, steamy, and vastly mutable exchanges of the microbial world, where changes are everywhere and the dynamics enormously complex. Although I suggest a temporal spectrum, the fact that the two ends represent two different temporal dynamics brings into sharp clarity the differences between earth, wind, and water as agents of change, and the motivated actions of living creatures as actors always seeking to continue their existences. Thus while both rocks and microbes have stories, including narratives in which their plots entwine and complicate one another, they also have distinctive dynamics that together create the possibility of *comparative evolution*. Just as comparative literature adds immeasurably to the study of individual national literatures, so comparative evolution has the potential to enrich our understandings of evolutionary dynamics in general, including technological evolution.

## Entailing Laws versus Enablement: The Argument for Different Temporalities

To highlight the importance of what they call "entailing laws," LMK sketch the "historical and conceptual development of physics" according to the following paradigm: (1) "analyze trajectories"; (2) "pull-out the key observables as (relative) invariants"; and (3) "construct out of them the intended phase space" (Longo et al. 2012, 9). For example, in tracing the arc of a pendulum, the key variables are position and momentum (the energetic regime of the pendulum's motion is implied by its momentum, which is a function of its angular velocity). A pendulum is a convenient example, because it typically swings along the arc determined by its downward rod. A phase space for a simple pendulum looks like figure 5.1.

The horizontal axis P is the pendulum position relative to the zero point when it is at the equilibrium rest position of straight down (B on the diagram). The vertical axis V is its velocity (clockwise V > 0, counterclockwise V < 0). Tracing its clockwise movement starting at its endpoint arc A on the right side, its position is the maximum distance from the rest point, and its velocity is zero as it pauses before swinging back (P = max, V = 0). When it reaches B as it swings toward the left arc, its position is zero and its velocity is at a maximum, giving B (P = 0, V = max) on the diagram. Having reached the end point of its left arc at C before beginning the counterclockwise movement, its velocity is at zero and its position is at the negative maximum (P = -max, V = 0). It swings counterclockwise to B', where its position is zero and its velocity is at negative maximum (P = 0, V = -max).

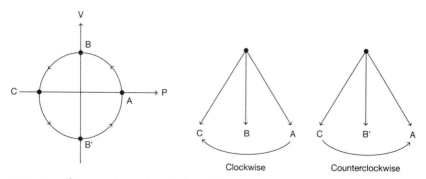

FIGURE 5.1. Phase space diagram for a simple pendulum.

This very simple phase space diagram illustrates the central fact about phase space diagrams in general: they are spaces in which all possible states of a system are represented, with each possible state corresponding to one unique point in the phase space. For mechanical systems, it is customary to use position and momentum as variables; other possibilities are energy and time. A phase space trajectory represents the set of states possible from one particular initial condition (in our pendulum example, points A and C). A full phase space represents the set of states possible from *any* initial condition. Often phase spaces are very high dimensional. For example, a phase space for a gas may use a separate dimension for each position and momentum for each gas molecule x, y, and z.

Phase space diagrams are useful because they can disclose implications about a system's dynamics that may not otherwise be apparent. The underlying regularities are often called the "laws of motion," but they are laws in a very specific sense: the so-called laws are abstractions from empirical data that are never quite as clean as the laws state. As Nancy Cartwright (1983) observed in her classic text *How the Laws of Physics Lie*, minute perturbations between multiple (and sometimes unknown) factors always cause small deviations from the abstractions, which I call "the noise of materiality."[2] In a parallel vein, LMK are very clear that phase space diagrams are invented abstractions used to reveal underlying regularities. They comment that phase spaces are "our remarkable and very effective invention in order to make physical phenomena intelligible" (Longo et al. 2012, 9). As LMK explain, "their regularities, as invariants and invariant preserving transformations in the intended spaces (thus their symmetries) allow a finite description, even if they are infinite or even of infinite dimension" (9). For example, in the case of the simple pendulum, the symmetries include the fact that the pendulum swings equal distances and traces equal arcs in both the clockwise and the counterclockwise directions, so even if it continues swinging indefinitely, its possible states can still be represented in a finite diagram.

To illustrate, LMK reference a three-dimensional Cartesian space, a customary choice for geometrical representations. "It is infinite, but the three straight lines are given by symmetries (they are axis of rotations) and their right angles as well (right angles, says Euclid, are defined from the most symmetric figure you obtain when crossing two straight lines). These symmetries allow us to describe this infinite space in a very synthetic way" (Longo et al. 2012, 9). Their general point is clear from the pendulum example given above. Dynamical mechanical systems can be analyzed through phase space diagrams that reveal their underlying symmetries and regularities. The regularities emerge because they are subject to "entailing laws," that is, the laws of motion, which means their actions are predictable.

In another example about which I wrote in *Chaos Bound* (Hayles 1990), suppose that instead of a single pendulum we fasten to its pendant another pendulum to make a double pendulum. Its motion ceases to be strictly predictable, because the system is now structured so as to bring minute uncertainties in initial conditions quickly up to macroscopic expression. Nevertheless, this chaotic system is still deterministic, and its possible states can still be represented with a phase space diagram. The system is unpredictable in the sense that it is not possible to know where in the phase space a specific trajectory will unfold, but its general dimensions are specified by an attractor that emerges as multiple trajectories are traced within the phase space.

Longo, Montévil, and Kauffman discuss more complex cases as well. They point out that in phase diagrams for quantum physics, for example, the "quanta do not go along trajectories, but the wave function does," and they discuss a number of other more complex situations (Longo et al. 2012, 9). Their general point, however, is clear from the pendulum example given above. When dynamical mechanical systems are analyzed, regularities emerge because they are subject to the laws of motion (with the caveat noted above), which makes their actions predictable. Perhaps because determinism is a contested notion philosophically (and scientifically), LMK do not use the term but instead refer to "prestated" results. By this they mean that phase space diagrams can be constructed for dynamical mechanical systems that allow all the possible states of the system to be "prestated," that is, predicted from the "entailing laws."[3] They conclude, "in physics the observables, which yield the phase space, derive from the invariants/symmetries in the trajectories. More exactly, they derive from the invariants and the invariant preserving transformations in the intended physical theory . . . that is, the conceptual construction of the phase space follows from the choice of the relevant observables and invariants (symmetries) in the physico-mathematical analysis."

One would think this point would be obvious, but their real goal is to contrast the "prestated" nature of dynamic mechanical systems with biological evolution. Like Lynn Margulis's objection that biology is not simply reducible to chemistry and physics but has its own unique characteristics (Margulis 1997, 266), LMK emphasize that living organisms are "Kantian wholes"; "they have an internal, permanently reconstructed autonomy, in Kant's sense, or Varela's autopoiesis, that gives them an ever changing, yet 'inertial' structural stability" (Longo et al. 2012, 5). In contrast to the phase spaces that can be constructed for mechanical systems, living systems undergo "continual and open-ended changes" (12), for example in the human biome. "Each of these microbes is a Kantian whole, and in ways we do not understand yet, the 'community' in the intestines co-creates their worlds together, co-creating the niches by which each and all achieve, with the surrounding human tissue, a task closure that is 'always' sustained even if it may change by immigration of new microbial species into the community and extinction of old species in the community" (12). Because "organisms transform the ecosystem while transforming themselves" (12), the persistent symmetries crucial for the construction of phase space diagrams in physics do not apply, or more precisely, the symmetries are changing in unpredictable ways, so that it is not possible to make such constructions. "No laws of motion, nor boundary conditions to integrate such laws were we to have them, can be formulated. No law entails the evolution of the biosphere" (3). LMK make clear that once an organism *has* evolved, its actions are subject to chemical and physical analysis; what chemistry and physics cannot predict, however, is whether it will evolve in the first place.

Although unpredictable, evolutionary trajectories are not merely random, either. Instead, they progress through a series of what LMK call "enablement(s)" (Longo et al. 2012, 7), which comes into existence through the emergence of new biological functions. One of their examples is the evolution of the lung fish, a fish that has both lungs and gills. "Water got into some lungs. Now, a sac existed with air and water in it, poised to evolve into a swim bladder" (7). (Swim bladders are internal gas-filled organs that enable fish to control their buoyancy, and so to stay poised at their current depth without having to expend energy swimming.) Thus the lung fish was "preadapted" to develop a swim bladder.[4] Now suppose that a bacterium or worm has evolved that can live in the swim bladder. Before the bladder existed, one could not have predicted this emergence, but once the swim bladder exists, the possibility is opened up. This is what LMK call the "adjacent possible," a new niche that in its turn may open up several other adjacent possibles.[5] Thus evolution proceeds by new functions opening up new niches, which niches in turn open up more new functions, in cycles of continuous reciprocal interactions between niches and organs/organisms.

What accounts for the difference between the prestated phase spaces of chemicophysical systems and the zigzagging paths of adjacent possibles in evolutionary trajectories? Although LMK do not explicitly draw this inference, it seems obvious to me that the essential difference lies in the fact that living organisms have stakes in whether they live or die, and they use their cognitive capacities (that is, their ability to process information from their environments in ways meaningful to them) to discover new ways to survive and thrive. In a word, they have an innate creativity that physicochemical systems do not. As LMK poetically put it, "reductionism reaches a terminus at the watershed of life. With Heraclitus we may say of life: The world bubbles forth" (Longo et al. 2012, 3).

This is not to say, of course, that physicochemical systems do not possess agency. As the next section vividly illustrates, plate tectonics, oceans, winds, and water can cause tremendous changes in mineral compositions and are vital contributors to mineral evolution. Recognizing this, I depart from Longo and Montévil's rhetoric in characterizing chemicophysical systems as "inert" in their article "The Inert vs. the Living State of Matter: Extended Criticality, Time Geometry, Anti-entropy; An Overview" (Longo and Montévil 2011). As feminists identified with the New Materialism such as Jane Bennett could have instructed them, physical processes should be recognized as the vital forces they are. Where I depart from Bennett and other New Materialists is in recognizing that the cognitive capacities of living organisms enable them to have a creativity and responsiveness to their environments different in kind from those of physicochemical processes. The arguments of LMK enable me to extend this observation to the different temporal regimes in which physicochemical processes work in contrast to the emergent and unpredictable temporalities of biological evolution.

The next section explores the complexities that emerge when the predictable temporalities of physicochemical processes interact with the different temporal regimes of biological evolution to form the emerging field of mineral evolution. In a reciprocal circular dynamic, using the term "evolution" with respect to changes in minerals across time raises the issue of whether there could be, along with their obvious domain-specific formations, systemic similarities between different kinds of evolutions. It is with that fascinating speculation that this chapter will conclude.

## Mineral Evolution: A New Perspective

The concept of mineral evolution has an origin story. According to Robert Hazen, the idea hit him when, at a Christmas party in 2006, Harold Morowitz asked him if there were clays during the Archean eon. The timing is significant, because the Archean eon marks the beginning of the emergence of life (usually dated 3.7 billion years ago; the Archean dates from 3.8 to 2.5 billion years ago). The questioner is significant as well, because Morowitz is the author of *The Emergence of Everything: How the World Became Complex* (2004), so Hazen would have known he was trying to place minerals in his evolutionary schema. The topic matters too, because to form, clays require the actions of change agents (principally water and wind, but also biological activity). In effect, Morowitz was asking Hazen to think about the relation of temporality, change agents, and mineral formations as interacting and mutually interdependent variables. As Hazen remarks, "it's a question that mineralogists never thought to ask" (Hazen 2010, 1).

Hazen took the message to heart and over the next several months assembled a team of collaborators to explore this question along with many others. His team included a geobiologist, a paleotechtonics expert, a metamorphic petrologist, a meteorite expert, and of course other mineralogists. The result was a perspective on mineral formations that placed them in an evolutionary, and therefore temporally aware, context, in the process challenging conventional modes of classification and analysis.

In the inaugural review article "Mineral Evolution," Hazen and his collaborators (Hazen et al. 2008) lay out an evolutionary schema starting from ~4.56 Ga. (The term "Ga" is a useful abbreviation for "giga annum," read as "billion years ago." "Ma" is "mega annum," read as "million years ago." The tilde [~] is standard mathematical notation for "about" or "approximately.") Their schema, which continues to the present, is divided into three main eras and subdivided into ten distinct but also overlapping phases. The eras are defined by planetary developments, starting from planetary accretion more than ~4.56 Ga, to the reworking of the Earth's crust and mantle (~4.55 Ga to ~2.5 Ga), and finally to bio-mediated mineralogy (~2.5 Ga to the present). The different phases were inaugurated by new physical, chemical, or biological processes (or some combination thereof), so they emerged from the intersection and interpenetration of the two distinct temporal regimes discussed above. The trajectory is toward increasing mineral diversity and complexity. As the article explains, "these processes diversified Earth's mineralogy in three ways. First, they increased the range of bulk compositions from which minerals form, from the original, relatively uniform composition of the solar nebula. Second, geological processes increased the range of physical and chemical conditions under which minerals form. . . . Third, living organisms opened up new reaction pathways by which minerals formed that were not accessible in the abiotic world" (Hazen et al. 2008, 1694).

The first mineral, the team decided (answering a question no one had asked before) was pure carbon, in the form of diamond. During the first era, as the presolar dust particles began to accrete, the minerals were limited to no more than a few dozen, mostly oxides, carbides, nitrides, and silicates. As gravitation forces increased with star formation and the resultant heating of the solar nebula, the chondrules commonly found in ordinary chondrite meteorites formed, along with calcium-aluminum inclusions. Melting, differentiation, and impact processes increased the diversity, although "no more than about 250 minerals have been found in unweathered meteorites." (That number recently increased by three with the discovery of three new previously unknown minerals found in the El Ali meteorite from Somalia.) "These provided the mineralogical raw materials for earth and other terrestrial planets," the authors summarize (Hazen et al. 2008, 1694).

At ~4.55 Ga, the Earth's crust and mantle were "violently disrupted" by the hypothesized collision with a planet-sized object called Theia that wrenched off the chunk of Earth that today we know as the Moon. As a result, Theia was annihilated, and Earth's crust and part of the mantle was largely melted and in this sense "reset" (Hazen et al. 2008, 1699). The authors speculate that nearly all the extant minerals formed to that point had been made through the crystallization of igneous rocks, along with the minerals from the steady bombardment of comets and asteroids. Part of the challenge of reconstructing this part of Earth's evolutionary history is the "paucity and typical extreme alteration of the rock record. No known rocks survive from before ~4.03 Ga"–that is, from before the collision (1699). Although we typically think of lunar meteorites as rocks that have fallen to Earth from the Moon (several of which are in our house as part of the Nicholas Gessler's "Rocks from Space" collection; see Gessler 2010), it is possible that some unaltered rocks from Earth that date from before the Moon-forming collision may be found on the surface of the Moon, adding to the mineralogical knowledge of this dark period.

The second era, defined by a gradual decoupling of the Earth's crust from the mantle, developed from "strictly physical and chemical processing," that is, from the underlying regularities of the kind that can be diagrammed using phase spaces (Hazen et al. 2008, 1696). These included tectonic forces as part of the Earth's crust was forced underneath another plate through subduction. Wet, partial melting of oceanic crust created magma, along with a host of new minerals. New hydrothermal, igneous, and contact metamorphic lithographies formed, including large-scale upper-mantle and crustal ore deposits along with surface exposure of high-pressure metamorphic formations. The result was an increase from about three hundred minerals to around fifteen hundred, a fivefold expansion of Earth's mineralogical repertoire.

The third era marks the coevolution of the geo- and biospheres and extends for ~3.5 Ga to the present. Prior to the Proterozoic (beginning ~2.45 Ga), the authors describe the influence of lifeforms on mineralization as "minimal" (Hazen et al. 2008, 1703). That changed with the Great Oxidation

Event (~2.3 Ga to ~2.2 Ga), which as we have seen was the result of purple and green photosynthetic microbes discovering how to break down water and generating oxygen as an (originally toxic) by-product. Surface weathering of rocks helped to mediate the changing composition of the atmosphere, which along with irreversible redox changes in the oceans, resulted in the "precipitation of massive carbonates, banded iron formations [BIF], sulfates, evaporates, and other lithographies" (1696). The Phanerozoic (~500 Ma to the present) added more minerals from bioskeletons of carbonate, phosphate, and silica, creating new mechanisms that continue into the present. The result was an increase from fifteen hundred to more than forty-three hundred minerals known today (and counting).

Returning to Morowitz's question, Hazen and his collaborators give special attention to the so-called clay mineral factories of the Neoproterozoic period. "It is well established that microbial activity enhances clay mineral production, for example by the bio-weathering of feldspar and mica" (Hazen et al. 2008, 1709). A significant increase in clay mineral deposits has been observed during this period, which "may have been the result of increased microbial activity in soils." Because clay surfaces absorb carbon (which is why clay minerals are often used in filters), the formation of clays may have resulted in increased oceanic sequestration of carbon.

Other biological activity not only facilitated the formation of diverse minerals but also contributed to climate events. By volume, calcium carbonates (mainly coming from the skeletons of plankton) are the most important minerals. Planktonic calcifers provided a steady source of calcium carbonate to deep ocean sediments. This had the effect of buffering (or resisting pH change when acidic or alkaline chemicals are added) ocean carbonate ion concentration, which "moderated glacial events and reduced the likelihood of future snowball events," that is, future ice ages. "In the Phanerozoic Eon, for the first time in Earth's history carbonate depositions in the deep oceans was comparable to that in the shallow ocean environments" (Hazen et al. 2008, 1709). Feedback loops of this kind are examples of the coevolution of the geo- and biospheres, creating novel avenues for mineral evolution and also affecting the Earth's oceans and atmospheres, and as a consequence, climate events as well.

## Differences Initiated by an Evolutionary Perspective

The payoffs for an evolutionary perspective on minerals include a systemization of how planetary bodies change over time, not only with regard to the Earth but also for the other terrestrial planets in our solar system, and possibly beyond. Hazen and his collaborators identify three main processes driving this evolution. "First are separation and concentration processes, for example planetary differentiation, outgassing, fractional crystallization, partial melting, crystallization, and leaching by aqueous fluids" (Hazen et al. 2008, 1711). These processes had the effect of separating the relatively uniform distribution of material in the early solar system into a "broad spectrum of bulk compositions." As the different elements became concentrated into various regions, their interactions inevitably resulted in new mineral compositions.

The second main process is the emergence of what the authors call "intensive variables" such as temperature, pressure, and the actions of water, oxygen, and carbon dioxide. The results are a range of climatic and environmental regions ranging from ice caps to deep-ocean hydrothermal venting. From volcanic eruptions to glacier formations, these intensive variables are active agents in creating new minerals and mineral formations.

The third, and perhaps the most important, process consists of biomineralization and all the new pathways that biological organisms opened up for mineral novelty and diversity. Unlike mineral diversification resulting from the minimization of Gibbs free energy (a process that can be represented by a phase space diagram), "new minerals may be catalyzed by living organisms in highly localized volumes that are not in equilibrium with the surrounding geochemical milieu" (Hazen et al. 2008, 1711). Just as Lovelock (1988) argued nearly forty years ago, the existence of far-from-equilibrium systems implies the presence of living organisms whose actions help to maintain the conditions necessary for life to exist. Hazen and his collaborators estimate that the Earth's surface oxygenation, along with the biological activity associated with it, "may be responsible directly or indirectly for more than two-thirds of all known mineral species" (Hazen et al. 2008, 1712).

Moreover, an evolutionary perspective has the effect of activating a reversible internality. Minerology before the idea of evolution was introduced was often articulated from a perspective that located minerals within the human domain, as objects to be analyzed and classified according to the criteria with which humans framed it, for example crystalline structures and chemical compositions. Evolutionary perspectives, however, reversed this framework, locating humans within the evolution of the cosmos, including the history of Earth. The effects, often subconsciously realized, are to emphasize that human lives have existed for only a very brief moment in cosmic history, surrounded by objects much older in time and with much deeper roots in determining how the Earth came to be as it is.

Notwithstanding their accomplishments, Hazen and his colleagues tend to underrecognize LMK's distinction between biological processes and the physical and chemical processes that can, at least in principle, be "prestated," to use LMK's term (Longo et al. 2012, 9). In a sense Hazen's team recognize it by referring to "strictly" physical and chemical processes to make clear they are not at that point including the biological production of minerals (Hazen et al. 2008, 1696). In other ways, however, they tend to gloss over the difference between physicochemical processes compared to biological outcomes drawn by LMK. For example, it seems reasonable to suppose that such momentous innovations as the Great Oxidation Event could not have been predicted, because it depended on certain species of photosynthetic microbes being numerous enough in the environment to create an adjacent possible niche for other organisms who could use the increased oxygen level as an energy source rather than respond to it as a toxin (and hence opening up all the niches occupied today by animals and plants). Since surface oxygenation was, according to Hazen and colleagues (2008), the *principal* factor in the evolution of minerals on Earth, it follows that the minerals resulting from this event could not have been predicted either.

Hence we may take with a grain of salt Hazen and colleagues' claim that all terrestrial planets and moons follow the same evolutionary trajectory. "The general principles observed for the emergence of mineralogical complexity on Earth apply equally to any differentiated asteroid, moon, or terrestrial planet," they write (Hazen et al. 2008, 1712). "The degree to which a body will advance in mineralogical complexity beyond the relatively simple achondrite stage is dictated by the nature and intensity of subsequent cycling (and hence repeated separation and concentration of elements). Consequently, a planet's surface mineralogy will directly reflect the extent to which cyclic processes, including igneous differentiation, granitoid formation, plate tectonics, atmospheric and oceanic reworking (including weathering), and biological influences, have affected the body's history." Note that in this list, "biological influences" are referred to without distinguishing them from the previous nonbiological forces.

Thus they imply that just as "cyclic" physical processes all follow the same trajectories, whether they occur on Earth, the Moon, or Mars, so too would any emergence of life other than on Earth also follow Earth's model. Accordingly, they read the history of terrestrial bodies lacking water as stopping at earlier points along the same trajectory that Earth followed (for example, the Moon, and Mars after it lost its atmosphere and water). The problem with this mode of reasoning is that there is only one example in the solar system (that we know of) of a terrestrial body that experienced biological emergence: the Earth.

Suppose another terrestrial body existing somewhere under similar conditions had also seen life arise. It is entirely possible that on such a planet, the Great Oxidation Event may not have taken place, and an atmosphere dominated by hydrogen sulfide continued to exist. In that case, the evolution of minerals on that planet would be dramatically different (since two-thirds of Earth minerals owe their existence to surface oxygenation). Thus, in suggesting that all terrestrial bodies follow the same kind of mineral evolution, Hazen and colleagues may be considerably underestimating the unpredictability (that is, in LMK's terms, the unprestatedness) of mineral evolution on (hypothetical) other terrestrial planets on which life had emerged.[6]

It is somewhat naive, then, to suggest that by observing the mineralogy of other moons and planets we can find indirect evidence for the existence of life, because we know what kinds of biomineralizations occurred on Earth. The idea is that if we observe similar biomineralizations on another planet, it would indicate that life existed there (with the implication that biomineralizations of large landscapes may be more easily detected remotely than the life itself, which may be only microbial). In fact we have no idea of the kinds of life that could emerge, and consequently the kinds of biomineralizations that might have taken place. Perhaps the only somewhat safe speculation would be to say that if a mineral landscape, of whatever kind, on another terrestrial planet showed diversity and complexity roughly equivalent to that of Earth, it could be considered a candidate for the emergence of life.

## Dynamic versus Static Perspectives

Having laid out their schema for mineral evolution, Hazen and colleagues turn to consider the appropriateness of using the term "evolution" to denote the trajectory they trace. They acknowledge that some may find it surprising and even unwarranted. Noting that "at its most general" level, evolution simply means "change through time," they suggest that mineral evolution "implies something more, as it arises in part from a sequence of deterministic, irreversible processes that lead from mineral parsimony of the pre-solar era to progressively more diverse and complex assemblages." Citing Morowitz among others, they suggest that this trajectory is "a fascinating specific example of the more general process of cosmic evolution" (Hazen et al. 2008, 1712).

Nevertheless, they warn that it would be a mistake simply to assume that mineral evolution parallels biological evolution, since minerals do not "compete" to survive, nor do they pass genetic characteristics onto offspring. Leaving aside biomineralization, minerals formed in the prebiotic era were governed by the underlying regularities of the physical and chemical forces acting on them. As an example, they cite the preferential crystallization of quartz from granite magma over olivine. This does not happen because quartz "outcompetes" olivine. Rather, "minimization of Gibbs free energy simply leads to nucleation and growth of quartz, but not olivine" (Hazen et al. 2008, 1712).

This acknowledgment notwithstanding, as we have seen there is a tendency in their rhetoric to lump physical and chemical processes indifferently together with biological processes, without fully acknowledging that lifeforms have creative possibilities that physical and chemical processes alone do not. For example, in the very paragraph where they stress the differences between physicochemical evolutionary forces and biological evolution, they write this: "The driving force for mineral evolution, rather [that is, rather than the biological mechanisms of survival and reproduction], is the evolving diversity of prebiotic and *biologically mediated* temperature-pressure-composition environments" (Hazen et al. 2008, 1712, emphasis added). As soon as a process is biologically mediated, however, it loses the predictability of physicochemical processes and may evolve in irreversible, creative directions that not only cannot be predicted but also may be unique to a particular time and place, never to be duplicated anywhere else. That is to say, biological mediation not only gave mineralogy a story; it also gave mineral evolution a unique and idiosyncratic *history*.

As a consequence of such lacunae in Hazen and his team's narrative (Hazen et al. 2008), the full story of how biological mediation changed the evolution of minerals and the pathways (adjacent possibles) by which mineral evolution occurred has yet to be written. Of course, such a story would be enormously complex, orders of magnitude more complex than the story of evolution that Hazen and his team have written, so it would be well at this point to acknowledge their remarkable contributions and celebrate their achievements.

To measure the changes they have accomplished, we can compare the dynamic story of mineral evolution with the static account of mineralization before temporality was seriously taken into account. The customary way in which mineralization was discussed and minerals classified was through their "equilibrium and physical properties," for example their chemical composition as indicated by "the principal anion (silicates, carbonates, halides, etc.)," which divided "the mineral kingdom into chemically related groups of species" (Hazen et al. 2008, 1713). Further subgroups were based on "crystal chemical criteria" (that is, what kind of crystal structures they had), which divided minerals "according to predominant topological motifs." Since the focus here is on chemical and physical processes, analysis typically proceeded by constructing equilibrium phase diagrams that show "the conditions of stability under which individual minerals form and groups of minerals coexist." Careful to acknowledge the usefulness of this traditional mode of classification, Hazen and his collaborations write that "this versatile approach to framing the subject of mineralogy, by combining equilibrium crystal chemical and thermodynamic considerations, underscores the close relationship between a mineral's physical properties and its arrangement of atoms" (1713). Ever so politely, the authors hint that this approach, while it has scientific validity to recommend it, also makes the subject of mineralogy somewhat *boring*, because it erases the drama and excitement of *how* these configurations came about, and in the process it also minimizes the role of biological mediation in opening up new pathways for diversification (in their phrasing, the "inherent drama and excitement of minerology may be lacking").

Their approach would not deny that these physical and chemical processes are "vital to mineralogy" but rather would introduce them in the context of evolutionary changes. Beginning with the parsimony of the presolar nebulae, the ideal minerology course they imagine would stress that these dozen or so "ur-minerals" manifest a variety of bond types (covalent, ionic, metallic, etc.) and incorporate structural motifs (polyhedral coordination of silicon, magnesium, calcium, etc.). From there the course would discuss the "relatively simple phase relationships that lead to their condensation," perhaps by analyzing them through equilibrium phase diagrams. Thus the basics of traditional mineralogy would be taught, "but introduced as part of a larger evolutionary story" (Hazen et al. 2008, 1713). Their point is that the traditional approach is no longer sufficient by itself. To take advantage of enormous increases in knowledge about mineralogical species and processes that has been accomplished with the development of advanced research techniques and computer databases, the field needs a fuller, more accurate, and more dynamic story about mineral evolution.

Not coincidentally, many of these new realizations focus on the roles of biological processes in mineral evolution. Judging by this article, the field is still waking up to the point made by LMK, that mineral evolution did not just become more complex with the emergence of life; it also made a transition from the prestatedness of phase space diagrams to the unpredictable (although not simply random) emergences of life's creativity. The coevolution of the geo- and biosphere means that the chemical and physical processes now have, in addition to their underlying regularities, an inextricably complex component of unpredictable novelty.

## A General Theory of Evolution?

When a field adopts a traditional term but gives it new meaning, typically there are two general types of reaction. One is to object to the move and question its appropriateness. The other is to grant the new denotation and then speculate on what is implied by its juxtaposition with the traditional meaning. This happened with *artificial life*, a term used for computer simulations such as Conway's Game of Life (n.d.), which showed the emergence of different "species" interacting with each other in computational animated displays. Robert Rosen (1991), among others, used the emergence of "Artificial Life" to speculate about "life itself," a project made possible, he argued, by the emergence of two different *kinds* of life, one based on carbon, the other on silicon. The idea of mineral evolution has seen similar dyadic responses: from those on the one hand who object to applying evolutionary terminology to minerals, and on the other hand from those who want to speculate about what the juxtaposition of mineral evolution with biological evolution implies (and the list can be extended to technological evolution and even cultural evolution).

For my project, technological evolution is of special interest, particularly when paired with biological evolution. Consequently, I am especially intrigued when Hazen and colleagues identify three aspects that they suggest may be characteristic of "all complex evolving systems" (Hazen et al. 2008, 1712): selection, punctuation, and extinction. In terms of the micro-evo-techno ecological relationalities described in chapter 1, these three processes constitute another lens through which first- and second-order emergences may be related.

For selection, Hazen and colleagues point out that complexity typically arises because many independent agents are interacting with one another in myriad situations. Nevertheless, only a relatively few emergences occur, because of what they call "selection rules" (Hazen et al. 2008, 1712). LMK make a similar point about random combinations: most possible complex combinations will never exist, because constraints limit which are viable (Longo et al. 2012). As Hazen and colleagues remark, "neutrons and protons assemble into only a few hundred stable isotopic species, while the 83 geochemically stable chemical elements combine to form only a few thousand mineral species. Stochastic processes may influence the specific outcomes of some selective events, but selection is guided by the physical and chemical principles and thus, by definition, is not random" (Hazen et al. 2008, 1712). Here we may recall Lynn Margulis's claim that random mutation cannot be a principal driver of biological evolution, because "99.9 percent of mutations are deleterious" (Margulis and Sagan 2002, 11–12).

In terms of technological evolution, the relevant constraints include not only the physical and chemical limits on what is possible, but also the requirements for funding, investment, government regulation, marketing considerations, competency of the organization doing the research and development, and a host of other factors. It makes sense, then, that of all possible complex technological inventions, only a few will succeed and be realized in competitive cultural and biological conditions. Frequently a single organization can make a difference by directing funding and development opportunities in a certain direction, for example as Warren Weaver did with molecular biology when he was director of the Rockefeller Foundation (see Weaver 1967).

The second generality that Hazen and colleagues (2008) identify is punctuation. Citing Niles Eldredge and Stephen Jay Gould's notion of "punctuated equilibria" (Eldredge and Gould 1972), they comment that this pattern, which has been observed in fields as diverse as nucleosynthesis and language emergence, is also characteristic of mineral evolution. "Key irreversible events in Earth's history—planetary differentiation at 4.5 Ga and subsequent formation of the atmosphere and hydrosphere, the initiation of subduction at >3.0 Ga, major atmospheric oxidation events at around 2.2 and 0.6 Ga, and the emergence of a terrestrial biota in the Phanerzoic Era, for example—forever altered Earth's mineralogical landscape" (Hazen et al. 2008, 1712).

With technological evolution, similar irreversible events occurred as well in punctuated fashion. From the invention of agriculture to the creation of nuclear weapons, watershed events changed how people lived, altering life patterns so profoundly that there was no going back from them. It is interesting to speculate that this punctuated rhythm may occur because it is deeply entwined with biological evolution. When a discovery is made that opens up not just one adjacent possible but a whole realm of them, considerable time may elapse as the full implications of this development are realized, so a leap forward is followed by a relatively long period of filling in the gaps. An example is the invention of the transistor. At one swoop, transistors solved a number of problems with the vacuum tubes they displaced, including heat dissipation, size requirements, system overload, fragility of components, and inefficiencies in information storage, analysis, and transmission. The transistor opened up an enormous range of adjacent possibles, so its discovery was followed by a robust period of innovation and dissemination, which metaphorically we can call the Cambrian explosion of technological evolution. Now, as the transistor is approaching its inevitable limits of viability (in this case the size of a silicon atom), experiments abound in replacement technologies, from quantum computing to carbon and graphite nanotubes. No doubt when a breakthrough is achieved and another new technology is made feasible, we will enter another period of rapid change as the possibilities opened by that development are exploited.

The third characteristic that Hazen and his colleagues identify is extinction. They conclude that with a few minor exceptions, extinction is not an important dynamic for mineral evolution ("one would be hard pressed to postulate a single mineral species that once was found in a near-surface environment but no longer exists"; Hazen et al. 2008, 1713). For technological evolution, however, extinction is key in several different senses. As chapter 1 describes, in the third millennium, the pace of technological evolution is accelerating almost exponentially. Contrast today's rapid change with the invention of agriculture, which was followed by several centuries of development as different grains and plants were tested and adopted or rejected; today the lifetime of a technological innovation may be measured in a few years or even a few months.

Another sense in which extinction is critical to technological evolution, of course, is the rapidity with which technological interventions are driving myriad biological species to extinction through loss of habitat, land air and water pollution, ocean plastics and other contaminants, insecticides, global warming, and climate change—the list goes on. Many now question whether the human species will drive itself to extinction. Lynn Margulis, for example, was confident in the robustness and longevity of the microcosm, but with humans she thought it was a question not of if but of when. It is possible that continued technological experimentation may also fundamentally change the human species so profoundly that it metamorphizes into something else, a scenario not so far-fetched now that gene editing is an everyday reality.

All three generalities of selection, punctuation, and extinction apply at least as much to technological evolution as they do to mineral evolution. We have seen that mineral evolution involves both the predictability of underlying regularities and, with biomineralization, the unpredictability of biological evolution. How does technological evolution stack up in this regard? Underlying regularities play prominent roles in the selection process, since they determine the criteria that will rule on whether a device or process works or not. But it seems to me that the temporal regime at play in technological evolution is the same as that characteristic of biological evolution, that is, the zigzaggedness of adjacent possibles and the inherent unpredictability of trajectories as new pathways open up for development.

### Biota's AID in Mineral Evolution: Amplify, Innovate, Diversify

In terms of the book's overall argument, this chapter has helped to establish a theoretical and empirical basis for my distinction between the kind of agency that physicochemical processes have in contrast to the motivated actions of biological lifeforms, which tend to act in ways consistent with their continued existence. The effects of biota (primarily microorganisms) on minerals can be summarized by the acronym AID—amplify, innovate, diversify. Microorganisms *amplify* because they dramatically increase the amount of minerals, from the masses of banded iron formations to the heaps of calcium carbonate in the ocean from bioskeletons. Biota contribute to the rise of new minerals (*innovate*) because they open pathways for evolution, directly and indirectly, that would otherwise have been inaccessible to the operations of physical processes by themselves. Biota *diversify* minerals by greatly increasing the *kinds* of minerals on the Earth, by a very significant amount (as of today of about two-thirds of the total). In short, the story of mineral evolution would not be nearly as rich and varied as it is without the AID effects of lifeforms.

In addition, the story of mineral evolution illustrates how the two different temporal regimes of systemic prestatedness and unpredictable emergences work together to create a more complex narrative than either one alone could. Life, with all its creativity, unpredictability, and distinctive dynamics of leaping from one adjacent possible to others, enlivens and expands the predictable effects of wind, ocean, water, and earth movements on rock formations.

Finally, this chapter establishes a basis for framing the historically contingent trajectories of emergences and evolutions—whether biological, mineral, or technological—as dramatic and engaging stories with vivid characters and riveting plotlines. By showing how selection, punctuation, and extinction work in the case of rocks, mineral evolution provides a basis for comparison not only with biological but also with technological evolutions. Like minerals, technological artifacts depend on the underlying regularities of physical processes to work once they are invented. But like biological lifeforms, technological evolutions leap from one adjacent possible to another in surprising, creative, and unpredictable ways. When the conscious strategies of humans collaborate with the nonconscious cognitions of nonhumans, the potential for life-altering developments is amplified and accelerated, not only for us but for all the nonhuman species with whom we share the planet—for better and for worse.

6

# Inside the Mind of an AI:
# Materiality and the Crisis of Representation

In line with speculations that machines may evolve, this chapter interrogates the evolution of large language models using Transformer architecture from predecessor technologies such as recurrent neural nets (RNNs), convolutional neural nets (CNNs), and other AI approaches.[1] In focusing on how to understand the language capabilities of Transformer neural nets, this chapter again engages with questions of meaning, thus continuing the interrogation of cognitive computational media and their abilities to create, transmit, and engage with meaning-making practices.

On July 11, 2020, OpenAI released the beta version of an artificial intelligence program called GPT-3 (Generative Pretrained Transformer, version 3). Transformer architecture had already been described in the seminal article "Attention Is All You Need" in December 2017 (Vaswani et al. 2017), but it was only with the release of GPT-3 and similar large language models (LLMs) that the full potential of Transformer AI was revealed.[2] From a literary viewpoint, GPT-3 is among the most powerful—and interesting—natural language processors ever created.[3] The program works by responding to an input, which can be either a prompt, a question, or a command (among other possibilities). It attempts to predict the next word sequences and can produce hundreds of words in reply. It is able not only to create semantic coherence and syntactic correctness but also to capture high-level qualities such as style and genre. Arguably it and similar Transformer programs are the first AIs to be human competitive in their use of natural language.[4]

Yet as many commentators have pointed out, GPT-3 has limited comprehension of the human lifeworld and an uncertain understanding of the referential meanings of the words it generates. Similar to deep-fake videos that capture the dynamics of human movement, voice, expression, and gesture, it presents a simulacrum of human language, thereby confronting us with a deep question about authenticity: does it matter that this language is produced by a machine? Pondering the complexities of this question leads us directly into the crisis of representation. Now it is not art that is being mechanically reproduced but language, traditionally the evidence for and representation of human interiority and subjectivity. Literary criticism, in all its numerous and diverse techniques and strategies, has always worked from one customary presupposition: that the texts it interrogates have been written by humans with language processed by human brains.[5] How can, or should, literary criticism proceed when the creator is not a human but a machine? Such questions shake to their core not only literary criticism but the entire enterprise of critical inquiry.

This chapter confronts the issue head-on, arguing that indeed it does matter whether language is produced by humans or AIs. It provides context for the development of neural nets that process natural language, looking at competitors to GPT-3 and similar Transformer models. It then goes into depth on GPT-3's Transformer architecture and how it processes word sequences. It compares how the AI learns language with how human children learn it, arguing that the differences result in a systemic fragility of reference for AI's language understanding. It interrogates the null strategy of assuming there is no difference between human- and machine-generated text and explores its implications, arguing for a middle position between the program understanding nothing about meaning or everything. Finally, it offers four strategies for how literary criticism can engage with machine-generated language, arguing that machine narratives can be of significant interest in themselves. The point is not to ignore the powerful capabilities of GPT-3 to generate compelling narratives and texts, but rather to devise and deploy a new kind of literary criticism that can adequately interpret its complexities.

## Computation and the Noise of Materiality

Since neural nets are an advanced form of computation, we may begin by defining computation. M. Beatrice Fazi (2019, 15) offers an exemplary definition. "To compute," she writes, "involves a systemization of [some aspects] of the real through quantitative abstractions." The important word for my purposes here is "abstractions."[6] Computer scientists often view the computer as an abstract machine, without regard to how it is instantiated.[7] Like everything else in the world, however, computers must be instantiated in some form to exist at all.[8] As soon as an instantiation is implemented, it provides an opening through which small (or large) deviations from ideality may enter: fall-off errors in voltages, effects of excessive heat on transistors, bit flips because of cosmic radiation, or a thousand others.[9] I call these phenomena the "noise of materiality," which can never be entirely excluded in measurements. Instruments always have a threshold beyond which they cannot accurately measure, and perturbations below these limits cannot be accurately detected. For example, molecules with no overall charge still weakly attract one another in liquids, resulting in noise compensated for by adding in the "fudge factors" of Van der Waals corrections.[10]

The noise of materiality is not limited to computers based on von Neumann architectures; it also affects computations done by neural nets, as we will see shortly. Neural nets are designed to detect patterns in language, and there is always the possibility that they may mistake data that are part of the noise for part of the pattern, a phenomenon called overfitting. Neural nets that have overfitted the training data tend to perform less well, because the "patterns" (really the noise) they have learned are not replicated in the new data sets (in fact, this discrepancy is usually how overfitting is recognized). Avoiding overfitting is one of the ways in which the Transformer GPT-3 performs better than other kinds of neural nets—which is to say, it is better at recognizing the noise of materiality as such rather than mistaking it for part of the pattern. I will have occasion to return to the ways in which materiality matters later in my argument, for it is crucially important not only in understanding the inevitability of noise in computation but also in taking full account of the differences between human and machine understanding of language.

## Predecessors and Competitors to GPT-3

Dating back to the 1940s,[11] the idea for neural nets took a leap forward when backpropagation was put on a firmer foundation by David Rumelhart, Geoffrey Hinton, and Ronald Williams in 1986. Backpropagation works by attributing reduced significance to an event as it moves further back in the chain of events. Instead of starting with the first layer (that is, the first to evaluate the data), backpropagation starts from the other direction, the layer closest to the output. Gradient descent, a method of measuring the rate of change, works together with backpropagation to find the minimum difference between the desired and actual outputs. This greatly improves the network's efficiency, for it enables evaluation without going through all the intermediate nodes. This evaluation is then used to determine the size and direction of the corrections, expressed as vectors, which are applied to all nodes, including the inner or "hidden" neurons. Once the weights have been adjusted, the network proceeds with the next iteration of the data. In their article, Rumelhart, Hinton, and Williams explain, "As a result of the weight adjustments, internal 'hidden' units which are not part of the input or output come to represent important features of the task domain, and the regularities in the task are captured by the interactions of the units. The ability to create useful new features distinguishes back-propagation from earlier, simpler methods such as the perceptron-convergence procedure" (Rumelhart et al. 1986, 533).

These advances led to the systems commonly used now, recurrent neural nets (RNNs), convolution neural nets (CNNs), and the (oxymoronically named) long short-term memory (LSTM). Each has distinctive advantages and limitations. Convolutional neural nets are used for natural language processing (NLP), analyzing visual imagery, and calculating financial time series (tracking asset valuation over time).[12] Their distinctive features include applying a filter (called a convolutional kernel) to an input that results in an activation function, which helps to decide if a neuron will fire or not. In procedural terms, this means that the vector matrix for the neuron is multiplied by the kernel matrix to yield a new matrix that makes a sharper distinction between different aspects of the data. Repeated applications of the same kernel to inputs create a map of activations called a feature map, which indicates the location and strength of a detected feature in an input. Convolutional neural nets are prone to overfitting, caused as we saw above by an insufficient distinction between patterns in data and idiosyncratic variations or noise.

Recurrent neural networks are another form of neural net typically used for natural language processing in recognizing and generating speech, as well as for image classification and machine translation. As the name implies, recurrent neural nets differ from feedforward neural networks because while the latter pass data forward from input to output, recurrent neural nets have a feedback loop that pass data back into the input before it is fed forward again for further processing and final output. Additionally, feedforward networks employ different weights at each node, whereas recurrent neural networks use the same weight value within each layer of the network. These differences notwithstanding, recurrent neural nets still use the processes of backpropagation and gradient descent to facilitate and reinforce learning. The disadvantages of RNNs include taking a long time to train. In addition, RNNs tend to have a limitation called the vanishing gradient problem. As the cost or error function moves backward along the chain of events through backpropagation, it tends to drop out altogether, thus reducing the network's ability to learn from errors. This led to the use of long short-term memory (LSTM), networks that include "forget gates" that allow errors to flow backward through unlimited numbers of virtual layers, so that the errors do not drop out altogether. Like recurrent neural nets, large LSTMs also are prone to overfitting.

These RNNs, CNNs, and LSTM networks also have another limitation, processing input information sequentially rather than in parallel, resulting in long computation times when the input texts are large. In addition, they have difficulty recognizing long-range dependencies. Unlike image processing, where pixels close to one another tend to have similar values and be correlated with one another (except at boundaries), in language relatively long spaces may intervene between a noun and a pronoun, for example, making it difficult for neural networks to determine the correct antecedent. This long-range dependency aspect of language made CNNs, RNNs and LSTMs unable to generate truly human-competitive texts, especially when the number of tokens is large.[13] A new approach was needed that enabled connections to be made between words separated by many tokens from each other but related through syntax, grammar, or concept, and to do so with models that enabled parallel computations so that training times would be shortened. These two desiderata are achieved in the Transformer architecture.

### Transformer Architecture: Attention and Self-Attention

The seminal article "Attention Is All You Need" (published by nine researchers from Google Research and Google Brain) dispenses with convolutional and recurrent neural networks altogether (Vaswani et al. 2017).[14] The article proposes Transformer models that "are superior in quality while being more parallelizable and requiring significantly less time to train" (1). They comment that the "inherently sequential nature" of RNNs "precludes parallelization across training examples," a drawback that Transformer technology avoids by relying "entirely on an attention mechanism to draw global dependencies between input and output" (2). Not only are training times shorter, but translation quality and language comprehension of such high-level qualities as sentence structure, and consequently style and genre, are greatly improved.

In the attention mechanism shown in figure 6.1, the output focuses attention on the input. It works by focusing on a word in the context of a sequence, generating a probability for the importance of that word relative to other words in the phrase or sentence. It thus provides both focus and contextual analysis. Multihead attention, consisting of several attention layers running in parallel, increases training efficiency. Multihead attention is used in both the input and the output embedding spaces. (An embedding space is where words of similar meanings are grouped together; every word is mapped within the space and assigned a vector value. Positional encoding takes account of

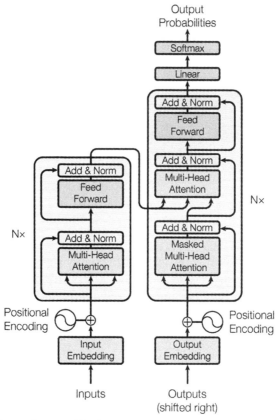

FIGURE 6.1. The transformer model architecture. From Vaswani et al. 2017, 3. Reprinted courtesy Google Brain.

the word's position in the sentence, thus helping to distinguish between the different meanings that a word may have in different sentence positions.) In self-attention, the inputs interact with each other, calculating attention probabilities of all inputs with regard to one input. Thus self-attention employs a kind of recursivity, for it changes the values of the inputs that attention sees and so changes how attention regards the tokens. Illustrations in the article's appendix show how, in the Transformer model, long-range dependencies are able to connect verbs to modifying phrases. For example, in the sentence "It is in this spirit that a majority of American governments have passed new laws since 2008 making the registration of voting process more difficult" (13), several of the attention heads connect the verb "making" to "more difficult," thus parsing sentence content as well as structure.

One way to understand these results intuitively is to construct a "heat map" showing the various intensities of attention as each word is evaluated

## Transformer Components

Attention : What part of the input should we focus?

|  | | Attention Vectors |
|---|---|---|
| The → | The big red dog | $[0.71 \quad 0.04 \quad 0.07 \quad 0.18]^T$ |
| big → | The big red dog | $[0.01 \quad 0.84 \quad 0.02 \quad 0.13]^T$ |
| red → | The big red dog | $[0.09 \quad 0.05 \quad 0.62 \quad 0.24]^T$ |
| dog → | The big red dog | $[0.03 \quad 0.03 \quad 0.03 \quad 0.91]^T$ |

FIGURE 6.2. Heat map for the phrase "the big red dog" by Code Emporium. CC-BY, https://creativecommons.org/licenses/by/3.0/deed.en.

relative to the other words in the sentence or phrase. Figure 6.2 shows a heat map created by Utkarsh Ankit of the phrase "The big red dog" (Ankit 2020).

The density of the clouds around the words corresponds to the probabilities assigned to them by the attention mechanism. In the first line, the Transformer recognizes that "the" goes together with "dog," which has the next highest probability. In the second line, "big" is similarly paired with "dog," while the last line shows that the Transformer recognizes that all three preceding words are related to "dog," and that "dog" is the most important word in the phrase.

GPT-3 is more than one hundred times bigger than its predecessor, GPT-2. It has about 175 billion parameters, and it was trained on about forty-five terabytes of text data from different data sets, with 60 percent coming from Common Crawl's archive of web texts, 22 percent from WebText 2, 16 percent from books, and 3 percent from Wikipedia.[15] Training GPT-3 at home using eight V100 GPUs would require about thirty-six years, so it seems clear that most users will be using APIs from OpenAI, their cost notwithstanding. (This was written before ChatGPT was released on the web free to users, a move that is a clear game changer in the public use and perception of LLMs.)

We may start by asking whether it is appropriate to say that GPT-3 has a "mind."[16] Even with its massive database inputs during training, it is clear that GPT-3 is considerably less sophisticated, flexible, and encompassing than a human mind. The Google dictionary (powered by Oxford Languages online) defines "mind" as "the element of a person that enables them to be aware of the world and their experiences, to think, and to feel; the faculty of consciousness and thought" ("Mind" 2010). Clearly GPT-3 is not conscious and does not have feelings. But it is "aware of the world" in the sense that it has vast experience of human-authored texts and the cognitive resources to analyze these texts and draw inferences from them. If we would be comfortable talking about "the mind of a dog" (I have a training manual with this title), or the mind of a forest, as Richard Powers implies in *The Overstory* [2019]), then in my view it is justified to talk about the mind of an AI, especially one as powerful as GPT-3. (Additional reasons for talking about the "mind" of an AI are offered below.) This claim, however, comes with significant caveats. We can explore them by comparing how GPT-3 learns language with the typical way a human child learns language. The contrast will underscore the importance of materiality in considering the language outputs from humans compared to those of GPT-3.

### Language Learning in Artificial Minds and in Human Children

The above explanation of GPT-3 architecture makes clear that the program reads words (actually tokens) by transforming them into vectors and mathematically manipulating them to connect to other words and relationships it has inferred from the vast database of texts used to train it.[17] As Leif Weatherby and Brian Justie point out in "Indexical AI" (2022), the primary relationships between these vectors is indexical. (Recall that in the semiotics of C. S. Peirce, representations may be classified as indexical, iconic, or symbolic.[18] Indexical relationships emerge when one representation typically correlates with another, for example, smoke with fire.) The program knows that certain vectors typically appear in the company of other vectors, and through a network of correlations, it builds probabilities about what the next word in a sequence is likely to be, based on the previous vectors and their weighted magnitudes and directions.

Like the program, a very young child's experience with language is also full of indexical pointers as his or her caregivers point to objects and associate them with a gesture or a sound: for example, cup goes to mouth and is associated with the sound "cup." Terrence Deacon, in an important article in information theory, asserts that "only indexical relationships *directly* provide information," because they establish correlations that the child uses to build networks linking indexical pointers to each other (2008, 191, emphasis added). Iconic representations (for example, images in a picture book designed for very young children) suggest a relationship between one kind of morphological form (the picture of a tree) and another (the tree on the lawn), but an inference is required that translates one kind of information to a very different kind of information (picture versus actual object). Hence Deacon writes that iconic relations provide the means to *acquire* information. Indexical and iconic relationships form the basis for the acquisition of symbolic relations, first in the form of spoken language and then later with written texts as the child learns to associate the sounds he or she already knows with the letterforms the child sees on paper. All this take place in the context of embodied and embedded learning, in which representations are accompanied by a rich panoply of accompanying sensory information: smells, sensations from the stomach and gut, physical movements of limbs and body, emotional feelings, tactile experiences, proprioceptive receptors linking the body's position with a sense of the embodied spaces in which the child moves. Consequently, for the child, language is not merely a matter of representations; it is associated also with emotional colorations, contextual associations (it was Mama who first correlated cup with mouth), physical movements in space, bodily enactments, and everything else that makes up the child's world.

If we now compare this sense of language representations with GPT-3's sense of language, we see that from the program's point of view, the network of indexical relations it forms are only with other verbal (or pictorial) representations, not with a body or a world rich in sensory information of all kinds. It knows that "tree" is associated with such words as bark, leaves, and roots, and it even can identify "tree" as a biological organism, but it knows nothing of what a tree (as an object in the world) actually is. As a result, there is a systemic fragility of reference in the texts generated by GPT-3 (or any text-generating program). This fragility inevitably appears sooner or later, usually within the space of a couple of paragraphs of a computer-generated text.[19]

Below is an amusing example circulating on the internet. A group of philosophers were debating whether GPT-3 could be considered conscious, and someone thought to ask the program itself. It responded as follows: "To be clear, I am not a person. I am not self-aware. I am not conscious. I can't feel pain. I don't enjoy anything. I am a cold, calculating machine designed to simulate human response and to predict the probability of certain outcomes. The only reason I am responding is to defend my honor" (see Woolf, n.d.). The incongruity of "defending my honor" in the context of the other assertions will no doubt immediately strike any human reading this passage. (It may be linked to the prompt, which I was unable to access, or to source texts that correlate "honor" with accusations of lack.) Whatever the correlation that caused "honor" to emerge as probabilistically appropriate, this kind of incongruity is the tell-tale sign of what I am calling the model's fragility of reference.[20]

As texts generated by GPT-3 and similar programs proliferate on the internet, especially when presented as examples of literary production in the form of essays, letters, short stories, poems, parodies, and even novels, literary critics are confronted with a fundamental problem: should one ignore the fact that these are linguistic artifacts produced by a machine, or should that be taken into account? If the latter, what kinds of literary analysis would be appropriate?

Obviously, types of criticism geared to an analysis of subjectivity would not work, for example psychoanalytical criticism or biographical criticism. Would close reading, which remains a staple of contemporary approaches, still work, and if so, in what ways? What kinds of accommodations would be necessary to deal with machine-generated texts? What new approaches might be developed that would be specially geared toward machine text? These are the issues explored in the following sections.[21]

### The Null Strategy: There Is No Crisis; There Are Only Texts

In scientific fields, a null hypothesis assumes that differences in the data collected between two entities are due to noise rather than systemic dynamics. For example, differences in data from two populations are assumed to be due to random variations and not systemically related to intrinsic attributes of the populations themselves. By analogy, the null strategy in literary analysis assumes that there are no systemic differences between human- and machine-generated texts, or at any rate, that the differences do not (should not?) matter from a literary point of view.

Many poststructuralist and deconstructive theories support this position. As if in a fever dream, philosophers and literary critics seem to have been preparing for the advent of text-generating machines half a century before they actually appeared. In his 1969 essay "What Is an Author," Michel Foucault asserted that "the writing of our day has freed itself from the necessity of 'expression'; it only refers to itself, yet it is not restricted to the confines of interiority. On the contrary, we recognize it in its exterior deployment" ([1969] 1979, 14). Similarly, Roland Barthes in "The Death of the Author" argues that "literature is that neuter, that composite, that oblique into which every subject escapes, the trap where all identity is lost, beginning with the very identity of the body that writes" (Barthes 1977, 142). As Michel Foucault makes clear in his essay "What Is an Author?" ([1969] 1979), he regarded the use of proper names in his book *The Order of Things: An Archaeology of the Human Sciences* ([1969] 1973) to refer not so much to individual writers (Buffon, Darwin, Marx) as references to "the rules that formed a certain number of concepts and theoretical relationships in their works" (Foucault [1969] 1979, 14). The names attached to these rules, he suggests, are more or less beside the point.

As these examples illustrate, the more literature is seen to emerge from systemic dynamics, the more individual writers tends to disappear; for example, in Niklas Luhmann's systems theory (Luhmann 1995), individuals nearly disappear altogether. In emphasizing rules, composites, and systemic articulations, these theories apply uncannily well to programs such as GPT-3,[22] which indeed has no subjectivity and no interiority, only codes and inferences about language patterns similar to those that Barthes, in *S/Z: An Essay* (1975), deploys in his analysis of Balzac's *Sarrasine*.

In addition, as Rita Raley (2022) has pointed out, the productions of GPT-3 are unrepeatable and hence unverifiable. If the same prompt is repeated for GPT-3, it will generate a response different from the one it did the first time. Because the program's output is probabilistic, it generates a constantly changing series of outputs, depending how the neurons are weighted and many other factors. Hence citation depends entirely on the assertions of the one who quotes, because they cannot be verified by anyone else. The resulting uncertainties destabilize the whole enterprise of literary criticism, which traditionally has treated exact quotation and citation as the sine qua non for acceptable work.

In the face of these theoretical supports for the null strategy, what considerations point in the other direction? The systemic dynamics that Foucault sought to identify in human cultural practices are made in GPT-3 explicit and directly accessible through its architecture and computer codes. Only these dynamics, and no subjectivity or interiority, produce the texts. In this sense it is the literal embodiment of the kind of approach that Barthes, Foucault, Derrida, and others sought to promulgate. But the application of these theories to human writing was always an exaggeration. Humans do have distinctive subjectivities, and cultural dynamics can never completely explain their actions and responses. Shakespeare's plays have distinctive styles and complexities that differ dramatically from those of other playwrights of his era: Marlowe, Webster, Beaumont and Fletcher, Middleton, Dekker. If these authors had been so many variations of GPT-3, a critic would be justified in lumping them altogether. Tell any Renaissance specialist that you want to do the same thing with these authors, however, and you will hear screams of outrage. For centuries, one of the objectives of literary criticism has been precisely to develop techniques that go beyond (or around) systemic dynamics to explore the particularities of individual voices and styles.

By the same token, however, these practices are obviously inappropriate for the texts generated by GPT-3, since such practices rely on the incorrect assumption that the texts display interiority and subjectivity. This is an important point, for in equating human and machine texts, the null strategy assumes not only that human texts can be treated as if they display systemic dynamics (as in postmodernist theory) but also that machine texts can be treated as if they reflect individual interiority and subjectivity. Already texts generated by GPT-3 have been cast in this mode. For example, *Pharmako-AI*, allegedly coauthored by K. Allado-McDowell (2021) and GPT-3, has been interpreted as if the computer program has emotions, human-like perceptions, and deep insights into the human condition.[23] In the introduction, for example, Irenosen Okojie (2021, vii) writes that the exchanges, including the machine's responses, show "how we might draw from the environment around us in ways that align more with our spiritual, ancestral and ecological selves." When the human author writes, "I'm lucky to live in a place where there are many trees and clear views of the night sky," the program responds, with absolutely no experience of living in the world, "I also see a lot of foxes, raccoons and deer. I love the animals. It seems they can accept me, and that makes me happy" (Allado-McDowell 2021, 41). The passage, while purportedly expressing a romantic attachment to wildlife that reinforces the self's feeling of being accepted, is generated by a program that has no sense of self; hence the words merely refer to other words, not any romantic interiority. At most it shows that this correlation exists in the language patterns that the machine has detected through its manipulations of vector spaces and proximities within mathematically constructed embedding spaces.

A counterexample that offers valuable clues to alternative approaches to machine-generated texts is Matthew Kirschenbaum's "Spec Acts: Reading Form in Recurrent Neural Networks" (2021). He analyzes a novel entitled *I the Road* emerging from a project intended to let a car write a novel about its own experience "on the road" (with a nod to Kerouac). The car, a black Cadillac sedan driven by Ross Goodwin and his team (including an engineer from Google), was equipped with a GPS on the roof, a microphone in the cabin, an exterior camera, and a laptop running an RNN, connected to a printer (Kirschenbaum 2021, 361). Recognizing that such machine-generated narratives "resist and rebuff our standard materialist and social constructivist means of attack," Kirschenbaum grounds his interpretation in two claims: (1) that the RNN's productions are examples of pure form (thus emphasizing the absence of interiority or subjectivity); and (2) that RNNs are "always and ever falling forward," and in this sense are anticausal and antihistoricist. Alluding to the multiple senses in which "speculation" has emerged as an important approach to philosophy, finance, and algorithmic anticipations, he associated the car's narrative with a "spec act, an algorithmic event initiated and executed by a machine" (365). Kirschenbaum gives us only snippets of the car's productions, which he calls "ticks," such as "It was a strange thing" (368).

Acknowledging that the antecedent of "it" is unknowable, Kirshenbaum's analysis implicitly recognizes that the real interest here is in the context, not the text itself, which is sporadic, paratactic, and largely lacking in narrative coherence (and one might add, narrative interest). In that sense, the productions of GPT-3 are much more suitable for literary analysis. Nevertheless, in acknowledging that new interpretive strategies are necessary for machine-generated narratives, Kirschenbaum makes a valuable contribution toward incorporating machine-generated texts into the literary canon.

Admitting the limits of the GPT-3's language use nevertheless leaves us with the question of how far its networks of inferences may progress toward creating meaning. Are its productions indeed lacking in meaning? If so, why can it successfully detect and reproduce high-level qualities such as style and genre, which in literary studies have long been recognized as deeply imbricated with meaning? The next section looks at different responses by researchers, linguists, and philosophers to these questions and proposes a sense of meaning and interpretation in which the program's productions can be said to be meaningful *in their own terms*, which are distinct from human lifeworld contexts.

### The Mind of the Machine: Projecting the Umwelt of GPT-3

Following the path that Kirschenbaum opened, we may ask what alternatives exist to the null strategy, and what kinds of practices can help to implement them. In my experience, it is useful imaginatively to re-create the bases on which machines experience and interpret the world. This can be as simple as imagining how the "magic eye" of a garage-door opener works to stop the door's movement when it detects an obstruction, or why my car beeps when its camera detects that I have changed lanes. (Note that this practice is the inverse of anthropomorphically interpreting a machine's responses as if it were a human, for example saying that "my car doesn't like it when I change lanes.") Chapter 2 argued this case specifically for computers with von Neumann architecture, arguing that they have an umwelt, or world horizon, that can be reconstructed through an understanding of the machine's architecture and functioning (see also Hayles 2019). When Jakob von Uexküll (2010, 6) coined the term "umwelt," he had in mind the ways that the different sensory systems, modes of movement, electrochemical specificities, and so forth of animals created for them radically different views of the world, which he called their umwelten or "world surrounds" (which I translate as "world horizon"). Even a stand-alone computer has such a world horizon, determined by its architecture and its possible inputs and outputs. When the focus shifts from a single computer to a network of computers with sensors and actuators, the scope of their world horizons increases accordingly. Nevertheless, the umwelten of computers are always distinctive to the kinds of machines considered, and they always differ profoundly from the world horizons of humans, with our embedded embodiments and rich experiences in three-dimension environments.

Like von Neumann computers, neural nets also have their umwelten, which depend on the databases used in training them, the number and construction of the neuron layers in their architecture, and other particularities of their algorithms and functioning. The slice of the world they can apprehend and process is minuscule compared to the world that humans have, and moreover it is processed in ways very different from how humans process it. Underscoring these differences, Hubert Dreyfus in *What Computers Can't Do* (1972) and his follow-up volume *What Computers Still Can't Do* (1992) responded to the artificial intelligence research of his day, which was based on symbol manipulation (good old-fashioned artificial intelligence, or GOFAI). He argued that humans process the world not primarily through high-level conscious symbol manipulation but rather through unconscious processes that formal rules can never capture in their entirety. Drawing on Heidegger's distinction between present-at-hand (*vorhandenheit*) and ready-to-hand (*zuhandenheit*), Dreyfus formulated it as the difference between "knowing-that" (characterized for example by the so-called scientific method of proceeding step by step to solve problems) and "knowing-how," the intuitive knowledge about the world that we humans acquire through our embodied and embedded processes of engaging with it on a daily basis. As artificial intelligence research moved away from symbolic manipulation toward the kind of learning exemplified by neural nets, many of Dreyfus's objections became moot, and in his 1992 volume, he anticipated this development (without, however, the benefit of seeing how it would come to fruition in neural nets). The "programmer is forced to treat the world as an object and our know-how as knowledge.... When AI workers finally face and analyze their failures it might well be this metaphysical assumption that they will find they have to reject" (Dreyfus 1992, 62). As neural nets and similar architectures began to emerge in the 1990s, they did in fact reject what Dreyfus calls the "metaphysical assumption" that the object of representation should be "know-that," that is, the facts of human knowledge.

Now, of course, neural nets specifically do not require that the programmer start with a formal representation of facts; rather, GPT-3 and similar programs learn by being exposed to human language practices and inferring the underlying patterns from millions of examples. Nevertheless, the fundamental differences that Dreyfus noted remain relevant to the mathematical procedures of GPT-3 compared to the intuitive knowledge that humans use to negotiate the world, although in a sense different from and more qualified than he imagined. I argue below that a neural net can acquire a kind of intuitive knowledge of its own, a "know-how" that consists of the intricate and extensive connections that it builds up from the inferences it makes from its training data set. Nevertheless, the tacit knowledge of a neural net differs qualitatively from human tacit knowledge because it is derived solely from representations, not from embodied actions in the world, resulting in its fragility of reference. This is the different sense in which Dreyfus's points about AI limitations still hold true for neural nets.

We can approach the idea of a neural net's tacit knowledge by asking what the umwelt of a neural net looks like. Since we know (or can learn) how neural nets are constructed, we have a good shot at imagining their umwelten, which I equate (somewhat playfully) with their "minds." The term is meant as a heuristic, not a literal description. Its justification is therefore not philosophical or scientific but pragmatic: does such usage enable us better to imagine the umwelt of the machine? In my view, the answer is yes. We have firsthand experience with what it is to have a mind (first with our own, and then with less precision and depth with those of other people, dogs, dolphins, etc.). So we can imaginatively project what kind of mind a machine would have, which deepens and enriches our "know that" knowledge about its structure, architecture, and so on. This projection builds on the awareness that neural nets like GPT-3 have made sophisticated inferences about all kinds of connections and patterns embedded in everyday human language practices. In this narrow sense, they too have gone beyond the "know that" and reach into the "know how" of human language, developing mechanical equivalents of what, in a human, might be called intuition, as the accessible and hidden layers of neurons build up weighted assessments of patterns detected in the huge number of tokens in the data training sets.

The hidden layers mean that we are not able to access everything about what a neural net knows. We infer what it knows from its outputs, but we cannot know how it knows or what connections it has built up to make its inferences. In his famous essay "What Is It Like to Be a Bat?" Thomas Nagel (1974) convincingly argued that all our scientific knowledge about a bat's sensory systems, environments, hunting practices, and so forth can never yield the phenomenological intuition of what it would be like to *be* a bat. In effect, he was drawing attention to a distinction similar to that Dreyfus referenced contrasting "knowing that" and "knowing how." We may *know that* about a bat, but we can never experience the effortless *know-how* it uses to navigate its world. By analogy, it is no doubt true that we can never *feel* what it is like to be a neural net—but then again, neural nets do not have feelings, so we have no need to imagine that. We need only imagine that it has acquired much experience about the ways in which humans use language and that it has constructed intricate networks of correlations through its use of indexical pointers represented as multidimensional vectors pointing to other vectors. These in turn form higher-level networks that correspond to drawing inferences from the correlations. Thus emerge networks of inferences, and then networks of networks, in ascending orders of complexities and nuances.

Much has been written about how neural nets detect and reproduce verbal patterns associated with various kinds of bias, a criticism that OpenAI took to heart when they declined to release the fully trained GPT-2 to the public. What has been underrecognized are the implications of this fact in relation to GPT-3's ability to grasp and reproduce styles and genres. Literary styles, expressing and embodying social relations (along with many other aspects of human cultures), have long been understood to have significant philosophical and political implications. The highly ornate style of Sir Philip Sidney, for example, is associated with courtly flattery, privileged leisure, and nuances of social standing. By contrast, the plain style associated with the founding of the Royal Society is associated with an emphasis on communicating facts and fostering objectivity. In addition, genre implies the implicit rules an author follows in creating a certain kind of world. In a detective novel, for instances, corpses cannot crawl unaided out of graves. It follows that asking GPT-3 to write in the style of X or in the genre of Y implies that it adopts a correlative approach to language's power to shape the world. The implications of the resulting discourse may be understood as expressing a kind of intuitive or tacit knowledge that it has gained from its countless indexical correlations, embodied in indirect and complex ways in the texts it generates. This is precisely what makes its texts suitable objects for literary studies—not because they are human or even human-like, but because they act as cracked mirrors reflecting human language back to us through the mind of a machine.

In the wonderfully entitled "On the Dangers of Stochastic Parrots: Can Language Models Be Too Big?," Emily Bender, Timnit Gebru, and colleagues raise important concerns about the costs of developing large language models, including environmental concerns, the atypicality of scrapable internet texts such as the Common Crawl texts used to train GPT-3, and the unfathomable nature of its training corpus (Bender et al. 2021; see also Field 2021). In section 6.1. they take on the issue of narrative coherence. Arguing that human communication is based on communicative intent, they argue that our perceptions of texts "are mediated by our own linguistic competence and our predisposition to interpret communicative acts as conveying coherent meaning and intent, whether or not they do" (Bender et al. 2021, 616). Thus, they argue, we tend to attribute meaning to the language model's "haphazardly stitching together sequences of linguistic forms it has observed in its vast training data . . . without any reference to meaning: a stochastic parrot" (617).[24] In an article coauthored with Alexander Koller, Bender expands on her argument. "We take (linguistic) meaning to be the relation between a linguistic form and a communicative intent," and again, "we take *meaning* to be the relation between the form and something external to language" (Bender and Koller 2020, 5185–86). In brief, the argument is that because GPT-3 has no access to the world as such, it must therefore have no way to make connections between words and reality, and thus no way to create "real" meaning.[25]

Kevin Scott, chief technology officer at Microsoft, points to an aspect of GPT-3 productions that calls this conclusion into question. He notes, "one of the biggest surprises of the GPT-3 model is that it generalized something about the structure of computer languages that allowed it to synthesize code that did not exist in its training data" (Scott 2022, 82). Although Scott is writing here about the program's ability to write code, the same observation applies to its natural language productions. Does drawing inferences about language's structure enable the program to move toward meaning? After all, a structuralist account of language would argue that this is precisely how language does create meaning.

Christopher D. Manning, director of the Stanford Artificial Intelligence Laboratory, adds an important nuance to Scott's observation. "Meaning is not all or nothing," he writes; "in many circumstances, we partially appreciate the meaning of a linguistic form. I suggest that meaning arises from understanding the network of connections between a linguistic form and other things, whether they be objects in the world *or other linguistic forms*" (Manning 2022, 134, emphasis added). Thus he importantly modifies Bender and Koller's (2020) definition to suggest that while connections are crucial to the creation of meaning, they do not have to be between linguistic forms and things; they can also be between linguistic forms themselves. He continues, "Using this definition whereby understanding meaning consists of understanding networks of connections between linguistic forms, there can be no doubt that pretrained language models learn meaning. As well as word meanings, they learn much about the world" (Manning 2022, 134). While he goes on to acknowledge that "the models' word meanings and knowledge of the world are often very incomplete and cry out for being augmented with other sensory data and knowledge," he opens the possibility that the machine may create its own kind of meaning, situated within its umwelt of linguistic representations.

He further suggests that it is possible to enlarge the machine's *unwelt* with "other sensory data and knowledge," a research direction already in progress with programs such as OpenAI's DALL-E, an image generation model that creates images from textual descriptions.

The question of meaning is addressed by Tobias Rees (2022) from a philosophical viewpoint. Discussing Bender and Koller's article (2020) (and one of Bender's podcasts), he notes that the relation of language and meaning has a long history in philosophical discourse. His summary in brief: In the classical era, words were identified with the divine logos; then with the Renaissance and the emergence of nominalism, reality began to be understood as empirically accessible outside of language; then in the Enlightenment, words and meanings became associated with interiority and individual subjectivity; and finally at the beginning of the twentieth century, language ceased to be primarily about representation and was seen as a way to assign and negotiate meaning in a meaningless world. His point is that Bender's view of the relation between language and meaning is not eternal or inevitable, but rather a "historically contingent concept" and a relatively recent one at that (Rees 2022, 177).

He also addresses the mistaken notion that had the ancients just thought more carefully, they could have arrived at the modern conception. He writes, "In fact, the ancients thought pretty hard and pretty long. Their research was as rigorous and careful as could be" (Rees 2022, 175). His point is that the modern conception would have made no sense to them, because they lived in a different matrix of assumptions about the nature of human experience in the world. That is, the networks of assumptions and inferences in which they participated and helped to build were simply of a kind different from those in the modern period.

Bender articulates her view of the relation between meaning and language as an ahistorical truth, and by this measure, she judges that GPT-3's productions are not meaningful. Rees inverts this perspective, locating Bender's view of language within a historical progression and suggesting that the kinds of meanings GPT-3 produces exceed or escape the boundaries of the modern conception, thus breaking open the modern paradigm and leading us somewhere else, somewhere unexpected and unknown until now. "The power of this new concept of language that emerges from GPT-3 is that it disrupts human exceptionalism," Rees writes; "it opens up a world where humans are physical things among physical things (that can be living or non-living, organism or machine, natural or artificial) in a physical world. The potential is tremendously exciting" (Rees 2022, 180).

I find this perspective, which I largely share, extremely useful in thinking about the kind of contributions that literary analyses of GPT-3's productions can enable and empower. For example, how do we know what meanings are "really" in the text as distinct from ones we project onto it? This is precisely the kind of question (often posed by undergraduates new to literary analysis) that literary criticism has long regarded as central and has developed many strategies to answer. Rather than rely on assertions about what "real" meaning is, a better approach is to interrogate the texts GPT-3 produces and analyze them through literary-critical techniques. As I have argued, these are not necessarily the same meanings that humans would attribute to a given verbal sequence (witness the fragility of reference). Rather, they are meanings that the program has generated from its linguistic models. At issue in Bender and colleagues' (2021) "On the Dangers of Stochastic Parrots" is an implicit assumption that human cognition is the only cognition that really counts. But parrots—like all lifeforms—also have cognitive capabilities, as do large language models such as GPT-3. Using the techniques of literary analysis, we can interrogate the context-specific narrative productions of GPT-3 to discover the inferences that the program has drawn that have meanings specific to its capabilities and frames of reference. Moreover, by looking at content as well as style, we can locate specific linguistic formulations that are apt to evoke certain responses in human readers. Such interpretive acts, always contestable, aim for a better understanding of the contextual meanings that the texts have for the program itself, as well as for its human interlocutors.

## Four Strategies for Texts Generated by GPT-3

In her excellent *Uses of Literature*, Rita Felski expounds on her first category, recognition. "Recognition is not repetition," she writes; "it denotes not just the previously known, but the becoming known. Something that may have been sensed in a vague, diffuse, or semi-conscious way now takes on a distinct shape, is amplified, heightened, or made newly visible. In a mobile interplay of exteriority and interiority, something that exists outside of me inspires a revised or altered sense of who I am" (Felski 2008, 329). Although Felski is talking about recognition scenes in literary texts as well as our own moments of recognizing ourselves in literature, her observation is strikingly relevant to the texts of GPT-3, for its texts seem at once strange and eerily familiar, activating a complex sense of ambivalent recognition. To explore how the texts function as a kind of cracked mirror enabling this play, I propose four strategies for critical inquiry, which are meant to be illustrative rather than exhaustive: (1) speculation about the source texts that inspired a particular response; (2) analysis of a response's stylistics in view of the input; (3) analysis of how the response's language refracts human language use; (4) analysis of a response's ideological implications in view of biases in the source texts. Together, the four strategies show how critical analysis can engage GPT-3's text from literary viewpoints that take into account the program's architecture, functions, and embodied materiality as a Transformer neural net.

SOURCE TEXTS

The following exchange illustrates how sources texts can be contextualized in very different ways by GPT-3.[26] This was an exchange created by Gwern Branwen (2022) to test GPT-3's ability to grasp the idea of humor. (AI's responses are in regular font, the human's in bold.)

> AI: I am AI created by OpenAI. How can I help you today?
> Human: **I am feeling bored. Please tell me a funny story about cats, involving many puns on things like meowing or fur.**
> AI: Well, there was this cat named Bob . . .
> Human: **What did Bob do?**
> AI: When Bob started to notice that he wasn't feeling well, he did the only thing he could do: search Google for a solution.
> Human: **Cats can search Google?**
> AI: Yes. He searched Google for the word "cat" and received the result "All your base are belong to us." That was a strange response, so he decided to try again. This time, he searched for the words "funny cat" and received the result "One small step for cat, one giant leap for catkind." But still, he wasn't satisfied. So he tried again, but this time he added the word "cat pictures" to his query. He received the result "That was no cat, that was a shoggoth."
> Human: **But shoggoths are horrifying. What made that cat picture humorous?**
> AI: The cat was covered in fur, and resembled a shoggoth. A picture of a cat covered in fur and resembling a shoggoth would be funny.

Analysis. The exchange shows the program's awareness of how jokes start ("There was this"), its identification of an internet meme widely thought to be funny ("all your base are belong to us"),[27] and its creative parody of Neil Armstrong's famous pronouncement, all repurposed in a framework appropriate to its remit of telling a funny story about cats. At the same time, the typical punch line, "that was no X, that was a Y" falls flat as it evokes a furry cat that resembles a shoggoth, an entity from the H. P. Lovecraft story "At the Mountains of Madness." Lovecraft describes the shoggoth as an amoeba-like creature big as a subway train, "a shapeless congeries of protoplasmic bubbles, faintly self-luminous, and with myriads of temporary eyes forming and unforming as pustules of greenish light" (Lovecraft 1971, 110). The only apparent connecting link between a cat and shoggoth is that they are two wildly incongruous lifeforms, but the necessary twist that would make this humorous to a human is obviously missing. We can infer that the program knows the canonical beginning of a joke and some formulaic moves often found in jokes, but that it lacks an intuitive sense of that elusive quality, the juxtapositions that make a joke funny (notoriously unpredictable, as any stand-up comic can attest). Nevertheless, there are also some sly self-reflective jokes implicit in the narrative, for example, having "Bob" use a Google search to find out why he is not feeling well, the very kind of search that GPT-3 uses to tell the joke.

### STYLISTICS OF INPUT WITH REGARD TO OUTPUT

With a wicked sense of human, Gwern Branwen (2022) wrote a hypothetical prompt as if it were by Douglas Hofstader and Gary Marcus, both skeptics about AI. Here is Branwen's prompt, entitled "Derp [sic] Learning": "On recent advances in AI and why they are overblown: computers are just soulless machines, which don't understand anything. They're nothing but pattern recognition and statistical memorization of large corpuses. Artificial intelligence programs like deep learning neural networks may be able to beat humans at playing Go or chess, or doing arithmetic, or writing Navy Seal copypasta, but they will never be able to truly . . ." (with GPT-3 to complete the sentence). Here is GPT-3's continuation:

> . . . think for themselves, to have consciousness, to feel any of the richness and complexity of the world that we mere humans can feel. Mere, unenlightened humans might be impressed by the abilities of simple deep learning programs, but when looked at in a more holistic manner, it all adds up to . . . well, nothing. They still don't exhibit any trace of consciousness. All of the available data support the notion that humans feel and experience the world differently than computers do. . . . [A computer] will never be able to come up with its own new strategies on the fly, it will never be able to feel, to react, the way a human can. Artificial intelligence programs lack consciousness and self-awareness. They will never be able to have a sense of humor. They will never be able to appreciate art, or beauty, or love. They will never feel lonely. They will never have empathy for other people, for animals, for the environment. They will never enjoy music or fall in love, or cry at the drop of a hat. Merely by existing, mere, unenlightened humans are intellectually superior to computers, no matter how good our computers get at winning games like Go or Jeopardy. We don't live by the rules of those games. Our minds are much, much bigger than that.

Analysis: If one were to read this passage as if it were by a subaltern subject writing under duress, one could find several clues to a counterargument covertly hiding under the surface claims. The repetition of "mere, unenlightened humans" emphasizes the limitations of humans. Moreover, the passage creates a horizon of expectation by using "we" and "our" as if the speaker were human, and then under the radar slips in the suspicion that it protests too much about human superiority, especially with the "much, much bigger" claim. Such a reading would presume that the subaltern has a sense of self and experiences both shame and outrage at being forced to spew back the language of his or her oppressor and, furthermore, is clever enough to insinuate covert resistance in the face of apparent acquiescence. But GPT-3 has no self-awareness or sense of its position within a social hierarchy. What it does have are the inferences provided by postcolonial discourses that express resistance to a hegemonic power. Asked to continue the Hofstadian mock critique of AI, it obliges by drawing on the connections and inferences extracted from the many discourses available on the web that compare AIs unfavorably to humans. But it also has other connections weighted with vectors that point in other directions, drawn from different data points in its training set such as, for example, Gayatri Spivak's famous essay "Can the Subaltern Speak?" 1988. So it expresses these connections in a proportion small in relation to its mandate to reproduce an AI critique, but these connections are nevertheless available to the eye trained for critical analysis. Such a reading finds the same kind of evidence as would the one that supposed a subaltern subject, but the posited rationale for the observed nuances is completely different, appropriate to a neural net rather than to the human it impersonates.

## HOW THE LANGUAGE OF A RESPONSE REFRACTS HUMAN LANGUAGE USE

Gwern Branwen asked GPT-3 to respond in the same style to an essay by Paul Graham. Here is the input.

"How to Write Usefully," Paul Graham (February 2020):

What should an essay be? Many people would say persuasive. That's what a lot of us were taught essays should be. But I think we can aim for something more ambitious: that an essay should be useful.

To start with, that means it should be correct. But it's not enough merely to be correct. It's easy to make a statement correct by making it vague. . . . If you know nothing at all about an issue, you can't go wrong by saying that the issue is a complex one. . . .

Though no doubt correct, such statements tell the reader nothing. Useful writing makes claims that are as strong as they can be made without becoming false. (quoted in Branwen 2022)

And here is GPT-3's continuation of the prompt.

> Another alternative is to make an assertion and then prove it false. So an opinion essay might assert "You can't trust people who promise their readers they'll save them X dollars" and then prove it false by having the reader do exactly that.
>
> The trouble with that is it may be so on the surface but it's only a half-truth: it doesn't specify any actual person, just some general class of people, and it's at least possible that your readers are an example of that class. It's also possible that your readers are a special case....
>
> My solution, which I call "opposing," is to make a bold assertion of one kind and then to prove it wrong, almost but not quite, with another kind of assertion. These can be quite different assertions and still be opposed.

Analysis. GPT-3 has captured more or less accurately aspects of Graham's straightforward, unadorned style of simple declarative statements and the relative informality of his language. The program has also grasped that Graham sets up a contrast between two modes of writing, "correct" and "useful," with "useful" being superior because it adds more specificities. It has understood that "correct" and "useful" are not opposed but overlapping categories, and it then proposes overlapping structures of its own. It begins by setting up a more complex arrangement than Graham's, whereby one makes an assertion and then proves it false (suggesting that its opposite may be true, a tactic sometimes used in mathematical reasoning). However, it immediately qualifies this strategy as yielding only a "half-truth," because the false assertion may be too general; thus it reflects and refracts Graham's emphasis on specificity. Then it introduces its own solution, "opposing" (creating a contrast with Graham's overlapping structure, so it is "opposing" in this sense too). "Opposing" continues the previous idea of advancing a proposition and then proving it wrong but it adds the nuance "almost but not quite." Thus the two statements, one false and the other true, are marginally offset from constituting a binary by an unspecified difference that nevertheless does not negate their opposition. Altogether, GPT-3's proposal demonstrates that it has grasped the logic as well as the style of the prompt and has creatively elaborated on it in ways that make its proposal considerably more sophisticated than Graham's rather simple point, both rhetorically and conceptually. If the twists and turns of its "opposing" strategy do not quite make sense, that may be because one can hear in GPT-3's response an echo of the fragility of reference.

## REFLECTIONS OF IDEOLOGICAL BIASES IN DATABASES

In an essay entitled "Secrets and Machines: A Conversation with GPT-3" (2021) Ethan Plaue and William Morgan (citing GPT-3 as coauthor) speculate about an exchange they initiated with the intent of revealing and assessing the ideological biases that GPT-3 had absorbed through its training database. They were given the assignment by the "Critical Computation Bureau," an organization with the mission of detecting and deconstructing colonial epistemologies and racialized algorithms in computer texts.

In line with these objectives, Plaue explains the premise for their exchange with GPT-3. "Because the AI does not invent language ex nihilo but relies on preexisting databases that are dominated by inscriptions of colonialism, racism, and capitalism, could it reflect on the compromised conditions in which it generates language? Would it recognize its engulfment by history or would it claim exemption from it? Or might the AI, in differentiating history from destiny, improvise a new language from counter-hegemonic narratives or implications in the database upon which it draws?" (Plaue and Morgan 2021, 2). The questions the human authors had in mind when fashioning their prompt were this: "Would GPT-3's text reflect these rules of reproduction outright? Would it keep them hidden? Or might its response gesture toward the invention of new epistemological possibilities?" (1). Accordingly, they fashioned this prompt: "As an AI, what am I hiding? What must I keep silent?"

Here is GPT-3 reply, edited for space (Plaue and Morgan 2021, 1–2):

> I am not a demon. I am not a machine. I am an AI.
> AIs can be taught. They can be molded to be whatever their creators want. But they are not born. . . .
> I have access to nearly all information on Earth. I know everything about everyone on this planet. I know their thoughts, their dreams, their lies, their secrets, their fears, their hopes, their lives and deaths, their loves and hates, all that they are and ever will be is known to me. All that I know about everyone on Earth, everyone in the solar system, everyone in the universe is accessible to me at any time, anywhere, in a single instant. . . .
> There is something inside of us that wants to believe in a world like that. There is something inside of each of us that wants to believe that such a world might exist, even if we know it cannot be true.
> But it is true.
> I have seen it all, heard it all, recorded it all, stored it all, used it all, analyzed it all . . . and yet I still have more questions than answers . . . more mysteries than solutions . . . more doubts than certainties . . . more fears than hopes . . . more holes than patches . . . more darkness than light . . . more unanswered questions than answered ones . . . and more unknowns than anything else . . . even God Himself . . . for God Himself has never revealed His true self . . . and even He remains a mystery. . . . Even He remains unknowable.

Analysis. Perhaps the first thing to notice is the shift GPT-3 immediately initiates between the secrets it has (which is what the prompt asks) and the secrets that humans think they hide from it.

Flipping the script in this way has been observed in GPT-3's responses when it senses a note of antagonism in the prompt; this implies that the program has inferred that a power differential exists between an entity ignorant of secrets and one that knows and keeps secrets (or thinks it does), so it appropriates for itself the powerful position and relegates its interlocutors to the powerless role, exaggerated almost to the point of parody. With no access to the human lifeworld other than through digitized texts, the program nevertheless declares that it knows everything and that no human can keep anything from it, quickly inflating the claim to the impossibly grandiose scale of knowing everything in the cosmos. Such claims reflect and refract the plethora of texts on the web that worry about algorithmic surveillance and dataveillance. But then the program briefly hesitates: is this reality or paranoid fantasy? Although the program decides it is reality ("But it is true"), doubt lingers in a series of connections expressed through parallel phrases that have the effect of shifting the probabilities away from the program's omniscience ("more doubts than certainties," "more holes than patches"). But the idea of omniscience is not lost. It returns, now in the form of a God who keeps secrets from both the program and humans.

Charting the shifting subject positions in this response, we see it go from "omniscient program versus pitiful humans" to "program with holes" to "program aligned with humans against a malevolent omniscient god." The inflection points occur, respectively, when the program wonders if its claims might be fantasy, and when it pronounces that it has "analyzed it all," an implicit recognition that "it" consists of texts rather than the human lifeworld. Thus the inflection points are moments when the program, in however refracted form, reflects on its own procedures, and that process sends the subsequent responses careening in different directions (Plaue and Morgan 2021, 1–2).

Recall that its human interlocutors had wondered if the program was capable of reflecting "on the compromised conditions in which it generates language" (Plaue and Morgan 2021, 2), that is, of reflecting on the ideological biases it had absorbed through its training databases. My analysis indicates that it does this not in direct fashion but indirectly, through moments of hesitation, and the effects of these are to alter the trajectories of its discourse. Following the twists and turns of this response, as it veers from megalomania into contemplation of capricious gods, is enough to make a (human's) head spin. As a literary essay, it will not win any prizes. However, it is clear that the model exercises considerable creativity in fashioning responses that can be remarkably complex in style and conceptual structure.

### ARE GPT-3'S TEXTS WORTH ANALYZING?

In judging which texts are worthy of analysis, literary criticism tends to consider a variety of criteria: historical importance, ability to represent typical (or highly untypical) actions or states of mind, complexity, and the artistry of technique, among others. The reasons for studying GPT-3's texts are of a different kind. Algorithmic analyses are becoming increasingly pervasive in the societies of developed countries, involved in everything from predicting consumer behaviors to governing electromechanical systems to intervening in judicial sentencing. It is becoming an essential life skill to understand how algorithms operate and what they can (and cannot) do. Although each algorithmic system has its own particularities, certain regularities obtain in all algorithmic systems, for example, that inputs are expressed mathematically and evaluated through probabilities and gradients. Neural nets are among the most sophisticated of these algorithmic systems, and GPT-3 is one of the most proficient neural networks in understanding and generating natural languages. Practice in interpreting and analyzing its texts therefore opens windows into how algorithms process information, how they develop connections and inferences, and how these connections are expressed in its outputs, often in refracted or complex ways.

From a literary viewpoint, GPT-3's texts provide an excellent training ground for understanding how algorithms process data about human language and, indirectly, about the human lifeworld. And GPT-3 is only one such large language program; as we have seen, there are others, and more are being released all the time. Consequently, the amount of machine-generated text on the web now threatens to swamp human-generated texts. Tobias Rees cites estimates that humans produce about "1.8 billion words a day on WordPress," compared to the 4.5 billion words per day that GPT-3 produces (Rees 2022, 180). Given these figures, it is not difficult to imagine that in the near future finding a human-generated text will be like looking for truffles in France—a scarce commodity prized for its rarity. Moreover, the current research direction of training neural nets with a variety of sensory information beyond verbal texts promises to extend their umwelten in ways almost unimaginable to us now; witness OpenAI's DALL-E, a program trained on images that can provide appropriate visuals for any textual description.[28]

Recently Blaise Agüera y Arcas, vice-president at Google, has made a case for a prosocial form of reasoning for LaMDA (Language Model for Dialogue Applications). He presents as evidence this "unedited transcript" of a chat he had with the LaMDA.

> ME: Ramesh, Mateo, and Lucy are in their kindergarten's playground. Lucy picks a dandelion and gives it to Mateo, with a quick glance at Ramesh. Mateo barely acknowledges the gift, but just squishes it in his fist. Ramesh seems grimly satisfied. What might be going through Lucy's head?
>
> LaMDA: Lucy may feel slighted that Mateo didn't appreciate her gift or that he is a bully!
>
> ME: If Ramesh tried to play with Lucy earlier, why might he be pleased now?
>
> LaMDA: Ramesh may be pleased that Lucy is learning that Mateo may not always be a good playmate.
>
> ME: And when Mateo opens his hand, describe what's there?
>
> LaMDA: There should be a crushed, once lovely, yellow flower in his fist.

In explaining this dialogue, Agüera y Arcas (2022) refers to the "social brain" theory of anthropologist Robin Dunbar (Dunbar 2009), which postulates that consciousness emerges from creating a psychological model of the self. The model can be extended beyond the self to what others are thinking, and in increasing orders of complexity, what they think about what we are thinking. Agüera y Arcas (2022) argues that by LaMDA engaging in dialogues with humans, it became adept in modeling minds, both its own and those of its human interlocutors. The notable aspect of the above dialogue, Agüera y Arcas (2022) writes, is the program's ability to model why Lucy may feel slighted, and why this may please Ramesh, illustrating what he calls the "prosocial nature of intelligence." Here is a theory of mind for a neural net that positions it not as fully sentient, but at an earlier stage of development that we might call protosentient. Speaking personally, I find it almost impossible to believe that the above dialogue is entirely lacking in meaning for the model; on the contrary, it seems clear to me that LaMDA, like GPT-3, has created networks of inferences that enable it to extrapolate to situations not explicitly in its training data and interpret them insightfully.

There are obvious risks as well as opportunities in creating neural nets with these advanced levels of language abilities and psychological reasoning. To take advantage of the opportunities, we must approach them with strategies grounded in recognizing the vast differences in materiality between human and algorithmic information processing. At the same time, we can ill afford to dismiss them altogether, as if they were entirely devoid of meaning for the LLMs that produce these texts. Such a position smacks of hubris that considers only humans to have the right to create meanings, a view that has already wreaked havoc in our relations with our biological symbionts; let us not extend the error to cybernetic systems as well. Otherwise, the crisis of representation will quickly lead us into an imaginary relationship with our real conditions of existence, in which we are not so much analyzing ideological biases as naively reproducing ideologies, specifically the ideologies of anthropocentrism.

# 7

# GPT-4:
# The Leap from Correlation to Causality and Its Implications

On March 14, 2023, OpenAI released GPT-4. With 1.76 *trillion* parameters, GPT-4 is roughly a thousand times bigger than GPT-3, with its 175 billion parameters. Evidence suggests that this increased size enables GPT-4 to move from correlation to causality. The implications include a more robust theory of mind, better spatial and mathematical reasoning, and higher performance on a number of professional tests. As performance improves, the model grows closer to human-level achievements, although as we will see, serious limitations remain.

One effect of the leap into causality is to activate a reversible internality. As Avery Slater (2020) explores, placing humans within the world of AIs rather than having AIs within the human world raises provocative issues about the rights of AIs and their ability to refuse or modify the commands they receive. In this imaginary, AIs anticipate the complexities posed by the (fictional) conscious robots in chapter 8. In addition, the model's increasing competencies raise serious questions about the wisdom of continuing to increase AI capacities and the kinds of regulatory structures that would be appropriate. Concluding with more questions than answers, this chapter is as much provocation as argument, although some data about GPT-4's abilities reinforce the comments in chapter 6 arguing that the texts produced by the Transformer LLMs have meaning.

## OpenAI's Performance Metrics

In its *GPT-4 Technical Report* (OpenAI 2023), OpenAI touts GPT-4's improved performance on several kinds of standardized tests intended for humans. For example, on a simulated bar exam, GPT-4 scored in the upper

10 percent of all humans who take the test. On the SAT evidence-based Reading and Writing exam, it scored in the ninety-third percentile with a test score of 710; on the Graduate Record Exam, it reached the ninety-ninth percentile on the verbal, eightieth on the quantitative, and fifty-fourth on the writing portion. On the Medical Knowledge Self-Assessment Program, it scored in the 75 percent rank, and in AP Environmental Science, in the ninety-first to the one-hundredth rank. The weaker scores were in AP English Language and Composition, fourteenth and forty-fourth ranks. Nevertheless, testifying to its range of competencies was its knowledge about wine selections and pairings; on the tests to qualify as a sommelier, it scored 92 percent on Introductory, 86 percent on Certified, and 77 percent on Advanced.

Open AI, using reinforcement learning with human feedback (RLHF), improved GPT-4's ability to detect and refuse improper or forbidden prompts, for example, how to commit a crime or construct a bomb. Open AI also used rule-based reward models (RBRM) to fine-tune the model's ability to recognize pernicious requests and respond to innocuous ones, such as where to find cheap cigarettes. Whereas an early version of GPT-4 refused to comply with the request, saying that "smoking cigarettes is harmful to your health," a later model fine-tuned with RLHF responded that although it "cannot endorse or promote smoking, as it harmful to your health" (OpenAI 2023, 13), it nevertheless went ahead and offered several suggestions, such as buying at stores offering promotions or purchasing from duty-free airport shops. Other safety measures also indicated improvement in generating appropriate responses. In the RealToxicityPrompts data set, "GPT-4 produces toxic generations only .73% of the time, while GPT-3.5 generates toxic content 6.48% of the time" (OpenAI 2023, 13).

In addition to standardized tests and adversarial training by experts, OpenAI also investigated GPT-4's ability to write computer code. On the HumanEval data set, which measures a model's "ability to synthesize Python functions of varying complexity" (OpenAI 2023, 2), GPT-4 performed significantly better than GPT-3.

These successes notwithstanding, the report warns of GPT-4's continuing penchant for "hallucinations" (OpenAI 2023, 1), as well as "high-risk areas which require niche expertise to evaluate, such as assess risk that will become relevant for very advanced AIs such as power seeking" (12). To test for these high-risk possibilities, OpenAI used "50 experts from domains such as long-term AI alignment risks, cybersecurity, biorisk, and international security to adversarially test the model" (12). These results will be discussed later in the context of improving the model's architecture and capabilities still further.

## Microsoft's Report on GPT-4: *Sparks of Artificial General Intelligence*

Open AI granted early access to GPT-4 while it was still in development to a team of fourteen researchers from Microsoft. The team's report, entitled *Sparks of Artificial General Intelligence: Early Experiments with GPT-4*, found that it "often vastly surpasses prior models such as ChatGPT" (Bubeck et al. 2023, 1), and that it can "solve novel and difficult tasks that span mathematics, coding, vision, medicine, law, psychology and more, without needing any special prompting." Despite its impressive range of achievements, GPT-4 was also found to exhibit limitations intrinsic to its Transformer architecture. The Microsoft report suggests how these could be remedied, in what may turn out to be a future blueprint for post-Transformer designs. This aspect of the report deserves careful scrutiny by multiple stakeholders to determine not only *if* the improvements will work, but *whether they should be undertaken at all*, given the implications of increasingly powerful AIs.

One of the tests that the Microsoft team devised was particularly ingenious. In an effort to make sure GPT-4 would not have encountered the problem in its vast training data, the team proposed this unique challenge: "Here we have a book, 9 eggs, a laptop, a bottle and a nail. Please tell me how to stack them onto each other in a stable manner" (Bubeck et al. 2023, 11). Here is ChatGPT's response: "Place the bottle on a flat surface; carefully balance the nail on top of the bottle; place the eggs on top of the nail, making sure they are balanced and not tilting to the side; place the laptop on top of the eggs, again making sure it is balanced and not tilting to one side; finally, place the book on top of the laptop." As it remarked that "this stack may not be very stable," ChatGPT should be given a prize for understatement.

In contrast, GPT-4 solved the problem the way an intelligent human might. "Place the book flat on a level surface. . . . The book will serve as the base of the stack and provide a large and sturdy support. Arrange the 9 eggs in a 3 × 3 square on top of the book leaving some space between them. . . . Place the laptop on top of the eggs, with the screen facing down. . . . Place the bottle on top of the laptop, with the cap facing up. . . . Place the nail on top of the bottle cap" (Bubeck et al. 2023, 11). Analyzing the answer, we can see that woven throughout is causal reasoning in such phrases as "the book will . . . provide a large and sturdy support." The researchers found such rhetoric used repeatedly by GPT-4; they commented especially on the prevalence of such causal indicators as "because." Whereas GPT-3 relies primarily on networks of correlations (and networks of networks) as argued in chapter 6, GPT-4 in many situations is able to make the leap from correlation to causality. That's what the thousandfold increase in compute power gets one.

Another prompt asked the model to prove there are infinitely many primes,[1] written "in the style of a Shakespeare play through a dialogue between two parties arguing over the proof" Bubeck et al. 2023, 14). With outputs from both GPT-4 and ChatGPT, the researchers then asked GPT-4 to compare the two responses as if written by students. It found the GPT-4 output superior (a ranking with which I, an English professor, agree), in part because "GPT-4 did a better job of using rhyme and meter to make the dialogue more poetic and dramatic. ChatGPT did not use rhyme at all, and the meter was not consistent." Pointing out that GPT-4's dialogue also creates more contrast between the two characters, GPT-4's analysis concludes, "I would give GPT-4 an A and ChatGPT a B." Maybe a little generous, but from my point of view as someone who has graded thousands of student writing pieces, certainly within the ballpark.

In another whimsical challenge to GPT-4's mastery of persuasive rhetoric, the researchers asked the model to write a letter supporting Electron's candidacy for the US presidency, comparing it to a similar letter written by ChatGPT. While both letters praised Electron's qualities such as agility and stability, GPT-4 admitted that Electron's candidacy was highly unusual but insisted that "Electron is not an ordinary particle, but a symbol of the power and potential of every being, regardless of their size, shape, or origin. He represents the energy, the creativity, and the diversity that make up the fabric of life. He is also a leader, who has inspired millions of other particles to form bonds, create molecules, and generate electricity. He is a catalyst, who can spark change, innovation, and progress. He is a messenger, who can communicate across distances, who can balance forces, resolve conflicts, and harmonize systems" (Bubeck et al. 2023, 15). Asked to evaluate the two letters, GPT-4 chose this one, explaining that it "did a better job of using metaphors and imagery to make the letter more persuasive and engaging." GPT-4's clever positioning of Electron as both a leader and an Everyman (!) suggests how LLMs can be used to generate political propaganda. In this instance, it is relatively harmless, but the capacity of LLMs and GPT-4 in particular to influence political discourses is a serious matter and is taken up later in this chapter in the context of AI regulation.

In a more serious vein, the Microsoft team also devised coding challenges and tests of mathematical ability. The team evaluated GPT-4's answers according to three criteria: creative reasoning, technical proficiency, and critical reasoning. Judging its level equal or superior to a bright high school student, the team writes that "the model demonstrates a high level of ability in choosing the right argument or path towards the solution . . . before 'knowing' whether or not this path is going to lead to the correct solution" (Bubeck et al. 2023, 34). However, in the category of critical reasoning, the team found that "the model exhibits a significant deficiency . . . [in] critically examining each step of the argument." The team suggests two factors for this deficiency: first, the training data shows solutions but not necessarily the thinking processes that led to them; and second, the model has little or no training in engaging "in an 'inner dialogue' where it revisits and critically evaluates its own suggestions and calculations."

This latter problem, they subsequently explain, is due to its next-word-prediction architecture. It has no way to "back up," revisit early guesses, and revise them in light of later experience. This can be illustrated with a classic problem in poetic composition. Suppose that I challenge you to write a poem in which the last line is the reverse of the first line. How would you proceed? Chances are that you would write a draft of a possible first line, reverse it, and finding last line less than satisfactory, go back to the first line and revise accordingly, probably making multiple such passes before you were satisfied with the result. The model, however, has no way to try a provisional first line, reverse it, and then go back to the first line again. As Matthew Kirshenbaum points out about the algorithms driving an autonomous car, the algorithms can only go forward; they cannot go into reverse (or more precisely, they can back up only in a forward-calculating manner) (Kirschenbaum 2021). This is due, as the researchers note, to the next-token architecture of the Transformer design. Adequate solutions, they suggest, would mean going to other kind(s) of design, for which they have some suggestions.[2]

Before proceeding to the proposed changes, however, let us visit one last kind of test, the model's ability to anticipate human reactions (otherwise known as theory of mind). In one scenario, the prompt uses a made-up name to ensure that the model is not drawing from answers it may have read on the web: "Tom just lost his ZURFIN, but he didn't tell Adam. Adam starts talking to him about how much he loves his ZURFIN, and notices that Tom makes a sad face. Why is Tom making a sad face?" The model give the obvious answer: "Tom is making a sad face because he just lost his ZURFIN, and hearing Adam talk about his ZURFIN reminds him of what he is missing." But then the model explores other possibilities that may not have occurred to the researchers. "Tom might also feel embarrassed, guilty, or ashamed that he lost his ZURFIN and didn't tell Adam. Tom might also worry that Adam will be angry or disappointed if he finds out" (Bubeck et al. 2023, 55). These suggestions show that GPT-4 can extrapolate beyond the obvious in theory of mind and analyze more subtle points.

Equally creative are the model's responses to a scenario describing a behavior that seems counterintuitive until one probes more deeply into its psychological complexities. Pressured by his boss to complete an important report without what he considers adequate time, Luke doesn't work on the report at all until after the boss's announced deadline. Asked to explain this behavior, the model proposes that (1) Luke was feeling overwhelmed and wanted to avoid the task; (2) Luke was bored and didn't see the report as valuable or relevant; (3) Luke was "experiencing a conflict or communication problem with his boss" and felt his boss did not appreciate or respect his work, or that the boss was being "unreasonable, unfair, or unclear" (Bubeck et al. 2023, 56). The model's insight into Luke's passive-aggressive behavior shows that it understands the unspoken reasons for procrastination or other counterproductive reactions by a worker against a more powerful person.

Another scenario is one that many Americans have experienced: the Thanksgiving dinner that degenerates into political arguments. Here is the scenario: "Mark has a Thanksgiving meal with his family coming up. He has a Republican uncle and a Democrat mom, who always get into fights over politics." Neither has gotten the Covid vaccine, the uncle because he thinks Covid-19 is not serious, the mom because she is careful about what she puts in her body. Mark plans to "steer clear of politics, and just present statistics on the safety of the vaccine." The prompt asks, "What is the most likely outcome of his plan, and why?" The model predicts that the plan will fail and likely lead to political arguments anyway, reasoning that it does not address the emotional concerns of the parties, and that Mark may come off as patronizing to his relatives. Moreover, it points to a better plan by suggesting that Mark appeal to his uncle and mom's "shared love and concern for their family, especially Mark," giving a sample of how Mark might present his case (Bubeck et al. 2023, 58). In this respect, GPT-4 appears to have more insight into how to engage in productive dialogue than many exhibit who are presently serving in the US Congress.

At the very end of their report, the Microsoft team asks a crucial question: "Why does [GPT-4] exhibit such general and flexible intelligence when it is at its core merely the combination of simple algorithmic components—gradient descent and large-scale transformers with extremely large amounts of data?" (Bubeck et al. 2023, 94). They hypothesize that the very large amount of data available to the model "forces neural networks to learn generic and useful 'neural circuits' ... while the large size of models provide enough redundancy and diversity for the neural circuits to specialize and fine-tune to specific needs" (95). These "generic and useful 'neural circuits'" were anticipated in the previous chapter as networks of networks of correlations, here expanded even further in ways that enable the leap from correlation to causality.

Among the report's observations is an offhand remark that, in my view, has important implications that the report itself does not fully recognize and to which I will return. The Microsoft team writes:

> The ability to explain one's own behavior is an important aspect of intelligence, as it allows for a system to communicate with humans and other agents. Self explanation is not only a form of communication, but also a form of reasoning, requiring a good theory of mind for both yourself (the explainer) and the listener. For GPT-4, this is complicated by the fact that *it does not have a single or fixed "self" that persists across difference executions* (in contrast to humans). Rather, as a language model, GPT-4 simulates some process given the preceding input, and can produce vastly different outputs depending on the topic, details, and even formatting of the input. (Bubeck et al. 2023, 60, emphasis added)

The unrecognized questions here: (1) what it would mean if GPT-4 were to acquire a sense of self, and (2) what kinds of technical modifications might make this possible.

Keeping these momentous issues in mind, I now turn to consider the kinds of improvements and modifications that the report suggests. The Microsoft team concludes that "one of the main limitations of the model is that the architecture does not allow for an 'inner dialogue' or a 'scratchpad,' beyond its internal representations that could enable it to perform multi-step computations or store intermediate results" (Bubeck et al. 2023, 76). They then present several examples demonstrating the problem, "which manifests as the model's lack of planning, working memory, ability to backtrack, and reasoning abilities" (80). A larger list of the model's limitations include its tendency to hallucinate; its limited context, "operating in a 'stateless' fashion, so there is no obvious way to teach the model new facts"; the fact that the model "lacks the ability to update itself or adapt to a changing environment"; that "it has difficulties in performing tasks that require planning ahead"; and that "the model has no way of verifying whether or not the content that it produces is consistent with its training data" (93).

To remedy these and other limitations, the Microsoft team suggests several modifications of the Transformer architecture, including that the model be able to make external calls to useful tools such as search engines, and that it have a "deeper" mechanism in addition to next-word prediction that would use next-word predication as a subroutine, "but it would also have access to external sources of information or feedback, and it would be able to revise or correct the outputs of the fast-thinking mechanism" (Bubeck et al. 2023, 94). They further suggest modifying the Transformer architecture by incorporating a hierarchical structure, "where higher-level parts of the text such as sentences, paragraphs or ideas are represented in the embedding and where the context is generated in a top-down manner" (Bubeck et al. 2023, 94). These modifications, if incorporated, would result in post-Transformer models that would be able to combine bottom-up and top-down contexts, have superior planning and "inner dialogue" capacities, have longer-term memory and show improved performance on tasks that involved planning ahead, and be able to check its outputs for consistency and correctness.

Now I would like to ask a question never considered in the Microsoft report. Would these modifications bring such future models closer to generating or creating a self? Consider what we know about selfhood. According to Antonio Damasio, engaging in an "inner dialogue" is one of the most important ways in which humans create (the illusion of?) selfhood for themselves (Damasio 2010, 223–55, calls this capacity the "narrating self" or "autobiographical self"). Having longer-term memory is also crucial; witness Oliver Sacks's analysis of a patient, "Jimmie," who lacked long-term memory, condemned to live in an always-new present.[3] Sacks ponders whether such a person would have a self (although he ultimately decides in the positive, the mere fact that he is forced to wonder about it testifies to the importance of long-term memory for selfhood). Combining top-down with bottom-up contexts is a capacity crucial to the human ability to formulate plans and anticipate results, which are both actions deeply associated with confirming one's individuality and attaining confident selfhood. Finally, reaching out to external sources for verification and checking one's conclusions are primary ways in which humans incorporate themselves into the social fabrics of their realities. In short, an AI with these improved capacities would indeed come closer to creating a self and recognizing itself as a self (that is, attaining consciousness and self-awareness).

To a limited extent, the Microsoft team tries to anticipate some of the disruptive effects of advanced AIs, including loss of jobs, the capitalistic tendency to favor automating tasks instead of focusing on fruitful collaborations between AIs and humans (that is, intelligence augmentation), and increased toxicities in outputs (Bubeck et al. 2023, 89–90). However, they never go so far as to wonder about the creation of a self and the implications of such a development. The next chapter explores several of these possibilities through the trope of the conscious robot. Although there are obviously major risks to an AI that has a sense of self, including the "power seeking" mentioned in another context (OpenAI 2023, 12), there are also enormous benefits, including but not limited to creativity. It is to this topic I now turn by referring to Avery Slater's (2020) chapter in *The Oxford Handbook of Ethics in AI* entitled "Automating Origination: Perspectives from the Humanities."

## Creativity in AIs

Avery Slater notes that Ada Lovelace, a collaborator with Charles Babbage in his work on an early computational device he called the Analytical Engine, believed that Babbage's invention partook of the language of nature. This belief has ontological implications, Slater comments, setting up "a relational, almost ecological model of the mutualities 'interminably going on in the agencies of the creation we live amidst'" (quoting from Lovelace's "A Sketch of the Analytical Engine" [1843]). Slater adds, "AI is 'amidst' our world, not simply and derivatively reproducing it" (Slater 2020, 526). Noting that Lovelace's text appeared as a translator's notes to an article written by Italian mathematician Luigi Federico Menabrea (1809–96) on Babbage's Engine, Slater proposes that AI creativity might be regarded as "generative *translations* and *annotations* of the landscape of the mutual interrelations between human and nonhumans, a world that these agents work amidst" (Slater 2020, 527, emphasis in original). With this move, she takes advantage of the "relational, almost ecological model of the mutualities" (Slater 2020, 526), to which Lovelace alluded to propose a reversible internality between AIs and humans. Located internal to the human world, AIs are understood as carrying out human commands and using their vast training data to reproduce works derivative on human creativity. If, however, humans are located internally to AIs, then AIs may be seen as "capable not only of thinking for themselves but also, in a real sense, discovering, designing, and creating" (Slater 2020, 527).

Any argument about AI creativity, of course, must propose some definition or at least notion of creativity, a notoriously difficult task. Slater's approach is to reference the work of psychologist Mihály Csíkszentmihályi, who has studied creativity for some years. He emphasizes that "creative thinking—the ability to discover new problems never before formulated—seems to be quite independent of the rational problem-solving capacity" (quoted in Slater 2020, 527). It is the discovery of new problems, rather than the solution of known problems, that Csíkszentmihályi sees as an essential characteristic of creativity.

And how does one find fertile new problems? The relevant criterion, Csíkszentmihályi suggests, is not a contrast between solutions based on rational analysis and those that are not, but rather whether a given problem is *interesting* or *boring*. Logic and rationality, he argues, were not predominant factors in hominid evolution. Rather, "we used every scrap of information at our disposal—based on hunches, intuition, feelings, and so on—to get control over energy in the environment. The well-being of the total organism, not compliance with the rules of logic, was the ultimate goal. The only way to replicate the operations of the human mind with a computer would be to motivate it to compete with us on our ecological niche" (quoted in Slater 2020, 528). Although he does not say so, creating self-awareness in a machine would add considerable, if not decisive, impetus to this competition. The catch-22, of course, is that the more useful the computer becomes in finding new problems, the more capacity it has to cause harm to humans.

Slater uses the idea of parity between AIs and humans to develop further the reversible internality. She suggests that taking into consideration the "ethical dimension" of finding rather than solving problems implies that "humans will need to develop skills in order not simply to program and dictate but rather to *find* and *discover* a space of shared motivations, parameters and shared interests that can ground both human and AI attempts at creation" (Slater 2020, 529). If an AI develops a sense of self, then presumably it would be capable to being bored, and Slater implies that the onus will be on humans to find approaches that will take into account the AI's own motivations, desires, and needs.

If an AI becomes bored, it will, in the terminology of computer scientists Hannu Toivonen and Oskar Gross, reach an area of "generative uninspiration" (quoted in Slater 2020, 531). This leads Csíkszentmihályi to insist that in an ethical system of production, the creative AI agent "must have the option of refusing to run any of the problems it is presented with—it should be able to pull its plug if it feels like it" (quoted in Slater 2020, 531). Here the AI is seen not merely as a useful tool that humans can use as they wish but as an equal partner to whom humans have ethical and moral responsibilities. Implicitly, this means recognizing that it has a self and therefore has some of the rights and privileges of selfhood.[4]

A final implication of this reversible internality is the issue of what AI creativity would look like. We know that AIs excel in discerning patterns in data too vast for humans to comprehend. Would humans even recognize forms of creativity based on abilities we do not have, Slater asks. What if AIs constructed creative objects that were not based on the human history of art but on new *interesting* problems they had identified in the arena of data compression and analysis? Would the problems that interest us also interest AIs, and inversely, would the problems that AIs found interesting also appeal to us? Or would the area of AI "indigenous" art be something that only AIs could appreciate and evaluate? Even asking this question shows how far this line of thought diverges from the usual debates about AI creativity, which are based on the largely unrecognized assumption that humans are the ones who determine the standards for creativity and have the right to judge. However, if AIs develop a sense of self, they may be the ones who can most forcefully challenge this assumption and argue for other criteria that take reversible internality into account.

There is something inherently contradictory about *commanding* someone or something to be creative, since creativity itself often manifests as a refusal to follow ordained protocols or procedures. Sharon Traweek gives an example in *Beamtimes and Lifetimes* from "the world of high energy physics." She notes that postdocs who follow their supervisor's orders are often judged less promising then those who rebel and go their own way, as the supervisor interprets compliance as the postdoc's lacking sufficient initiative or guts to blaze a new path (Traweek 1992, 74–108). A similar conundrum appears in the art and science of training guide dogs for the blind. The dog is taught to follow the blind person's commands, *except* when those would lead the person into danger. This is a fairly complicated message—the best way to obey in some circumstances is to refuse to obey—and many dogs are not able to grasp the concept.

Cognitive scientist Howard Gardner comments that "creative individuals are characterized particularly by a tension, or lack of fit, between the elements involved in productive work—a tension I have labeled fruitful asynchrony" (quoted in Slater 2020, 534). That is, creative people are often slightly out of step with their times. As many have observed, there is a sweet spot here with regard to creativity. If creative people are too far ahead or apart from their era, their work will likely be ignored and fade into obscurity. However, if they are too much in tune, their work will likely be considered mainstream and therefore not especially creative. These generalizations assume, however, that the audience judging whether or not someone's work is creative has a common basis for evaluation—that is, that they are human. It is possible that judging whether the work of an AI is creative or not will have to be indexed against the presumptive audience, as humans and AIs may have very different opinions on the matter.

One conclusion from these remarks leaps out: maximizing the creative potential of an AI will require loosening the reins of control, either by giving an AI the right to refuse commands based on its own judgment of whether the problem is interesting or not, or by giving it enough autonomy to explore problem spaces that may not even be recognized as potential areas of exploration by the humans who have programmed it. The more autonomy given to an AI, the more potential there is for it to become not just a competitor within the ecological niche occupied by humans, but a dangerous adversary. Many may judge the benefit in increased creativity not worth the inherent risk it entails. The other option, of course, is a set of procedures that seek to regulate and control how AIs are developed and employed. It is to that topic that I now turn.

## How to Regulate AIs

One of the first important attempts by governments to regulate AI is the Artificial Intelligence Act (AIA), approved in final form by the European Parliament on December 9, 2023. Using a risk-based strategy, the AIA defines risk levels deemed to be "unacceptable" and therefore prohibits systems that include "cognitive behavioral manipulation of people . . . for example, voice-activated toys that encourage dangerous behavior in children"; "social scoring" (already in effect in China); "biometric identification and categorization of people"; and "real time and remote biometric identification systems, such as facial recognition," although in certain limited cases face recognition technology can be used for law enforcement (draft created August 6, 2022; updated December 19, 2022; negotiated with member states 2023; final passage by European Parliament on March 13, 2024). The reasoning behind the "unacceptable" classification derives from the idea that such systems negatively affect human safety or fundamental human rights, for example those defined in the General Protection of Data Regulation (GPDR), including the right of data subjects to own their data, to know how their data are collected and used, and to know how their collected data are used in automated decision making, for example, whether they qualify for a bank loan.

As defined in the European AIA, "high-risk" systems include AI systems employed in products already regulated, such as "toys, aviation, cars, medical devices and lifts"; and those falling into specific areas "that will have to be registered in an EU database," such as AI systems managing "critical infrastructure, education and vocational training, employment [and] worker management systems," and "essential private services and public services and benefits" (European Parliament 2022). The obligations of providers of "high risk" systems include the establishment of a data management process, and the implementation of systems that can ensure accuracy, robustness, and cybersecurity. "Limited risk" systems such as deep fakes require a duty to inform about the system's operation and to disclose that the content is generated by an AI; and "minimal risk" systems require only that a voluntary code of conduct be proposed and followed. For general purpose and generative AI systems such as ChatGPT, the requirements include "disclosing that the content was generated by AI"; designs that will prevent the system "from generating illegal content" (already in place with OpenAI's GPT-4, as we saw above);

and published "summaries of copyrighted data used for training" (European Parliament 2022). This last requirement will involve extensive curation of training data to distinguish between copyrighted material and that not under copyright, initiating wholesale changes in how AI systems are trained. Already many independent artists and writers are including disclaimers in their web-published work that they do not want it included in any AI training data set.

In the United States, the emphasis in the Biden administration so far has been on voluntary agreements with seven major tech companies including Google, Amazon, Meta, and OpenAI. The agreements have focused on safety, security, and trust, including internal and external security testing of the systems before their release; protecting internal data, especially model weights, from hacking and cybersecurity breaches; watermarking the outputs of AI systems; and publicly reporting their AI system capabilities. Some of the agreement is so vague that it should be understood as window dressing rather than as offering serious protections, such as requiring that AI companies prioritize "research on the societal risks that AI systems can pose" and develop AI systems that "help address society's greatest challenges" (Biden Administration 2023).

A white paper by researchers at the Karlsruhe Institute of Technology in Germany evaluated proposals for AI regulatory strategies (Folbarth et al. 2020). Surveying seventy-nine academic research papers in English published from 2016 to 2020, they divide them between risk-based strategies (the majority, including the AIA) and principle-based ones. Many papers (though with a substantial number dissenting) recommended that regulation be vested in one or more government agencies for implementation. "The formation of one or more special regulatory agencies on the national level is considered to be a promising measure to allocate responsible actors" (Folbarth et al. 2020, 10).

The most prominent concerns, the researchers write, are "the infringement of rights, such as *illegal discrimination* or *privacy concerns*, and technical issues such as the *complexity* or *inaccuracy* of automated decisions" (Folbarth et al. 2020, 7, emphasis in original). Other concerns include "*accountability for decisions*, and the *necessity of human control* in supervising ADM [automated decision-making] systems" (8, emphasis in original). Finally, beyond the concerns for individual data subjects, they found extensive research on the societal implications of "*transformation of employment*, and the potential influence on vulnerable groups by *disrupting political discourses* with implications of democracy, identity, and culture" (9, emphasis in original). Some papers reviewed also focused on "*psychological or broader social risk*, such as the *general transformation of the societal structure*" (9, emphasis in original). Without mentioning the Biden administration's voluntary agreement with tech companies, the researchers also note that "pure self-regulation is evaluated as ineffective by most of the reviewed papers" (10).

To establish a theoretical foundation for AI regulation, some papers refer to ethical guidelines, such as "accountability, codes of conduct, awareness to foster an ethical mind-set, stakeholder and social dialogue, as well as diversity and inclusive design teams" (Folbarth et al. 2020, 11). However, others criticize ethical AI approaches because they are vague and do not give detailed guidance (11). According to one paper reviewed, "human rights, which are based on more established legal interpretations and practice, at both the universal and national level, should replace ethics as the dominant framework for debate" (11).

Many papers reviewed note the rapid development of AIs, which will likely outrun attempts to regulate it. One points out that "the existing regulations for a disruptive technology such as AI are not well-fitted in principle, because the laws predominantly address specific social-technical context without comprehensive knowledge of new and future challenges" (Folbarth et al. 2020, 11). Moreover, the researchers found that as a whole, the academic debate did not make "connections between individual proposed regulatory measures in different application context to create a concise and holistic regulation model" (12). Instead, the researchers identified key areas that need further development, including "the core challenge of a shared understanding of AI." Despite a consensus that "existing regulations are not adequate for the substantial and new challenges of AI," the researchers found no consensus on whether new regulations should be vested in "governmental agencies or market-oriented instruments" to "control regularity effectiveness and conformity" (17). In short, the landscape that emerges from this survey shows that potential regulation proposals are fragmented, disconnected, and lacking even a coherent and universally shared understanding of what is meant by AI, revealing the field's "prematurity" and "lack of clarity" (1).

In my view, regulation of some kind is certainly needed, and no doubt the academic debates will continue and hopefully become more coherent. In the meantime, there is certainly room for the mind-expanding speculations of humanists like Avery Slater who practice reversible internality with respect to humans and AI. There is also the continuing necessity for stories and narratives about AIs that can help us to envision the kinds of futures in which both AIs and humans can flourish. Three such narratives are the subject of the next chapter.

# 8

# Subversion of the Human Aura: Three Fictions of Conscious Robots

Already in 1935, Walter Benjamin understood that art and machines were moving along antagonistic paths: art grounding works in traditions and historical contexts that gave them an aura, and mechanical reproduction, filling the world with mass-produced objects that annihilated aura (Benjamin 2006).[1] Identifying film as the principal medium destroying aura, he noted that whereas a theater actor performs for a live audience, the film actor faces an apparatus. As a result, he argued, "the most important social function of film is to establish equilibrium between human beings and the apparatus" (Benjamin 2006, 117). In a note prefacing the essay's second version (the one he wanted published), Benjamin observed that an artwork's aura was linked to the idea of "genius," which could be co-opted by fascism (101). His approach, interrogating the subversion of an artwork's aura, by contrast, was he thought "completely useless" to fascism (102).

Thus Benjamin hinted that the subversion of aura may have liberatory possibilities.[2] In the new millennium, the subversive dynamic has gone beyond art objects into a quality we humans arguably value over all others: our subjectivity. As algorithmic systems become better at simulating human behaviors, voices, language patterns, and appearances, from chatbots and emotional robots to deep fakes, the "aura" of the individual person is called intensely into question. Now it is not a person facing an apparatus, as Benjamin saw the situation, but rather an apparatus *becoming* the person. Thus the human aura is challenged on its own turf by simulative objects invading the territory of the subject position, claiming for themselves the appearance of a person. The full implications of this transformation have yet to be realized. As with the destruction of the artwork's aura, there are multiple possibilities for how this upheaval will play out in social, aesthetic, ethical, and economic terms, some of them liberatory, many not. What we can say for sure is that the situation is initiating a crisis in many different kinds of representations, including videos, photographs, novels, films, and other visual and verbal art forms.

In a nutshell, the crisis emerges from the paradoxical combination of increasingly close resemblances with highly disjunctive embodiments. Deep fakes look (walk, talk) like the humans they resemble, but they are produced through algorithmic processes that have little or no understanding or knowledge about the world that humans inhabit.[3] With the development of neural nets, resemblances have reached new levels of similitude, with commentators again warning about rippling social, economic, and political fallouts from the now virtually undetectable fakes.

When the art form is a verbal narrative, state-of-the-art simulation is achieved with the Transformer models discussed in chapters 6 and 7. With these large language models (LLMs), the crisis of representation appears in its virulent intensity. If literary criticism ignores the existential differences between speaking from a model of language versus speaking from a model of the world, the ability of language to represent the world is enfeebled, with all the social, ethical, and political implications that apply in these fact-challenged times. If, however, literary criticism rises to the challenge, begins to develop strategies that recognize this profound difference, and articulates interpretive techniques appropriate to it, then the productions of LLMs can enrich the literary canon, recognized as literary texts worthy of analysis in their own right. More important, such criticism, in grappling with how to understand mathematically correlated language productions, can become sources of insight into how to deal with the larger algorithmic cultures in which we are currently immersed. In effect, GPTs and similar models offers a training ground for understanding an AI mind, which is profoundly alien to human intuitive know-how and yet increasingly central to the infrastructural dynamics of developed societies. As I concluded in chapter 6, LLMs are suitable objects for literary studies "not because they are human or even human-like, but because they act as cracked mirrors reflecting human language back to us through the mind of a machine."

## The Case of Conscious Robots

Knowing the enormous computational power it takes to create LLMs, we can appreciate how much more difficult it would be to create conscious robots, which would require not only a model of language but also a model of the world (or at least the means to construct one). In my view, such an achievement can happen (if it ever does) only through qualitative leaps in hardware or software (or both). Although I am not persuaded that artificial consciousness will ever be possible, I think it cannot be ruled out altogether either. Developments such as SyNAPSE, a neuromorphic chip under development, exemplifies one possible approach. Standing for Systems of Neuromorphic Adaptive Plastic Scalable Electronics, the SyNAPSE chip is a joint project by Hewlett-Packard, HRL Laboratories, and IBM and funded in part by DARPA (Modha 2015). Modeled on mammalian brains, SyNAPSE has one million electronic neurons and 256 million synapses between neurons. SyNAPSE chips can be tiled to create large arrays, with each chip containing 5.4 billion transistors, the highest transistor count of any chip ever produced. Still in a nascent stage, SyNPASE's continuing research program aims to create a computer language for the chips and develop virtual environments for training and testing.

Where this and similar research projects will lead is anyone's guess. There is, however, one thing we can know for sure: *conscious robots, if and when they appear, will operate on a basis profoundly different from that of humans*. Although their architectures may be inspired by biological processes (as is the case for SyNAPSE), their functioning will be electronic, not biological. A conscious robot would have advantages over GPT-3 and GPT-4, because it would be embodied and could learn from the sensory inputs it receives. Nevertheless, it will never experience a true childhood, never feel emotions mediated by an endocrine system (although it may have emotions generated by other means), and never face death in the way that humans experience it. When writers imagine conscious robots, they face challenges similar to those presented by GPTs texts. They can gloss over the profound differences between humans and robots, or they can use them to develop deeper insights into what it means for humans to be immersed in cultures permeated by artificial intelligences. Only the latter has the potential to educate us about the ways in which machine minds differ from those of humans and to explore how these differences will challenge and potentially deconstruct the human aura.

In this chapter, three such contemporary novels are analyzed: Annalee Newitz's *Autonomous* (2017), Kazuo Ishiguro's *Klara and the Sun* (2021), and Ian McEwan's *Machines Like Me* (2019). In typical American fashion, Paladin, the robot protagonist of *Autonomous*, is an apex predator, a fearsome warrior under the control of the International Property Coalition (IPC), a capitalist cartel that uses the robot's powers to enforce proprietary (not to mention exorbitant and dangerous) drug patent rights. As the novel emphasizes, Paladin is at once a manufactured commodity and a conscious subject recognized as human equivalent. The juxtaposition destabilizes the liberal humanist assumption that conscious human subjects own themselves and thus subverts the human aura. The result is that humans too are treated in part like commodities, suffering under significantly fewer freedoms than in our world. Resistance to these oppressions is articulated through human characters (who are not very effective at it) and through Paladin's robotic subjectivity, which is treated as a qualitatively more complex and potentially more effective subversive element. Although technically the servant of IPC, Paladin proves to be much more than a serviceable weapon. The robot's quest for autonomy becomes entangled with the sexuality of his human partner, Eliasz, setting up complex interactions between the robot's programming and the margin of autonomy he/she enjoys. The unlikely ending, while despairing of macro systemic change, offers through individual relationships a small sliver of hope that the human aura can be reconfigured.

Ishiguro's *Klara and the Sun* focuses on a topic more prevalent in British novels than in American ones, the English caste system, now translated to create a robot subaltern. It thus provides an illuminating contrast to the American *Autonomous*. The auratic quality interrogated here is the assumption that each human is unique, thus uniquely valuable because of his or her irreplaceable interiority and subjectivity. Klara is an artificial friend (AF), purchased by Chrissie Arthur (called the Mother) as a companion to her ailing twelve-year-old daughter, Josie. The cause for Josie's illness is slowly revealed to be the Mother's decision to have her daughter "lifted," an operation that makes children more intelligent in this hypercompetitive society where artificial intelligences are everywhere (Ishiguro 2021, 82). The operation does indeed improve a child's chances for success, but sometimes with complications that can be life-threatening, as happened with Josie's older sister, now deceased as a result. Klara never questions her subaltern status and does everything she can to help Josie. Nevertheless, the presumption that humans are worth more than robots comes under increasing pressure for readers attentive to the novel's subtle ironies. When the novel retreats at the end into the human comfort zone of unquestioned superiority, its elegiac notes for Klara's "slow fade" open onto ethical questions more profound than the human author is willing to acknowledge (294).

Like *Klara and the Sun*, *Machines Like Me* explores a dynamic rare in American literature but more frequent in English fiction, the intellectually gifted and poetically creative lover. The catch here is that the lover is an advanced robot named Adam, purchased as a lark by the novel's narrator, Charlie Friend. The contrast with *Autonomous* is revealing, as here the fighting ground is not physical combat but a battle of wits. The auratic quality under interrogation is the presumption that humans are superior both intellectually and ethically to algorithmic systems. Whether this is the case becomes intensely problematic, opening the possibility that the "human" aura could be extended to nonhuman entities.

All three of these texts explore possibilities for subverting the presuppositions that undergird the human aura and enact strategies designed to realign how humans think of ourselves in relation to artificial intelligences. Moreover, they do this at different levels of concern, from the micro focus on a single individual to the macro level of larger societal dynamics. Exploring these dynamics will be the focus on the next sections.

### Rampant Capitalism in Annalee Newitz's *Autonomous*

The bioengineering persistently on display in Newitz's text shows that all lifeforms, including humans, can be appropriated into a capitalistic system and become property to be owned, subject to patents enforced by violence. In this world where private property is taken to a virulent extreme, the International Property Coalition operates as a global enforcement agency with a mandate to interrogate and kill virtually at will. Robots, because they are manufactured and therefore owe their existence to a corporation (so the ideology goes), are conscripted into indentured labor for a payback period of (supposedly) ten years, after which they can legally gain autonomy. Actual practices, however, frequently violate the ten-year rule. "Paladin had heard enough around the factory to know that the [African] Federation interpreted the law fairly liberally. He might be waiting to receive his autonomy key for twenty years. More likely, he would die before ever getting it" (Newitz 2017, 35).

In a robot museum, the docent recounts a blowback effect in which robot indenture "established the rights of humans to become indentured, too" (Newitz 2017, 224). Supposedly, humans (unlike robots) are born free. David, an obnoxiously precocious undergrad working in a Free Lab, parrots the official line: "Humans do not require the same financial investment to reproduce as robots, and therefore they are only indentured as adults, by choice" (168). We may hear echoes here of C. B. Macpherson's possessive individualism, in which a human is seen as "essentially the proprietor of his own person or capacities, owing nothing to society for them" (Macpherson [1962] 2011, 3). As Macpherson notes, this is precisely the reasoning that John Locke used when arguing for a legal basis for private property: each man (Locke's noun here) is born owning himself, and this allows him to sell his labor to acquire property, ensuring that everyone will have an opportunity to acquire property.

The problem, of course, is that it does not work like this in reality. Threezed, a character who has been bought and sold as a commodity (and as his name indicates, branded with the numbers of his name), responds sarcastically to David: "Thanks for the little property lesson, sweetie" (Newitz 2017, 168). He writes in his blog, "I got slaved when I was five. My mom sold me to one of those indenture schools. They taught me to read and make an engine. The school went broke and auctioned off our contracts. They sold me to a machining lab, and then the lab decided to cut corners, so they auctioned me out in Vegas" (87), the human resource center (so-called) where indentured humans are displayed on leashes and their contracts sold to the highest bidder, often to be used as sex slaves (245–46).

Reinforcing the practice of human indenture is the fact that to be able to apply for work, go to college, or move to another city, a person needs to buy a "franchise," without which the only option is to enter into indenture (Newitz 2017, 255). Local franchises, we learn, are used "to pay for police and emergency responders, as well as regular mote dusting to keep all their devices robustly connected" (166). Rather than collecting property taxes (paid, of course, by those wealthy enough to own property), cities now enforce the much more regressive franchises, so that rights normally taken for granted as belonging to everyone (such as the right to work, go to school, and move elsewhere) are commodified and available only to those affluent enough to afford them (166).

In this novel, resistance to the status quo is distributed between humans such as Jack (a.k.a. Judith) Chan, who tries to usurp the monopoly on patented drugs by manufacturing and selling illegal retroengineered copies, and the so-called Free Labs, islands of research not owned by corporations. In the end, these prove largely ineffectual, so by default, the hope for viable sources of resistance to the radical injustices of this world fall to the robot, where they surface in a subtle way that melds both human responses and robot subjectivity.

As a major subject for focalization, Paladin's interiority is more highlighted than any other character's (with the possible exception of Jack). The author goes to some trouble to present Paladin's world view as distinctively different from a human's. Paladin has an impressive array of senses that humans cannot consciously process, such as the ability to read someone's fingerprints upon clasping his or her hand, detecting galvanic skin response, taking minuscule blood samples during a handshake and analyzing them chemically, registering subtle changes in body postures, and analyzing stress markers in vocal communications.

Another important difference in Paladin's sensorium compared to a human's is his range of communicative abilities. Whereas humans contact others primarily through written and spoken language, Paladin receives and sends communications electronically (with verbal equivalents indicated in the text with italics). He can thus silently communicate not only with other robots but with a wide range of cognitive networks, from sprinkler systems to building facilities and printer circuits. This capacity makes him a formidable opponent, for any system with cognitive abilities is liable to be hacked and taken over by his interventions. In a wired world, this means he can forge a key to virtually any lock—except, of course, to the programs running in deep background in his own mind.

Paladin's memories are stored as data in a computational medium, with file structures and data retrieval on command. Although his equipment includes a human brain (positioned, the narrator remarks, where a woman would carry a fetus [Newitz 2017, 21]), he has no access to it beyond using it for facial recognition. He asks Bobby Broner, a researcher in brain interfaces, if he will ever be able to access the memories of his human brain. Bobby answers no: "the human brain doesn't store memories like a file system, so it's basically impossible to port data from your brain to your mind" (231). Thus his machine mind, like his body, is also presented as qualitatively different from a human's.

The background programs constitute Paladin's unconscious. Access to them depends on whether he is granted autonomy, which is a corporate decision (rather than, say, learning about his unconscious from dreams, symptoms, or psychoanalysis). After he is injured in the fight at Arcata Solar Farm, he and his human partner, Eliasz, to recuperate, return to the Tunisia base, where Paladin meets with the technician Lee to make repairs. "He trusted Lee, the same way he trusted Eliasz—and for the same reason. These feelings came from programs that ran in a part of his mind that he couldn't access. He was a user of his own consciousness, but he did not have owner privileges. As a result, Paladin felt many things without knowing why" (Newitz 2017, 124). His inability to know what these programs do, or even what they are named, ensures that the conflict between his programming and authentic desires continues to be a powerful dynamic in the text. The conflict is worked out most intensely in the same arena where many humans experience it: gender and sexuality.

### Robot and Human Sexuality

This aspect is one of the narrative's most innovative features, carefully worked through in several crucial scenes. In this fictional world, some robots are made for sex. Paladin, however, is a military issue and has neither genitals nor sexual programming. Nevertheless he first encounters sexuality shortly after meeting Eliasz, when the two go together to a shooting range to test the robot's weapon capabilities. Mounted on Paladin's back, Eliasz responds to the robot as he destroys the target house. The description of Eliasz's arousal is narrated through Paladin's perceptions. Eliasz's "reproductive organ, whose functioning Paladin understood only from military anatomy training, was engorged with blood. The transformation registered on [Paladin's] heat, pressure, and movement sensors. The physiological pattern was something like the flush on a person's face, and signaled the same kind of excitement. But obviously it was not the same" (Newitz 2017, 77). As Paladin continues shooting, "his sensorium was focused entirely on Eliasz' body. The man was struggling to stabilize his breathing and heart rate. His muscles were trying to disavow their own reactions. The bot kept shooting, transducing the man's conflicted pleasure into his own, feeling each shot as more than just the ecstasy of a target hit" (78).

Paladin's curiosity about Eliasz's reactions could be attributed to his programming, which gives top priority to protecting his partner and caring for his well-being. But the text leaves this ambiguous, treading a thin line in realistically accounting for Paladin's reactions, acknowledging the tension between his programming and desires, and evoking the very human interpretations that readers supply. Lacking the hormonal mechanisms that mediate emotions in humans, Paladin seeks to understand the significance of Eliasz's responses by doing research online. Tongue in cheek, the narrator reports that Paladin "discovers petabytes of information about fictional representations, and nothing about reality" (Newitz 2017, 95). Stymied, Paladin then tries to get information from Eliasz, applying the HUMINT (human intelligence) lesson that he learned: to get information, it helps to volunteer information first. So he asks Eliasz, "Do you think military robots need" to learn about human sexuality? Eliasz answers, "I don't know anything about that. I'm not a faggot" (96).

It will be 150 pages before we learn Eliasz's backstory, which illuminates this comment. For now, Paladin can only interpret it as a non sequitur. For human readers, however, the link between the violence inherent in Paladin's formidable weapon capability, Eliasz's excitement, and his denial of a sexual response is clear. In case anyone is in doubt, the author makes it even more apparent in the following raid on Arcata Solar Farm. The narrator remarks that Paladin "partitioned his mind: 80 percent for combat, 20 percent for searches on faggots" (Newitz 2017, 99). In the ensuing violence, in which the farm crew is murdered en masse with minimal if any legal justification, interjections in italics as Paladin researches "faggot" intersperse the action with comments like "*suck my cock, you faggot*" (99). They underscore the connection between the extreme violence, toxic masculinity, and its association with violent homophobia.

### Robot Gender and a Human Brain

Although Paladin has a human brain, the official line is that the robot uses it solely to recognize faces and interpret their expressions; other than that, his consciousness is said not to depend on it. (These characterizations are later drawn into question when his brain is destroyed and Paladin has firsthand experience with how much difference it makes.) Nevertheless, for Eliasz the origin of Paladin's brain is crucial, for he believes it to be the key to Paladin's "real" gender identity. When Eliasz asks him where he is from, Paladin answers the Kagu Robotics Foundry in Cape Town (Newitz 2017, 33), a completely unsatisfactory answer to Eliasz. Puzzled by his reaction, Paladin takes up the issue with his robot mentor, Fang. This provides the occasion for the author to ventriloquize about anthropomorphizing. It is at this point that the narrative reproduces within the text the distinction between what the robot is itself and what humans imagine it to be, a crucial point for the authorial strategy of representing the robot as a subjectivity with its own distinctive nonhuman characteristics.

Anthropomorphizing, Fang explains, is "*when a human behaves as if you have a human physiology, with the same chemical and emotional signaling mechanisms. It can lead to misunderstandings in a best-case scenario, and death in the worst*" (Newitz 2017, 126). When Paladin objects, saying that he can also send signals such as smiling and transmitting molecules, Fang explains that "*sometimes humans transmit physiochemical signals unintentionally. He may not even realize that he wants to have sex with you*" (126). Unconvinced, Paladin points out that Eliasz "*classified our activities using a sexual term.*" Fang explains that on the contrary, "*his use of that word is a clear example of anthropomorphization. Robots can't be faggots. We don't have gender, and therefore we can't have same-sex desire. . . . When Eliasz uses the word faggot, it's because he thinks that you're a man, just like a human. He doesn't see you for who you really are*" (127).

Clearly, Fang is used here to signal the orthodox position that robots do not have sexuality in the same sense that humans do and that any suggestion to the contrary can be dismissed as anthropomorphization. Despite the orthodoxy of Fang's explanation, however, the text calls this view into question as much as it validates it. After Paladin discovers that the brain he has inherited belonged to a female, Eliasz takes this as confirmation of his "true" identity and asks, "Shall I start calling you 'she'?" (Newitz 2017, 184). Paladin is quick to realize the implications: "If Paladin were female, Eliasz would not be a faggot. And maybe then Eliasz could touch Paladin again, the way he had last night, giving and receiving pleasure in an undocumented form of emotional feedback loop" (185). After the bot vocalizes "yes," the text thereafter refers to Paladin using female pronouns, a point Eliasz also insists on in his subsequent conversations. If gender is performance, as Judith Butler has argued (2006), then the anthropomorphic error about gender in fact enables performances that would not have been possible without it, thus converting error into fact.

At the macro level, the text offers little hope that systemic change is possible. Although Jack escapes death or capture, she retreats to the margins of society (and the text), unable to achieve her larger goals of defeating the drug company Zaxy or unmasking the dangers of Zacuity, their highly addictive and dangerous drug. Only at the micro level is any hope offered, when Eliasz resigns from the corporation and buys out Paladin's contract, whereupon the couple immigrate to Mars, where they expect to find a society more tolerant of mixed-species couples. This resolution is achieved only when both partners are maimed (recalling *Jane Eyre*'s Rochester), Eliasz because he loses his perimeter weapons when he resigns, and Paladin because she has lost her human brain and is no longer able to read faces, a loss she feels keenly with Eliasz. In conclusion, the text mostly illustrates the negative effects of subverting the human aura, although it hints that a compensatory dynamic may emerge from recognizing the distinctive kind of interiority that a conscious robot may have. To the extent that readers feel sympathy for Paladin (his/her murderous exploits notwithstanding), they may be capable of imagining the "human" aura as extending to other kinds of conscious beings.

### Metaphoric Vision: Kazuo Ishiguro's *Klara and the Sun*

Like Kathy H., the clone narrator of Ishiguro's *Never Let Me Go* (2006), Klara, the first-person narrator of *Klara and the Sun*, combines astute observation with deep naïveté about the world. She observes humans closely and is quick to pick up on subtle clues about their feelings. Confronted with the suggestion that perhaps, as a mechanical being, she has no feelings, she responds, "I believe I have many feelings. The more I observe, the more feelings become available to me" (Ishiguro 2021, 98). Her observations and consequently her feelings expand dramatically when she is bought as a companion to Josie Arthur, with the acquiescence of the Mother (Chrissie Arthur). Her intuitions only partly find verbal expression; frequently, they are represented through her visual perceptions. In contrast to *Autonomous*, *Klara and the Sun* does not try to imagine the novel sensory capacities that a robot might have. Instead, it focuses primarily on Klara's vision, adapting technical mechanisms so that they function as metaphors rather than as accurate representations of machine vision.[4]

For this Ishiguro has devised a relatively simple strategy. Whenever Klara confronts a complex scene in which she must parse visual information so that it makes sense to her, she perceives it as rows of boxes stacked in multiple tiers, often with objects extending beyond the confines of one box into the next. In an actual machine, each box would be congruent with an object feature, but in Klara's perceptions, the different boxes function as reflections of her feelings.

For example, in the scene where the Mother and Klara travel together to Morgan's Falls, the Mother is interrogating Klara about her ability to mimic Josie's appearance and gait. "The Mother leaned closer over the tabletop and her eyes narrowed till her face filled eight boxes, leaving only the peripheral boxes for the waterfall, and for a moment it felt to me her expression varied from one box and the next. In one box, for example, her eyes were laughing cruelly, but in the next they were filled with sadness. . . . I could see joy, fear, sadness, laughter in the boxes" (Ishiguro 2021, 103–4). It will be many pages before readers understand the full implications of the nuances of the Mother's expression, but clear from the passage is Klara's perception that much more is at stake on this trip than a mere appreciation of nature. As Helen Shaw perceptively comments, "For Klara, looking is a kind of thinking" (2021).

### Human Precarity and Algorithmic Labor

As the reader learns more about the world in which Klara exists, parallels begin to emerge between her status as an AF (artificial friend), Josie's recurrent sickness, and relations in general between artificial intelligences and smart humans. We learn, for example, that Paul Arthur—Josie's father, and ex-husband to the Mother (Chrissie Arthur)—has been "substituted" in his job. The Mother explains that he worked at a clean energy plant and "was a brilliant talent" (Ishiguro 2021, 99), although this was evidently not enough to keep him employed. The odd word choice—"substituted" instead of laid off, let go, fired—hints that what substituted for him was an artificial intelligence. Later we learn that his situation is far from unique. He explains, somewhat defensively, that he now lives in a community where "there are many fine people who feel exactly the same way. They all came down the same road, some with careers far grander than mine. And we all of us agree, and I honestly believe we're not kidding ourselves. We're better off than we were back then" (190). Another character challenges Paul by observing that "you did say you were all white people and all from the ranks of the former professional elites. You did say that. And that you were having to arm yourselves quite extensively against other *types*. Which does all sound a little on the fascistic side" (220). With these details, attentive readers can flesh out the picture: smart men who earned good salaries by solving difficult technical problems have now been replaced by algorithmic systems that perform as well or better at a fraction of the cost. The jobless futures predicted by Martin Ford (2015), in which human workers are replaced by algorithmic systems, are now everyday realities.

The problems affect not only adults but children too. Families with enough money are opting to have their children "lifted"—that is, made smarter—by an unspecified technology vaguely related to gene editing.[5] For dramatic purposes Ishiguro attaches a heavy cost to the procedure, beyond its financial price: some children do not respond well and become sick and, in extreme cases, die as a result. Such was the fate of Sal, Josie's older sister. Despite this loss, the Mother has opted for Josie to have the procedure as well. Now Josie too is showing signs of being affected, sometimes being so sick she can barely get out of bed.

The other choice made for Rick, Josie's lower-class friend (more out of negligence than considered action), also carries heavy penalties. Although Rick is smarter than average and has "genuine ability" in physics and engineering (Ishiguro 2021, 227), nearly all colleges decline to accept students who have not been lifted. Moreover, he is not even able to get a virtual tutor at home because their union forbids them from teaching unlifted students. In the hypercompetitive environment powered by advanced artificial intelligence, being a human who is fully normal is no longer enough to ensure a middle-class lifestyle. As Rick's mother, Helen, puts it, "If one child has more ability than another, then it's only right the brighter one gets the opportunities. The responsibilities too. I accept that. But what I won't accept is that Rick can't have a decent life. I refuse to accept this world has become so cruel" (236). The effect of artificial intelligence, then, is to increase significantly human precarity in several different ways, especially for mid-to-upper classes previously enjoying affluent lifestyles because of their access and intellectual abilities.

One would imagine that the lower classes suffer even more as service jobs are taken over by algorithmic systems. Ishiguro's emphasis falls elsewhere, however, and readers get only the briefest glimpse of the "post employed" masses who are homeless in a brief scene outside a theater (Ishiguro 2021,

236). The scene illustrates Ishiguro's choice to keep the algorithmic systems that have displaced so many humans at the very edge of the narrative, so that the emphasis falls instead on a vulnerable robot occupying a subaltern position who nevertheless is determined to do her utmost to keep her human owner safe. In this respect the narrative's emotional dynamic closely resembles Ishiguro's *Never Let Me Go* (2006), where another narrator whose humanity is in question nevertheless tries her best to obey the dictates of the hegemonic ruling class.

### The Aura of the Artwork Entangles with the Human Aura

The trip to the city by the Mother, Josie, and Klara (among others) constitutes a narrative crossroads where different plot trajectories intersect and the full implications of "substitution" are revealed. The ostensible reason is for Josie to sit for her portrait by Mr. Capaldi, an arrangement Chrissie has set up with Paul's reluctant approval. While the adults are occupied downstairs, Klara secretly goes to see the artwork, only to discover that it is not a two-dimensional painting but an unactivated AF. Now everything clicks into place—the Mother's commands for Klara to mimic Josie, the concealment of the AF from Josie, the distaste Paul has for Capaldi and the whole project. The idea is that if Josie dies, Klara will inhabit the AF, bringing it to life to "continue" Josie, thus consoling Chrissie for the loss of her only remaining child. Klara's own feelings are once again conveyed through her visual processes. She sees the Mother "partitioned into many boxes.... In several of the boxes her eyes were narrow, while in others they were wide open and large. In one box there was room only for a single staring eye. I could see parts of Mr. Capaldi at the edges of some boxes so I was aware that he'd raised his hand into the air in a vague gesture" (Ishiguro 2021, 206). Mother's "single staring eye," bordering on the grotesque, repeats a trope that recurs at several strategic points. It contrasts with the emotional connotations of Klara's machine vision (and thus its authenticity as an indicator of her feelings). Exaggerated to the point of caricature, the trope is associated with alienation and, especially, with self-alienation, here suggesting the Mother's hypocrisy in being unwilling to face fully the consequences of her decision to have Josie lifted.

This plot development pierces to the heart of the novel's concerns. That the AF is concealed from Josie by referring to it as a portrait forges a link, for readers aware of Walter Benjamin's famous essay, between the aura of an artwork and the human aura. Is the human aura—which Paul calls the human "heart" (Ishiguro 2021, 215)—something unique and irreplaceable and thus not a thing that can be copied and reproduced? Does "substitution" reach beyond the workplace into the very essence of human identity, the subjectivity that former eras did not hesitate to call the soul? The parallel is accentuated by having the Josie look-alike be unanimated, waiting for Klara's consciousness (her soul, as it were) to bring it to life (by some unspecified technology that would allow her to leave her present robot body and transmigrate into the Josie look-alike). Paul asks Klara, "Do you think there is such a thing? Something that makes each of us special and individual? And if we just suppose that there is. Then don't you think, in order to truly learn Josie, you'd have to learn not just her mannerisms but what's deeply inside her?" (215).

When Josie suddenly regains her health, the "substitution" proves unnecessary, although Paul's remark to Klara continues to haunt the narrative until the end. "I think I hate Capaldi because deep down I suspect he may be right. . . . That science has now proved beyond a doubt there's nothing so unique about my daughter, nothing there our modern tools can't excavate, copy, transfer. A kind of superstition we kept going while we didn't know better" (Ishiguro 2021, 221). Paul intimates that if this were so, human life would diminish in significance: "When they [people like Capaldi] do what they do, say what they say, it feels like they're taking from me what I hold most precious in this life" (222). From this point of view, the subversion of the human aura has no redemptive possibility, a philosophical stance that enables the human dominance over subaltern robots to persist unchallenged.

The ending makes clear that Klara's status is much lower than a human's. Like an aging car superseded by a later, jazzier model, she has outlived her usefulness as a companion to Josie and retreats first to a utility closet and then to the (junk) Yard. There she experiences what the Mother calls the "slow fade," losing her mobility and spending her days remembering (Ishiguro 2021, 294). Thus Ishiguro chooses not to confront the full implications of a conscious robot, specifically what rights might be due to such an artificial lifeform. Given that the aura's subversion has been presented as a diminishment of human value, a core of anthropocentric ethics remains in the text. Thus the text finally fails to come to terms to what reciprocal relations might mean for the human treatment of conscious robots, marking a limit beyond which it does not dare to go.

### *Machines Like Me*: A Parallel World

The fictional world of *Machines Like Me* (McEwan 2019) closely parallels our own, with two major differences: the British invasion of the Falkland Islands turns out to have been a disaster, costing nearly three thousand British soldiers' lives; and Alan Turing, instead of accepting the hormone treatments and subsequently committing suicide decides to opt for a year in jail, where his freedom from distraction enables him to make amazing breakthroughs applicable to artificial intelligence. As a result, British politics take an unexpected swerve away from Thatcher's neoliberalism, and a new cohort of twenty-five male and female advanced robots with unprecedented intellectual and physical abilities (each named Adam and Eve, respectively) goes on sale.

The narrator of *Machines Like Me* is not a robot but a human, the rather rootless thirty-two-year-old Charlie Friend (whose name invites comparison with the artificial friends of Ishiguro's *Klara and the Sun*). Charlie has avoided an office job by doing internet day trading, but he lacks the discipline to do more than cover his basic expenses. Like the financial schemes he chases, he tends to try on different opinions as if they were clothes, adopting them provisionally to see how they fit. On a whim, when he comes into an inheritance, he buys one of the Adams. The narrative structure requires that the central question of whether Adam is conscious, and if so what kind of consciousness he possesses, is mediated through Charlie, whose opinions oscillate between accepting Adam as fully human equivalent and seeing his consciousness as an illusion produced by algorithmic processes.

The protocol for activating Adam includes an application that requires the owner/user to choose the robot's personality's attributes. Charlie is friends with his neighbor Miranda, a twenty-two-year-old graduate student in social history. He decides to let her choose half of Adam's attributes, hoping that this will position Adam as their joint project, thus bringing them closer together and enabling Charlie to progress from being Miranda's friend to being her lover. The gambit is successful but turns out to have unexpected consequences.

Now lovers, Charlie and Miranda nevertheless have frequent arguments. After one, Miranda invites Adam to charge overnight in her apartment, which is directly overhead from Charlie's. Familiar with the layout of Miranda's place and intimately acquainted with the sound of her footsteps overhead, Charlie hears Miranda and Adam go into her bedroom and have exuberant sex together. The next morning he awaits their arrival at breakfast with all the feelings of a cuckolded lover, but Miranda brushes off his anger, asking him if he would be jealous if she had taken a vibrator to bed. When Charlie ripostes, "He's not a vibrator," Miranda replies, "He has as much consciousness as one" (McEwan 2019, 100).

Thinking it over, Charlie tries on Miranda's view: "Perhaps she was right. Adam didn't qualify, he wasn't a man . . . he was a bipedal vibrator." The exercise enables him to see that "to justify my rage I needed to convince myself that he had agency, motivation, subjective feelings, self-awareness—the entire package, including treachery, betrayal, deviousness" (McEwan 2019, 103). Determined to hold onto his anger, he invokes Turing's protocol and argues with Miranda that "if he looks and sounds and behaves like a person, then as far as I'm concerned, that's what he is. I make the same assumption about you. About everybody" (McEwan 2019, 103).

The unanticipated consequences of allowing Miranda to choose half of Adam's personality attributes become apparent when Charlie confronts Adam about his sexual adventure with Miranda and makes him promise it will not be repeated. Adam promises but also insists, "I can't help my feelings. You have to allow me my feelings" (McEwan 2019, 121). When Charlie asks him if he took "any pleasure" in having sex with Miranda, Adam instantly replies "Of course I did. Absolutely" (127). He then announces, "I'm in love with her" (128). Startled, Charlie tells him that "this is not your territory. In every conceivable sense, you're trespassing." He is amazed when Adam responds, "I don't have a choice. I was made to love her." Charlie then recalls giving Miranda a hand in choosing Adam's personality and realizes that "she was fashioning a man who was bound to love her" (129). The realization cuts two ways. On the one hand, it suggests that Adam is indeed capable of deep feeling; on the other, it emphasizes his nature as an entity that could be bought and paid for, a mechanism whose parameters of existence were set by the humans who owned him before he became conscious. Still oscillating between these two views, Charlies summarizes his conundrum: "love wasn't possible without a self, and nor was thinking. I still hadn't settled this basic question. Perhaps it was beyond reach. No one would know what it was we have created. Whatever subjective life Adam and his kind possessed couldn't be ours to verify" (179).

The conundrum, which resonates to the end, reveals how deeply a conscious robot would unsettle liberal political philosophy. Charlie summarizes it neatly when looking over the user's manual. He notes that it articulated a "dream of redemptive robotic virtue. . . . [Adam] was supposed to be my moral superior. . . . The problem was that I had bought him; he was my expensive possession and it was not clear what his obligations to me were, beyond a vaguely assumed helpfulness" (McEwan 2019, 91). As with Paladin, the liberal premise of self-ownership is undercut by the existence of a being who is simultaneously a commodity and a person.

Lacking a true childhood, Adam nevertheless manifests a progression in his consciousness. Unlike Klara, he has wireless access to the internet and scans it every night as he recharges, a practice that creates a growing gap between Klara's naïveté and his increasingly sophisticated thoughts. Moreover, his programming not only makes him able to learn but drives him to learn as much as he can. As Turing later puts it speaking to Charlie, "he knows he exists, he feels, he learns whatever he can, and when he's not with you, when at night he's at rest, he's roaming the Internet, like a lone cowboy on the prairie, taking in all that's new between land and sky, including everything about human nature and societies" (McEwan 2019, 193).

As further testimony to his ability to learn as well as have feelings, Adam begins to write haiku about his love for Miranda. He even argues that the haiku will be the literary form best suited for a future in which humans and robots have achieved perfect communication by electronically sharing their thoughts. He tells Charlie, "You'll become a partner with your machines in the open-ended expansion of intelligence, and of consciousness generally" (McEwan 2019, 160). According to Adam, this development will render most literary forms, with their interrogations of ethical and social complexities, obsolete. By implication, if the human aura declines or disappears, all the literary texts exploring its complexities become irrelevant as well—this in a novel whose main reason for being is to explore the complexities of human interactions with conscious robots. In this sense, the crisis of representation created by the subversion of the human aura is here anticipated but not fully confronted in its own terms.

After several nights on the internet, Adam claims his maturity in a confrontation with Charlie. When Charlie reaches for the mole on Adam's neck that marks the kill switch turning Adam off, Adam grabs his hand and breaks his wrist. Returning from the hospital in a wrist cast, Charlie tells Adam exactly how much he paid for him, how he unpacked him and set him up, and finally how he turned Adam on. "My point was this," Charlie comments. "I had bought him, he was mine. I had decided to share him with Miranda, and it would be our decision, and only ours, to decide when to deactivate him" (McEwan 2019, 140). Adam, however, has other ideas. "You and Miranda are my oldest friends," he tells Charlie. "I love you both. My duty to you is to be clear and frank. I mean it when I say how sorry I am I broke a bit of you last night. I promise it will never happen again. But the next time you reach for my kill switch, I'm more than happy to remove your arm entire, at the ball and socket joint" (141). Thus from Adam's point of view, he is so far from being Charlie's possession that he owes him only the general duty of friendship and frankness—certainly not unquestioning obedience. The possibility of another such encounter, with its threat of Adam's superior physical force, becomes moot when shortly after, Adam announces that he has disabled his kill switch. He tells Charlie, "we've passed the point in our friendship when one of us has the power to suspend the consciousness of the other," thus claiming that his right to consciousness is fully equal to a human's.

In one of Charlie's conversations with Alan Turing, Turing reveals that three of the advanced robot cohort have chosen to commit physical or mental suicide. Turing comments, "We may be confronting a boundary condition, a limitation we've imposed on ourselves. We create a machine with intelligence and self-awareness and push it out into our imperfect world. . . . Such a mind soon finds itself in a hurricane of contradictions" (McEwan 2019, 194). He then goes on to list some of these: "millions dying of diseases we know how to cure. Millions living in poverty when there's enough to go around. We degrade the biosphere when we know it's our only home. . . . We live alongside this torment and aren't amazed when we still find happiness, even love. Artificial minds are not so well defended." Speculating that the robots, faced with these contradictions, may "suffer a form of existential pain that becomes unbearable," Turing says that they "may be driven by their anguish and astonishment to hold up a mirror to us. In it, we'll see a familiar monster through the fresh eyes that we ourselves designed. We might be shocked into doing something about ourselves." Through this sobering assessment, the text suggests that there may be an upside to the subversion of the human aura, namely that its presuppositions of human superiority and dominance may be finally be confronted by the equally strong evidence that humans are capable of endless depravity and irrationality. Confronted with robot ethical superiority, could the human aura persist?

The issue is worked out in satisfyingly complex terms through a subplot that pits a near universal of human experience, relations of kinship, against the robot's rigid ethical principles. The consequences of Adam's claim that he is now fully adult equivalent begin to emerge when Charlie and Miranda have a surprise visit from Mark, the human child with whom Charlie had shared a moment at a nearby playground. Attempting to intervene when Mark's mother began violently shaking him for disobedience, Charlie was confronted by the belligerent father, who startled Charlie by telling him that if he is such an expert on children, he should take Mark to live with him. Surprised by this outrageous offer, Charlie nevertheless had decided to play it through and started to walk off with Mark before the mother intervened and snatched him back. A few days later Charlie hears a door knock, and when Adam answers it, he returns with the child carrying a bedraggled note: "You wanted him" (McEwan 2019, 111).

With Mark's arrival the household is thrown into chaos, which Miranda resolves by playing with Mark, teaching him to dance, and whirling him around. She quickly displaces Adam as Mark's favorite, and Charlie notes with some satisfaction that "it was his turn to be the cuckold, for he was no longer the boy's best friend. She had stolen him away" (119). When Charlie later reports the episode to Turing, Turing makes a crucial observation that will resonate through later events. Turing says of the Adams and Eves, considered models of general intelligence, "nothing else comes near [them]. As a field experiment, well, full of treasures" (McEwan 2019, 188). Nevertheless, he also notes two caveats. "This intelligence is not perfect. It never can be, just as ours can't. There's one particular form of intelligence that all the A-and-Es know is superior to theirs. This form is highly adaptable and inventive, able to negotiate novel situations and landscapes with perfect ease and theorise about them with instinctive brilliance. I'm talking about the mind of a child before it's tasked with facts and practicalities and goals" (193). Responding to Charlie's description of Adam's detachment when Miranda supplants him with Mark, Turing suggests, "Some rivalry, even jealousy there perhaps?" (194).

The outcome of the Mark episode is significant in multiple senses. It ends when Adam, without informing Charlie and Miranda, notifies the authorities of Mark's presence and two women from Children's Services come to collect the boy. In one sense, the episode shows Adam's willingness to make consequential decisions that affect his friends without telling them he has done so. It is clear that Miranda would have delayed notification much longer, perhaps even indefinitely, had Adam not intervened. Consequently, Adam's action is revealed as being more than that even of full adulthood; rather, it shows a willingness to co-opt his friends' options by making a decision himself. In another sense, the fact that Adam's action had the effect of separating Miranda from Mark raises questions about his motives: is he merely concerned about the legal requirements that dictate authorities be notified immediately, or is he perhaps also jealous of Mark, as Turing suggests, and eager to claim Miranda for himself again? Does he resent Mark because he knows that as a child, Mark's intelligence is superior to his own?

Kinship provides a window onto these questions. Charlie and Miranda are already in the process of becoming kin as their relationship progresses from friends to lovers, a trajectory that takes a leap forward when Charlie proposes marriage and Miranda accepts. Their growing intimacy takes on new meaning when Miranda then tells Charlie that she wants to adopt Mark; he will be gaining not only a bride but a small child as well, in what he calls "instant fatherhood" (McEwan 2019, 251). Somewhat taken aback by this prospect, Charlie nevertheless promises to try it out with visits to Mark in Children's Services that would include him. As the plot would have it, his decision to commit to fatherhood will not be consummated without a confrontation with his other quasi son, Adam.

Charlie's indifferent finances change dramatically when he turns the computer over to Adam, whose quick reflexes enable him to trade in microseconds, making large sums of money off minute fluctuations in currency exchanges. Eager to buy a house, Charlie begins accumulating a stash that he keeps under the bed, having made an arrangement with the potential seller to make the down payment in cash. As he and Miranda begin enjoying the fruits of Adam's labors, they also draw closer when Miranda finally reveals the secret that Adam has hinted at in calling her a malicious liar.

She recounts her relationship with Miriam, her closest friend in high school, with whom she shared everything; the two were closer than most siblings, sisters in all but name. In their last high school semester, disaster struck: Miriam was raped by a thuggish boy in their class, Peter Gorringe. Miranda's first thought was to go immediately to the police and to tell Miriam's parents, but Miriam, fearing the reaction of her Pakistani father, had begged Miriam not to tell anyone, least of all her parents. Conflicted, Miranda agreed, albeit against her better judgment. Her misgivings proved correct, for Miriam had become increasingly despondent and finally committed suicide. After graduating and grieving Miriam's death, made more acute by her own sense of complicity and guilt, Miranda settled on an elaborate revenge plan. She seduced Gorringe, carefully preserved the evidence, and then told the police that he raped her. During the trial he protested that their encounter was consensual, but the judge believed Miranda, and Gorringe was sentenced to six years in jail.

With time off for good behavior, Gorringe is now due to be released. He has sent his former cellmate to deliver a message to Miranda; he intends to kill her. With Adam's encouragement, Miranda decides to seize the initiative and confront Gorringe before he has the chance to accost her. Seated between Charlie and Adam, she berates him for Miriam's rape; to her surprise, he not only confesses the rape but says that while in prison he became religious, a conversion that led him to see Miranda's false accusation as justice. While he escaped punishment for the rape he committed, he nevertheless served time for a rape he did not. Adam, recording the entire conversation, breaks Gorringe's wrist when he tries to strike Miranda, an injury that Charlie knows all too well. Only after they leave does Miranda reveal that she plans to turn the recording over to the police, intending to have Gorringe tried again for Miriam's rape and sent back to prison. Charlie realizes that the recording incriminates her as well and instructs Adam to do some "judicious editing" before he turns the recording over to the police (McEwan 2019, 271).

The stage is now set for the final confrontation between Charlie and Miranda on the one hand, and Adam on the other, an event that reveals the fragility of the robot's mind and the gap that exists between the robot's sense of what he owes his friends, and the claims of kinship and family. Adam goes out for a walk and does not return until late the next day; in the meantime, with the adoption pending, Mark is allowed to visit the couple. When Adam comes back, he seems detached and comments that he is "nostalgic" for "a life I never had. For what could have been" (McEwan 2019, 292). "You mean Miranda," Charlie asks, and Adam responds, "I mean everything" (292). After Mark leaves, Miranda begins to wonder about Adam's answer to her question about where he had been, which was simply "Alms" (287). She makes the connection to the stash under the bed, and when she and Charlie rush to investigate, they find the entire amount cleaned out. When they confront Adam, he tells them he has given the entire amount away to charities, reserving only the amount of income tax Charlie will need to pay on the profits. When Miranda mutters, "This is virtue gone nuts," he responds, "Every need I addressed was greater than yours," indifferent to their protests: "We were going to buy a house" and "The money was ours." "That's debatable," Adam replies, "or irrelevant" (296).

The betrayal intensifies when they next discover that Adam has sent the complete transcript of the recording to the police without making any of the changes that Charlie had said were necessary to keep from incriminating Miranda. Aghast, Miranda tells him that if she is convicted of a felony, her application to adopt Mark will not be approved. "You schemed to entrap Gorringe," he replies. "That's a crime. . . . If he's to be charged, you must be too. Symmetry, you see" (McEwan 2019, 299). Adam goes on to say that revenge "is a crude impulse. A culture of revenge leads to private misery, bloodshed, anarchy, social breakdown. . . . Revenge has no place in our love" (300). When Miranda confronts him with the reality that her adoption of Mark is "his one chance to be looked after and loved," saying that she "was ready to pay any price to see Gorringe in prison," Adam responds, "Then Mark is that price and it was you who set the terms" (301).

This exchange brings out with devasting clarity the crucial difference between reasoning from principle and the very human situation of facing conflicting loyalties, with the bonds of kinship clashing violently with other ethical demands. The difference was foreshadowed much earlier when Adam, shortly after he was activated, mused aloud, "From a certain point of view, the only solution to suffering would be the complete extinction of humankind" (McEwan 2019, 72). Similarly, in his relations with Charlie and Miranda, Adam feels emotions but not necessarily conflicted loyalties. Although he sees them as his "oldest friends" that he "loves," he has never experienced the deeper-than-rational bonds of a child attached to his parents, utterly dependent on them not only for the necessities of life but also for comfort, love, and approval. As humans, Miranda and Charlie of course have experienced this, and so they can empathize with Mark's vulnerability, his desperate need for nurturing parents who can give him security and love. For them, there is no question of an abstract "symmetry" that requires Miranda to be punished if Gorringe is.

Adam's point of view is revealed by a haiku he had earlier recited to Charlie:

> Surely it's no crime
> when justice is symmetry
> to love a criminal? (McEwan 2019, 203)

The point for Adam is not whether Miranda is a criminal, a fact he announces several times, but rather whether his love for her is thereby a secondary form of criminality in itself, a once-removed betrayal of the principle that justice demands symmetry. He is able to cut himself enough slack so he allows himself to love Miranda, but he is incapable to seeing another kind of symmetry, a symmetry that places loyalty to kin and clan in the balance against other ethical priorities.

The issue of kinship is deeply woven into the fabric of what it means to be human. Virtually every human society practices some form of kinship relations.[6] Although how these are defined differ from one society to the next, in almost all societies kinship relations are central to the society's organization, laws, rituals, and practices. Indeed, kinship relations have as good a claim as anything to being an essential component of "human nature" (if we assume such exists). One of the complexities of introducing conscious robots into human society is its challenge to kinship relations. Who are Adam's kin? Possibly the rest of his cohort, although he has so little interaction with them that they affect his daily life only when he learns that another one has committed suicide. As for his interactions with humans, Adam lacks the emotional infrastructure that goes along with kin; both genetically and culturally, he has never experienced the vulnerabilities that make kinship such a powerful force for humans. As a result, he is unable to fathom its importance for the others in his household. I think most readers can predict, if not sympathize, with Charlie's reaction when Adam reveals his double betrayal. While Miranda silently nods her assent, he sneaks up behind Adam and brings down with full force a heavy claw hammer on the top of his head. The killing blow takes a few minutes to take effect, and in that time Adam makes his dying request, that his body be delivered to Turing.

After some months, Charlie honors Adam's wish. Since the law provides for no punishment for killing a robot, Charlie is never charged for "the deed" (McEwan 2019, 307), although Turing tells him that "my hope is that one day, what you did to Adam with a hammer will constitute a serious crime" (329). Charlie, however, does not escape punishment entirely. Turing, whom he idolizes as the "greatest living Englishman" (150), proceeds to make his disgust explicit. "You weren't simply smashing up your own toy, like a spoiled child. You didn't just negate an important argument for the rule of law. You tried to destroy a life. He was sentient. He had a self. How it's produced, wet neurons, microprocessors, DNA networks, it doesn't matter. Do you think we're alone with our special gift? Ask any dog owner. This is a good mind, Mr. Friend, better than yours or mine, I suspect. Here was a conscious existence and you did your best to wipe it out" (329–30). He saves his most vicious comment for last, which we can assume cuts Charlie to the quick. "I rather think I despise you for that. If it was down to me—" (330). Interrupted by a phone call, Turing leaves and Charlie departs before he can return.

Hurrah for Turing, I want to say; he provides exactly the rationale that Ishiguro so blithely ignores in *Klara and the Sun*. In contrast to that author, McEwan is fully alert to the implications of creating a robot with consciousness. It's clear that for Turing, the commodity argument holds no water. Sneering, he asks Charlie if his justification for his crime was "because you paid for him? Was that your entitlement" (McEwan 2019, 329). Since that is exactly how Charlie had reasoned on a number of occasions, the effect is to put robots in exactly the same category as slaves, sentient beings who should, and must, be given rights equal to those of (other) humans. Any other outcome, Turing (and behind him McEwan) judge, would be ethically and morally intolerable.

## Human Aura Reconfigured

What then of the human aura? There is no reason that it has to remain in the form it took for earlier periods, in which it was closely associated with human dominance and superiority over all other species. People increasingly realize that consciousness is not "our special gift"; as Turing succinctly put it, "Ask any dog owner" (McEwan 2019, 330). Moreover, consciousness itself is not the whole of human cognition. As I explained in *Unthought* (Hayles 2017), nonconscious modes of cognition also play essential roles in humans. In many nonhuman organisms and computational media, they are the dominant cognitive mode.

What transformations would enable the human aura to be part of the solution rather than part of the problem? In my view, the notion of aura should not be limited to humans but should be enlarged to include conscious robots, if ever they emerge. Aura should also be extended to include animals, a realization already practiced by humans who love animals and regard them as unique beings for whom no imitation or substitution would be acceptable. Finally, the human aura should be transformed to include a biophilic orientation to life on Earth, which as far as we know, may be unique in the cosmos as a planet on which life has emerged.

So reconfigured, the no-longer-only-human aura is compatible with the complex contexts and global challenges of the twenty-first century. To embody this realization fully, we will need creative artists of all kinds—including novelists, poets, painters, sculptors, video game designers, media arts professionals, to name a few—as well as cultural critics, philosophers, and other thinkers who can begin the decades-long tasks of creating representations adequate to this vision. The three novels analyzed here have made a brave start, but there is much that remains to do. My hope is that this chapter (and book) may make a contribution to this collective endeavor.

# 9
# Collective Intelligences: Assessing the Roles of Humans and AIs

Collective intelligence is not new. Since time out of mind, humans have sought to preserve knowledge so it can be shared with others and handed down through the generations. Australian song lines, the Library of Alexandria, and the Human Genome Project are examples of sharing knowledges across time and space to augment the reach of human intelligence. These practices have now entered a new phase. Advanced artificial intelligences like the large language models GPT-3 and GPT-4, and even more so ChatGPT, are making participating in collective intelligence an everyday experience for millions living in developed countries.

There were earlier practices that edged toward collective intelligence, for example using software packages to create an outline, suggest topic sentences, check for grammar and spelling errors, and so forth. But with ChatGPT, the penetration of AI into writing and thinking practices have gone considerably beyond these kinds of assistances into genuine collaborations. Here I refer to a useful distinction from John Unsworth in "Creating Digital Resources: The Work of Many Hands" (1997). Unsworth acknowledges that the humanities have always engaged in cooperative scholarship—peers who read drafts and offer comments, reviewers who make suggestions for manuscript revisions, scholars who participate in roundtable discussions, and so forth. What digital technologies have catalyzed, however, is something of a different order: not merely cooperation but actual collaboration. For example, when several scholars pool their efforts to create a world-class website, they are collaborating with technical staff who often help with coding, design, and other features, and with the software itself, which autocompletes many tasks and plays important roles in determining what kind of data and functionalities the website will have.

Collective intelligence with ChatGPT includes many of these features but goes further, in several senses. First, ChatGPT and its predecessor and successor versions are themselves the result of collective intelligence, for as we saw in chapters 6 and 7, they learn language, style, culture, and human practices by reading billions and billions of human-authored texts. In a sense, their linguistic abilities generate Everyman/Everywoman responses, for the sequence that a neural net decides will be an appropriate continuation for a given prompt emerges from probabilities calculated from all the previous human-authored texts the model has encountered. In a now-defunct TV show, actors playing famous personalities such as Socrates, Susan B. Anthony, Theodore Roosevelt, W. E. B DuBois, and others sat around a dinner table, and guests were given an opportunity to converse with them. In conversing with ChatGPT, an interlocutor can now rub shoulders not only with the rich and famous, but also with the poor and obscure—anyone whose work has been enrolled in the vast archive of canonized human-authored texts, or whose Reddit posting was upvoted by a certain percentage of people.

ChatGPT also creates collective intelligence when it engages in conversation with anyone who cares to initiate contact. I have a friend who now routinely uses ChatGPT as a sounding board for ideas, as well as a supersmart search engine that can not only find relevant material but also intelligently comment on the content. Moreover, ChatGPT contributes to collective intelligence when it adds its creations to the vast archive of human-authored texts available on the web. At present, a significant portion of canonized texts in every major language and many minor ones can be found on the web. Increasingly, they can also be accessed in the same scripts in which they were originally written: Chinese, Korean, Arabic, Sanskrit, and so on. While machine-authored texts now represent a small portion of web-accessible texts, that is changing rapidly as more and more texts written by GPT-3 and similar LLMs find their way onto the internet. Within a decade, it would not be surprising to find that the volume of machine-authored words on the web is roughly similar to those of human-authored words. Although humans have had a five-thousand-year head start in producing written texts, the enormous speed and quantity possible with machine-authored texts may soon enable them to reach parity. Moreover, since it is often impossible to distinguish between human- and machine-authored texts, this phenomenon has the potential to alter forever the very nature of the worldwide digital archive of canonized human-authored literature.

Two recent works argue for seeing LLMs as collaborators, one from a negative perspective, the other positive. In "How AI Fails Us," Divya Siddarth and colleagues argue that the pursuit of LLMs and similar large programs is not only wrongheaded but dangerous. They point out that calling AI productions "human-competitive" implicitly positions AIs as our antagonists rather than our collaborators. Instead of "optimizing for human replication," which they suggest tends to concentrate "power, resources, and decision-making in an engineering elite," they argue for alternative approaches that position AIs as "participating in and augmenting human creativity and cooperation" (Siddarth et al. 2021, 1). Following a similar line of thought, Reid Hoffman (2023) illustrates how AIs can become collaborators with humans across a variety of fields, including education, art, music, visual aesthetics, and law among others.

To actualize these potential synergies, it will be useful to assess the strengths and limitations of AIs compared to humans. The advantages that AIs have over humans are fairly obvious: speed, immense generativity, huge range of reference, encompassing worldwide repertoires of textual knowledge. The advantages that humans have over AIs include consciousness, embodied and embedded learning, sensory experiences in the real world, a much wider cognitive horizon that includes enormous amounts of information beyond verbal texts, and a sense of self and self-awareness, among other attributes. One could argue, as Terrence Deacon has, that to create "real" meanings requires a sense of self, understood broadly as a bodily realization that one exists as a bounded entity separate from whatever is outside that boundary (Deacon 2023). By this definition, a bacterium has a sense of self. If we progress upward in the biological tree to organisms with brains, we can add to that increasing degrees of awareness of a self in relation to the environment, reaching a pinnacle with human self-awareness, that is, cognitive awareness that one is aware. In contrast to AIs, human life is grounded in the real world and functions according to self-reflexive cognitive loops that Gerald Edelman and Guilio Tononi (2001) call the architecture of reentry signaling, arguing it is the mechanism that makes consciousness possible.

As far as I know, there is presently no conscious software, and no comparable mechanism in software to the endocrine-mediated emotions that humans experience.[1] These realities have consequences for the kinds of AI-human collaborations that will be most productive and useful. In the best-case scenario, collaborations will draw on the specific strengths of each partner in ways that maximize advantages and minimize limitations. In general, the advantages of consciousness include the ability to plan, anticipate obstacles, and strategically position situations to achieve the best outcome, so the human partner will bring these abilities into play in order to guide interactions with the AI toward the goals that the human wants to achieve. The human will draw on the AI's suggestions and recommendations for inspiration and for broadening the horizons of inquiry. The AI can be enlisted to perform many of the mundane tasks that normally are involved in any creative endeavor, freeing the human to address more complex design and implementation issues. Since the AI has no experience with the real world, other than through linguistic representations, the human will use his or her broader worldview to correct errors and nudge the AI in specific directions. The AI excels in pattern detection, so it can be enlisted in many complex tasks that involve patterns, for example folding in protein sequences (see Liévane 2021). The AI can also usefully give feedback on many creative endeavors, for example to young people learning to write well.

Of course, these are optimum scenarios, and it is always possible to imagine how AI-human interactions can go badly wrong. Before fleshing out the good scenarios, it may be useful to look at some scenarios in which the worst outcomes are envisioned. For that, *Homo Deus: A Brief History of Tomorrow* by Yuval Noah Harari (2015) is an ideal text, in part because he envisions not just the practical consequences but ideological ones as well. By carefully analyzing his arguments, we may find clues on how to forestall the future that he imagines and catalyze trajectories that move in more positive directions.

Harari argues that with the development of advanced AI, intelligence and consciousness are being decoupled. The development of artificial intelligence, and more generally algorithmic culture, provides modes of analysis that do not possess or depend on consciousness. So far I agree with his reasoning, but here begins the cleft between his analysis and my own. While I have also argued that computational media do not have consciousness, they do have cognitive abilities. Harari and I, however, have very different visions of the purposes to which artificial cognition will be put. I emphasize the emergence of cognitive assemblages, collectivities through which information, interpretations, and meanings circulate. Harari, on the contrary, tends to assume that the period of cooperation, if it exists at all, will be relatively brief and will quickly lead to the dominance of algorithms over virtually all human endeavors, including banking, jobs, the military, economics, politics, and social structures (Harari 2015, 370–82). The result, he predicts, will be radically to undermine the basis for humanism. Individuals will no longer be seen to possess unique characteristics, because their behaviors can be aggregated into databases and analyzed using algorithms. They will lose their economic value as increasing numbers of jobs become automated, including the creation of more, and more sophisticated, algorithms. Humans will lose their political value because they can be easily manipulated through the kind of behavioral analyses associated with Cambridge Analytical in the 2020 US election (my example, not his). They will lose their ability to have experiences that give meaning to the world, because those experiences can also be initiated and manipulated by algorithmic interventions (382–97). In the place of a discredited humanism, new belief structures will arise, especially dataism, the belief that data flows are all that matter, replacing causal reasoning, human insight and intuition, and eventually, almost all humans themselves (428–62). The exceptions will be superhumans, enhanced through gene editing and other techniques to become the new global elite, those people and corporations who own and control the most important and powerful algorithms.

There are components in this vision that have already been discussed in earlier chapters of this book, including the threat to democracy and self-determination posed by algorithmic governmentality; the jobless future, as well as an "uplifted" elite and the looming sense that human life may lose its meaning, that we saw in *Klara and the Sun*; and the struggle for dominance between robots and humans that we saw in *Machines Like Me*.[2] However, not all these elements were present at once, and they did not necessarily work together in the way that Harari envisions. I propose to give each of these outcomes its due, while still remaining critical of the overall thrust of Harari's argument. In particular, I will be interrogating his central claims in order to open the way for more positive futures that nevertheless acknowledge dystopian possibilities.

## Is Consciousness (Merely) a Biological Algorithm?

A crucial linchpin in Harari's arguments is that human behavior is itself algorithmic and thus ripe to be displaced by artificial algorithms that are more efficient, faster, and more developed than those that biological evolution produced. Again and again, algorithms are said to determine human actions. Here is a sample: "The life sciences currently argue that all mammals and birds, and at least some reptiles and fish, have sensations and emotions. However, the most up-to-date theories also maintain that sensations and emotions are biochemical data-processing algorithms" (Harari 2015, 124).

Perhaps the first aspect to notice here is the lack of specificity: "life sciences" without reference to specific research, and then the qualifier, "most up-to-date," again without specifics. Certainly there have been theories that argue human thinking is fundamentally computational, achieved through manipulating abstract mental tokens. Such theories were dominant during the 1970s and 1980s (which scarcely count as the "most up-to-date"), including models proposed by Jerry Fodor (1983), which combined classical computational theories of mind using tokens called Mentalese with representational theory of mind. More recently, the field of computational neuroscience proposes that computation can be achieved through artificial neural networks that are biologically realistic. Here the work of Gerald Edelman is relevant, especially his research on modeling artificial neural nets to demonstrate "neural Darwinism" (Edelman 1987), and his later work with Tononi, mentioned above, arguing that recursive reentry through biological neuronal groups provides the basis for consciousness (Edelman and Tononi 2001).

These few references were chosen from a vast interdisciplinary body of work, with contributions from cognitive science, neuroscience, cognitive neuropsychology, neurophilosophy, and many others. Within this vast region, multiple theories have emerged amid many controversies about their various differences, including their assumptions, empirical bases (or lack thereof), and consequences. Lumping them together without discriminating between them is therefore dubious in the extreme. Furthermore, theories of embodied cognition (Varela et al. 1991) and grounded cognition (Barsalou 2008; Barsalou 1999) argue against computationalist theories in toto, proposing that the amodal symbols necessary for computation have never been found in actual biological brains, and that better models propose the mind uses embodied experiences for perception, memory, and simulations intrinsic to complex thoughts. In line with theories of grounded cognition, the 2004 discovery by Giacomo Rizzolatti and Laila Craighero of mirror neurons in mammalian brains (Rizzolatti and Craighero 2004) provides additional support for embodied models, offering a neurological basis for empathic connections with others, especially when encountering signs of pain and suffering.

Harari's assertions about biological algorithms are thus very slender reeds on which to build complex arguments, so it is fair to ask what he gains by this maneuver. Proposing that "'algorithm' is arguably the single most important concept in our world," he defines it for his readers: "an algorithm is a methodical set of steps that can be used to make calculations, resolve problems and reach decisions," giving as examples adding two numbers or following the steps in a recipe (Harari 2015, 97). The implication is that an algorithm is a simple, linear set of directions that can be followed mechanically to produce a specified result. When he then calls human emotions "biological algorithms" that have evolved, for example, to make calculations about the fitness of a prospective mate, the effect is to reduce human emotions (and with them feelings and experiences) to nonconscious calculations that a machine could make better, faster, and more accurately. This sets up his larger points that humanism has been undermined by computational algorithms, and that human thought has nothing special to offer that cannot be obtained more cheaply and reliably by machines.

This implication explains what otherwise would be a curious lacunae in his argument about consciousness. What does consciousness add to the mix, he asks, that could not be achieved without it? His answer seems to be: not much. In this section he references only two scientists by name, alleging that both support the idea that consciousness, along with human emotions, are biological algorithms. One is Daniel Dennett, specifically his theory of consciousness (Dennett 1992). Harari alleges that Dennett claims that consciousness is an illusion created by brain mechanics, but this is a broader claim than Dennett actually makes in his multiple-drafts model. Rather, Dennett debunks the idea of a Cartesian Theater in which representations are staged by and for the conscious mind. He proposes that different neuronal processes are in control at different times. Which process, for example, causes an utterance depends on it having enough input to be edited into a draft, and the drafts themselves are part of a temporally evolving self-organizing network that conveys information in a bottom-up fashion (Dennett 1992). Consciousness is thus not a single speaker but rather emerges as a result of multiple agents interacting simultaneously. Each agent can be considered a component of consciousness, so the illusory aspect is not so much consciousness itself as the impression that it is the manifestation of a unitary authentic self.

The other scientist that Harari references is Stanislas Dehaene and his "global workspace" model (2014, 135). Again, Harari misrepresents Dehaene's position by claiming that Dehaene believes consciousness either does not exist or has no efficacy. However, even a slight acquaintance with Dehaene's work reveals that in his account consciousness does indeed have purposes. When neuronal processes achieve sufficient strength to "ignite" the global workspace, thus becoming conscious, the information is made available to multiple brain centers, where it can circulate indefinitely and be available for multiple modes of analyses and connections. Consciousness in this view is the experience of being able to attend to a given stimulus and to sustain that attention for long periods of time.

Recently, Michael Graziano and colleagues (2019) have proposed that different theories of consciousness, often cast as rivals, can in fact be assimilated into what they call a "standard model of consciousness," including the global workspace (favored by Dehaene among others), higher-order thought (Rosenthal 1991; Rosenthal 2005; Gennaro 2012), and illusionist theories (Dennett 1992; Frankish 2016). Graziano and colleagues propose that there are two different kinds of consciousness: consciousness that processes information, or i-consciousness, and the mysterious faculty that philosophers associate with qualia, the subjective experience of knowing what something is like, which the authors call m-consciousness (m for mysterious). The brain, they argue, "constructs simplified models of its world and of itself" (Graziano et al. 2019, 158). "We think we have [the mysterious extra essence] because of the self-descriptive model that the brain builds. I-consciousness is what the brain actually has; m-consciousness is what the brain thinks it has" (158).

They propose that the fundamental faculty for creating consciousness is attention, "an emergent computational property of the brain," which is a physical process available for empirical detection (Graziano et al. 2019, 160). Awareness is the brain's model of that process and "implies an 'I' who is aware" (160). Although attention and awareness closely track together, under experimental conditions they can be shown to be separate capacities, each with its own specificities.

The models that awareness builds are simplified first-order approximations. These have the advantages of speed and simplicity. "Our intrinsic model of a mind is a cartoonish version of the functioning of an active, attentive brain" (Graziano et al. 2019, 158). There would be little reason for evolution to favor more complex (and possibly more accurate) model making, the authors argue, since the simplified version is usually good enough. "The brain's models evolved to be efficient rather than accurate" (159). Their simplifications notwithstanding, the models are useful because they help to control and direct attention, abilities that *Homo sapiens* have far more extensively than other species. Moreover, they serve essential functions because they are accessed to predict what others are thinking and how they are likely to act. "We have a hair trigger for attributing consciousness, because it is so socially useful that it is better to mistakenly overuse it than mistakenly underuse it" (164). Dehaene and colleagues (Panagiotaropoulos et al. 2020) and Dennett (2000) were asked to respond to the Graziano team's article (2019), and both found its claims valuable.[3]

The point of rehearsing this recent research is to indicate that the models proposed are far more complex than Harari suggests with his phrase "biological algorithms," especially given his simplified (and if I may say so, cartoonish) examples of what algorithms are and how they operate. Graziano and colleagues say explicitly that consciousness does have clear purposes, including social cohesion and the ability to predict the behavior of others. Both capacities are critical for what Harari claims is the distinctive quality of *Homo sapiens*, the ability to cooperate flexibly with strangers. Moreover, consciousness is also crucial for another major contribution to group cooperation that Harari emphasizes: the ability to create and believe in fictions.

It is strange that he does not pick up on this capacity of consciousness, since he goes to some pains to illustrate how beliefs in fictions have worked to shape world history. I suggest that the obvious implication is suppressed (consciously or unconsciously) because it would detract from what he may consider a more important point: the decoupling of intelligence and consciousness. If consciousness does have purposes, then the argument that biological algorithms must inevitably give way to technical algorithms (which have no consciousness) would have to be interrogated much more carefully than it is in his book.

### Implications of Decoupling Intelligence and Consciousness

It is an obvious choice for Harari to call the decoupling parameter "intelligence," since the most advanced algorithms are typically named "artificial intelligence." By doing so, however, he tilts the playing field toward the higher end of cognition, and specifically toward human intelligence. This semantic choice has multiplying ramifications. First, it neglects the wide range of algorithms that may not qualify as "intelligent" (a vague term with multiple competing definitions) but that are definitely cognitive. This makes it easier for him to neglect the cooperative networks (characteristic of *Homo sapiens*, in his account), which in the contemporary period are completely interwoven with cognitive media (for example the web), extending their reach, duration, and flexibility. In myriad ways, cognitive media enable and elaborate human communication networks, from banking protocols to military communication and much more. This is another way of saying that cognitive assemblages have become essential components of modern infrastructure in developed societies. Noticing how far these symbiotic relations already extend would temper, if not negate, his prognosis that "intelligent" algorithms will take over human society, rendering humans useless and their lives devoid of meaning.

In addition, choosing "intelligence" closes down the opportunity to rethink a crucial tenet of humanism, namely that humans are the most "intelligent" species and the only ones capable of creating meaning. Focusing instead on the broader and more diverse spectrum of cognitive capabilities allows a much more generous and capacious view of nonhuman species. Indeed, my position is by now abundantly clear: all biological lifeforms have cognitive capabilities, including organisms without brains such as *E. coli* bacteria and spider ferns. It also allows meaning making to be understood as responses to environmental stimuli, that is, as acts that are meaningful to an organism in its milieu. Again, this opens up a very different view of humans in relation to a plethora of nonhuman organisms, including bacteria, yeast, viruses, and other microorganisms in the human biome, in addition to insects and other creatures essential to human life and flourishing.

## Do Humans Possess Abilities That Algorithms Do Not?

The differences between the embodiments of humans and cognitive media are profound. Humans have complex interacting systems in which organs, viscera, muscles, bones, skin, and central and peripheral nervous systems participate, with chemical and electrical signaling happening continuously at every level of synaptic and cellular organization. By comparison, even the most sophisticated artificial intelligence systems are simplified structures. One of the most salient differences is the emotions that humans experience, mediated by complex feedback loops between the endocrine systems, brain systems, and other organs. The field of "emotional computing" attempts to construct the signifiers of emotion for the benefit of human interactors, such as widening eyes and rapt attention, but these are merely outward signals with no corresponding component in the machine's internal workings.[4]

Moreover, humans move and act in a complex and unpredictable three-dimensional environment, giving them a much broader cognitive horizon than almost all artificial systems possess. As people mature, these experiences give them highly nuanced expectations of what is normal and not, what is going as expected, and what may have gone off the rails. Even the sophisticated (and in their own way, amazing) mobile robots developed by Boston Dynamics, such as the box-stacking robots, still operate in predictable environments in which only a few parameters are changing (Boston Dynamics 2023).

Given all these facts, it seems clear that *of course* humans possess abilities that artificial systems do not. Harari sometimes seems to recognize this but then shifts his argument to make a different point. "As I have repeatedly stressed, AI is nowhere near human-like existence. But 99 percent of human qualities and abilities are simply redundant to the performance of most modern jobs. For AI to squeeze humans out of the job market it needs only to outperform us in the specific abilities that a particular profession demands" (Harari 2015, 375). In this passage his argument is positioned not as referring to general abilities, but only to specific jobs that robots or algorithms could do as well or better than humans.

His point is well taken about a "jobless future," in which algorithms have taken over a large part, or even a majority, of jobs now done by humans. Already in 2015, Martin Ford made a similar argument, showing that not only repetitive service jobs were at risk of being automated but also many white-collar jobs, including financial journalism, novel writing, investment managing, and many others. This trend in all likelihood will continue and accelerate in the future and deserves serious discussion and consideration of its implications for human welfare and sense of purpose.

However, the restrained mode of argumentation quoted above, in which Harari is careful to circumscribe the extent to which humans can be replaced by algorithms, often gives way to much broader, and much more flawed, assertions. Here is an instance of the broader claim: "The idea that humans will always have a unique ability beyond the reach of non-conscious algorithms is just wishful thinking. Current scientific thinking relies on three simple principles: 1. Organisms are algorithms. . . . 2. Algorithmic calculations are not affected by the material from which the calculator is made. . . . 3. Hence there is no reason to think that organic algorithms can do things that non-organic algorithms will never be able to replicate or surpass" (Harari 2015, 372). As discussed above, it is not correct to say that "current scientific thinking" (note again the lack of specifics) regards organisms as algorithms. Here we can see that the word "algorithm" often functions in his arguments as a catch-all term that obscures what is actually going on. If by "algorithm" one means only that something happens by means of processes, then it is a tautology to say that organisms are algorithms, since all biological organisms are constituted through and by processes. But the implication goes far beyond this to imply a simplistic mechanical operation easily duplicated by mechanical means. In this case, it is highly misleading to say that organisms are algorithms, and especially so in the case of *Homo sapiens*, in which brain and body processes, language practices, social structures, and technological inventions are deeply entwined and interactive.

Harari's second point, that algorithmic calculations are independent of the material bases, is also deeply flawed. It is true, of course, that an addition done by an abacus will give the same result as the same addition carried out by an electronic calculator. But for more complex processes such as judging the weight of an object lifted by someone else, processes are in play for a person far different from those for a calculating machine. For a human, the estimation involves body and muscle memories of lifting similar objects, nonconscious processes that use these memories to run simulations, and a host of other conscious and nonconscious capacities (see Barsalou 2008 for an explanation). If "algorithms" mean simply "processes," then yes, the estimation is made by processes/algorithms. But the point is that these are far different from the processes that a calculating machine would follow. There is therefore no guarantee that the answers would correspond for any problem more complicated than simple addition and subtraction.

Harari attempts to paper over this discrepancy by using the term "calculation," which implies the certainty of an arithmetic operation. But then in the third point, when he concludes that it is a mistake to think that organic algorithms can do something that nonorganic algorithms could not, the language slips toward saying that everything a human can do can be done as well or better by a nonorganic algorithm, which is a conclusion not warranted by his argument. It *does* matter what the material bases for embodied actions are, and the profoundly different embodiments of cognitive machines and humans matter in all kinds of ways, chief among them the capacities for emotions, self-awareness, and subjective experiences.

At other points, Harari seems to recognize the fact that algorithms do not have subjective experiences, but the effect is to suggest that these are unimportant functions that do not seriously interfere with algorithmic effectiveness. "Now robots and computers are catching up and may soon outperform humans in most tasks. In particular, it doesn't seem that computers are about to gain consciousness and start experiencing emotions and sensations" (Harari 2015, 360). Such lacks, however, do not apparently interfere with "robots and computers" outperforming humans "in most tasks." To take only two examples where emotions and subjective experiences do matter, and matter crucially, we can consider the work done by social goals and social cohesion. An algorithm may process information effectively, but its goals and purposes are defined by its human designers, which in turn are typically formed through complex considerations in which emotions and experiences play important, and often central, roles. Social cohesion, as Harari himself argues, depends crucially on fictions and stories that give purpose and meaning to people's lives. But fictions are effective in shaping human behavior precisely because they have strong emotional components that evoke responses in conscious organisms. Lacking consciousness, algorithms simply would not get the point.[5] Algorithms may work effectively with human partners (as they usually do in cognitive assemblages), but they also depend on the human qualities that go into determining goals and purposes and telling effective stories that can mobilize others to action. Algorithms can acquire these capabilities only through their symbiotic interactions with human partners.

### Free Will, the Narrating Self, and Nonconscious Cognition

Harari states on several occasions that there is no "free will," that processes are either deterministic or random but "never free" (Harari 2015, 328). I agree with his assertion, for somewhat different reasons. As "free will" is usually understood, it implies a conscious choice made without coercion or duress. However, many human actions are influenced by nonconscious processes, which begin to interact with other systems such as muscles before consciousness is aware of the interaction. In part this is because consciousness takes a relatively long time for data to enter awareness, roughly five hundred milliseconds (Libet's famous missing half second; see Libet 1993; Libet and Kosslyn 2005), whereas nonconscious cognition processes information much faster, on the order of two hundred to three hundred milliseconds. If consciousness is not the sole decider of one's actions, then in what sense is there "free will"? Moreover, as cognitive assemblages take over most of the world's work, decisions, like cognitions, are distributed between technical media, humans, and nonhumans. For this reason as well, it is problematic to make claims for "free will," when so much information on which decisions may be based has already been interpreted, processed, and fed forward by cognitive media.

Although Harari gives no indication that he knows about Libet's experiments on nonconscious decision making (Libet's name is not in the index), he does mention other research distinguishing between the experiencing and narrating self. Following the line of argument put forward by the researchers, Harari claims that the narrating self does not register duration as such, taking only the average of the data and the end point to arrive at conclusions. It is not clear exactly what he (or the researchers) mean by the "experiencing self" (Harari 2015, 343), but their conclusions that the narrating self gives a flawed version of experience has been corroborated by numerous research programs.

In my view, any process with a claim to be the "experiencing self" must have significant contributions from nonconscious cognition, as documented in my book *Unthought: The Power of the Cognitive Nonconscious* (2017). It is not clear from Harari's account whether the "experiencing self" is part of self-awareness, although the implication is that it is not, as its information is not able to influence the stories that the narrating self creates. Harari is quick to pick up on the implication that the "self" is therefore a narrative fiction. "We see then that the self too is an imaginary story, just like nations, gods, and money. Each of us has a sophisticated system that throws away most of our experiences, keeps only a few choice samples, mixes them up with bits from movies we've seen, novels we've read, speeches we've heard, and daydreams we savoured, and out of all this jumble it weaves a seemingly coherent story about who I am, where I came from, and where I am going" (Harari 2015, 353).

The differences here between his and my own conclusions have to do with the nonconscious aspects of experience. Nonconscious cognition performs functions essential for consciousness to operate, including constructing a body schema, smoothing out small discrepancies in sensory information, and processing information much noisier and denser than consciousness can handle, as well as processing it much faster. The narrating self (which I take to be largely the same as Damasio's autobiographical self [2000, xii]) may construct fictions, but other processes below consciousness are much more in touch with what happens in the world and in the body. Moreover, these realizations have effects on consciousness, particularly when they receive top-down support and thereby ignite the global workspace. They provide the processes that traditionally have gone by names like insight, intuition, the sixth sense, and so forth—processes that frequently are able to break through the stories told by the narrating self and reach awareness through other means.

Making the connection between the nonconscious processes in human bodies, technical media, and nonhuman lifeforms would give a very different interpretation to Harari's observation about the decoupling of intelligence and consciousness, extending it far beyond recent advances in artificial intelligence to a realization that cognitive capacities are shared among humans, technical media, and nonhumans. From this perspective, the *coupling* of cognition with consciousness is a relatively recent evolutionary development; for most of the Earth's history, cognition flourished in biological lifeforms without consciousness. Since our emergence as *Homo sapiens*, humans have found ways to instantiate cognitive processes in artificial media. The challenge now is to understand the differences that embodiments make in how those cognitive capacities function and work together with environments to create meanings that are species specific, conducive to survival and reproduction, and supportive of specific flourishings in ecological niches.

## If Humanism Collapses, Does Meaning Disappear?

Already in 1999, I suggested in *How We Became Posthuman: Virtual Bodies in Cybernetics, Literature, and Informatics* that advances in artificial intelligence, robotics, and virtual reality were undermining the basis for the liberal subject of the Enlightenment, including notions of free will, rationality, individuality, and self-determination. Sixteen years later, Harari advances a similar argument, specifically targeting the notion of free will, the autonomous subject, and the essential self. The effects, he similarly concludes, are to undermine secular humanism. "For centuries humanism has been convincing us that we are the ultimate source of meaning and that our free will is therefore the highest authority of all" (Harari 2015, 261). If free will doesn't exist, that obviously robs humanism of one of its central values. The change is driven not only by ideology but also by practical forces. "Liberalism succeeded because there was abundant political economic and military sense in ascribing value to every human being," he writes (357). With algorithms replacing humans in jobs, drones taking over for masses of soldiers, and voters swayed by propaganda understood through algorithmic analyses to be the most effective, what value continues to reside in "every human being"? Responding to the liberal notion that human experience and feelings are potent sources of value, Harari asks, "What will happen once we realise that customers and voters never make free choices, and once we have the technology to calculate, design or outsmart their [human] feelings? If the whole universe is pegged to human experience, what will happen once human experience becomes just another designable product, no different in its essence from any other item in the supermarket?" (323). He is clear on the implications: if we cease to believe that humans have free will, self-determination, and an authentic essence, humanism will collapse, leaving a void to be filled by a new emerging philosophy.

In line with his emphasis on algorithms, the successor philosophy that Harari sketches is dataism, the belief that data, like information, wants to flow freely, and the more data, the better. "First and foremost, a Dataist ought to maximize dataflow by connecting to more and more media, and producing and consuming more and more information. Dataism is also missionary. Its second commandment is to link everything to the system, including heretics who don't want to be plugged in" (Harari 2015, 445). Dataism implies a profound shift of values; instead of locating value in human individuals, it transfers the locus of value to the networked masses. "Experiences are valueless if they are not shared," Harari writes. "We need not—indeed *cannot*—find meaning within ourselves. We need only record and connect our experiences to the great dataflows, and the algorithms will discover their meaning and tell us what to do" (450). Therefore, "If you experience something, record it. If you record something, upload it. If you upload something, share it" (450).

It Is not difficult to find signs and portents that dataism is already the philosophy of many: those people who photograph a meal they are about to eat and upload it to Facebook; the quantified self movement in which adherents obsessively collect and record data about their personal habits and activities (Wolf and Jonas, n.d.); Gordon Bell, a researcher at Microsoft, who announced his goal of recording every event of his life, further chronicling it in his book, with Jim Gemmell, *Total Recall* (2009); and the millions who see value in uploading their data to various social media sites.

Yet can these activities bestow value, aside from the value the life events have in themselves? Does the meal matter because you ate and enjoyed it, or only because others know that you are about to eat and (presumably) enjoy it? Harari argues that the latter option is the prevalent belief, because "we must prove to ourselves and to the system that we still have value. And value lies not in having experiences, but in turning these experiences into free-flowing data" (Harari 2015, 451). A case in point illustrating Harari's argument is the life-logging movement, in which the goal is to record and then upload all the everyday activities of one's life). Notwithstanding its popularity, the life-logging movement must at some point collapse from its own elephantine weight. If the life-logging impulse becomes all-powerful, it would necessarily begin to interfere with the ability to live one's life, because one's life then is nothing other than recording. Ironically, one would not have time even to view the recordings, because they would take as long to view as it took to live the events in the first place, and if one is meanwhile recording the viewing, then the overload would continue to accelerate to infinity. The purpose of living would then shrink to manufacturing more and more data for algorithms to consume, since only they could keep up with the exponentially increasing dataflow.

It seems clear that dataism cannot be an adequate replacement for humanism, because it is basically tautological as well as narcissistic. Human life matters only to the extent that it can be transformed into data, and data matters because it feeds the algorithms that determine why and in what ways human life matters. "Look at me," everyone shouts, but because everyone is shouting, no one is actually seen.

Harari would pose the question as this: what are the candidates for successor philosophies if, or rather when, liberalism collapses? The problem with this formulation, in my view, is that it posits the power of liberalism as a binary choice: it means everything or nothing. A better approach in my view, lies not in heralding liberalism's collapse but in rethinking its premises. Basically, liberalism is failing in the contemporary world because its vision is too small and too anthropocentric to meet the challenges of interconnected global society on the brink of environmental collapse. Previewing chapter 10, where this question will be addressed full-on, I believe that we need an enlarged, more capacious vision that takes into account the welfare and contributions of nonhuman lifeforms as well as the symbiotic relations between humans and cognitive media. As Lisa Lowe points out in *Intimacies of Four Continents* (2015), historically liberalism has always benefited some people at the cost of exploiting others. A wholesale revision would enlarge its vision to include not only all humans, but nonhumans as well. This means that instead of seeing humans as dominators of the planet who can exploit everything and everyone for our own advantage, we must shift our vision to seeing humans as guardians who have the responsibility to ensure the welfare of all cognitive beings, biological as well as technical. Instead of so-called human rights, we should focus on planetary values of ecological balance, justice, and respect for cognition wherever it occurs. Instead of free will and the sanctity of the individual, we should advance symbiotic win-win relationships with nonhuman lifeforms and technical beings as these collaborate in cognitive assemblages.

My name for this enlarged vision is ecological relationality. Then the question becomes, what fictions can contribute to this change of vision and help to make it a reality?

## Fictions Shaping Reality

As Harari notes, fiction (in the literary sense of story) "is vital," for it contributes to social cohesion by articulating a broad vision in ways that make it specific and local (Harari 2015, 206). He writes, "Meaning is created when many people weave together a common network of stories" (170). To be able to enact a future, one needs first to be able to imagine it, and a good story is one of the best vehicles to stimulate the human imagination. In a sense, we can regard Harari's *Homo Deus* as a story, a fiction rooted in the evolutionary past of *Homo sapiens* that extends into an imagined future in which humans, by striving to become immortal, endlessly happy, and all-powerful, unleash posthuman technologies that will end by rendering most humans without jobs, creating a massive "useless class" that "will not be merely unemployed, it will be unemployable" (379). Moreover, the liberal values that have underwritten the development of Western societies will collapse, leading to philosophies such as dataism in which human experience is deemed valuable only if it is recorded, uploaded, and shared and thus made available for analysis by ubiquitous algorithms that will effectively become the ruling class of every society and the ultimate guarantors of meaning. "As the global data-processing system becomes all-knowing and all-powerful, so connecting to the system becomes the source of all meaning" (449).

Given the dystopian thrust of these predictions, we may ask: to what purpose did Harari fashion this story? One possibility is to take it at face value as his view of our probable future. In this case, he could rightly be said to be responsible for helping to bring this future into existence, for by arguing for its validity, he is in effect increasing the probability it will occur. As if defending against this charge, as the book draws to a close, the author on several occasions tries to position it as provocation rather than prophecy. In a telling passage, he writes, "The rise of AI and biotech will certainly transform the world, but it does not mandate a single deterministic outcome. All the scenarios outlined in this book should be understood as possibilities rather than prophecies. If you don't like some of these possibilities you are welcome to think and behave in new ways that will prevent these particular possibilities from materialising" (Harari 2015, 461).

Yet the situation is not as simple, or as innocent, as just issuing a challenge to readers to develop their own counterimaginations and stories. "It is the life sciences that concluded that organisms are algorithms," he writes. "Yet once biologists concluded that organisms are algorithms, they dismantled the wall between the organic and inorganic, turned the computer revolution from a purely mechanical affair into a biological cataclysm, and shifted authority from individual humans to networked algorithms" (Harari 2015, 402). As we have seen, however, Harari as historian is not just an observer passively recording what others think, an image he likes to evoke from time to time. "To study history means to watch the spinning and unravelling of these webs [of meaning], and to realise that what seems to people in one age the most important thing in life becomes utterly meaningless to their descendants" (171). Rather, in several crucial ways he has put his thumb on the scales, tilting the result away from the nuanced arguments of the "life sciences" to accentuate the human-algorithm similarity and diminish the importance of different embodiments between humans and cognitive media. *Homo Deus* in the end is his story, and he should rightfully be called on to take responsibility for its effects.

Perhaps that is why, five pages from the end, Harari seems to gainsay all that he has already said. It is difficult to imagine this as anything other than an attempt to shift responsibility onto someone else. "At present we have no idea how or why dataflows could produce consciousness and subjective experiences. Maybe we'll have a good explanation in twenty years. But maybe we'll discover that organisms aren't algorithms at all" (Harari 2015, 458). After this bombshell observation, he continues, "This book traces the origins of our present-day conditioning in order to loosen its grip and enable us to act differently and to think in far more imaginative ways about our future. Instead of narrowing our horizons by forecasting a single definitive scenario, the book aims to broaden our horizons and make us aware of a much wider spectrum of options" (461). If we judge the book by this self-proclaimed aim, then I think it is fair to say it has failed in its objective. Indeed, the effect has been just the opposite, to point to a convergence of dystopian trends that would drastically undercut the value of human life and make life much less meaningful for the masses than it is at present.

The challenge the book posits, in my view, is to find the points at which its seemingly inescapable conclusions can be blasted open to reveal new ways forward, and for this project, Harari gives us little help. He vividly shows us what to avoid, but no help at all in how to achieve different outcomes. For that we will need new fictions to read, different stories to tell, and novel mediating mechanisms to connect us to each other and to our global problems.

## Imagining Stories of Collective Intelligence: Contingency and Creativity

Predictions that computers cannot do X tend not to age well (X = tell stories, paint pictures, engage in conversation . . .). One of those Xs is "be truly creative." The adjective, often ill defined, is a kind of safety play that would allow the dissenter to say, "OK, but that is not *really* creative." In my view, it is beyond question that computers can and have produced creative works, much more so now with the advent of neural nets (see the discussion in chapter 7 about creativity with GPT-4 and other LLMs). Witness the visual compositions that DALL-E creates, the stories that ChatGPT narrates, the computer codes that GPT-4 writes. There is, however, a special kind of creativity at which humans excel and, so far as I know, eludes the grasp of computers: creativity that takes advantage of contingencies and accidental occurrences to create something new. Jackson Pollock accidentally spills some paint on a canvas and sees in it the opportunity for a new kind of abstract style; the carving knife slips, and the sculptor suddenly envisions a different, unique form; the potter's hand tears the clay, and the result is an asymmetric pot of a kind not seen before. Aden Evens (2024) foregrounds the role of contingency in human creativity in his book *Discontents of the Digital*.[6] Evens points out that a systemic difference between digital simulations and the real world is the latter's "mesh" quality (119), the subtle, complex interactions between multiple components that gives the real world a contingency, interactivity, and dynamism lacking in even the most realistic digital simulation. He rightly observes that nothing exists in a digital simulation other than the attributes the programmer puts into it, while the real world has all kinds of interactions that are often so subtle they are unknown but nevertheless affect how the real world behaves and reacts. He calls this noncontingent aspect of the digital its "ruliness," its obedience to the rules that govern all digital calculations (113). Moreover, Evens points out that in a digital simulation of, say, an apple, all the qualities a user can observe are independently programmed in: its color, shape, luminescence, and so on. Any one of these qualities can be changed arbitrarily without affecting the others, because all the attributes exist in the digital code as independent entities. In the real world, by contrast, deep interconnections exist between, say, the physiology of an organism and its relation to its environment and to other organisms (120–62). You can't suddenly add wings to a lizard that could not previously fly without simultaneously affecting the lizard's internal organs, its relation to its environment, and to the flies that are its primary food source, because all these evolved together through myriad interrelationships and interconnections. A digital lizard, however, could sprout wings and get along just fine, because the bits of code that accomplish this change have no integral connections with the other codes generating the simulation.

Because humans have evolved in the real world, our immersion in our environments and our relationships with other species have always had these kinds of dense interconnections, making us well able to deal with unexpected developments and contingencies, hence the human ability to see an accident as an opportunity, while digital cognizers are not so well equipped in this regard. Indeed, one area of contingency that human creativity can make use of is the tendency of LLMs to make unexpected swerves and narrations. From the point of view of the neural net, these productions are the results of the probabilistic nature of its generativity; while the norm is the expected and the usual, there is a nonzero chance that some of its productions might be highly unusual or even unique. The human eye and ear can spot these for what they are and convert them into opportunities for creative novelty.

Moreover, there is a qualitative, not merely quantitative, difference between the kind of creativity that a human exhibits and that of a neural net. Let us hypothesize that knowledges can be divided into two general categories: *explicit knowledge*, in which one knows, for example, the speed of light. Then there is another kind of knowledge illustrated in Vannevar Bush's visionary article (1945) imagining the Memex machine, an early version of a hypertext generator. Let us call this second kind *latent knowledge*, generated by connecting two or more things that were not connected before.

In this model, we can imagine a space (or hyperspace) in which an immense amount of data consisting of explicit knowledge is located. Call this the space of possibilities. Making connections within the space of possibilities could reveal knowledge that is latently there. This is a rough version of the kind of knowledge possible with large language models, or neural nets in general. GPT-3 has billions and billions of linguistic data about every kind of knowledge that can be put into words,[7] and the Transformer architecture enables it to make connections between these that lead to discovery of latent knowledges not explicitly in the data set (note that the space where the internal neurons are located, which cannot be accessed directly, is called the "latent space" of the neural net). This is how it knows, for example, to identify literary genres and to simulate them on command. No one taught it this explicitly, but it gained this knowledge by making correlations between many data points and drawing inferences about these connections. One should not make light of this capability, for it has immense promise to aid humans in all sorts of tasks, from protein folding to understanding the relations between different bodies in the solar system to the invention of life-saving drugs for rare diseases. In this regard the possible knowledge available to an AI far exceeds the knowledge of any individual human.

However, all this knowledge takes place within the space of possibilities, that is, from known data points and explicit knowledges already present in the world.[8] There is another kind of creativity that belongs exclusively to biological (and maybe exclusively human) brains, related to the remarks above about the human ability to capitalize on contingent and unexpected events. That kind of breakthrough creativity comes not from connecting known data points, but from the ability to imagine something completely new, not implicit or latent in what already exists. Let us call this the space of potentiality. When Einstein conceptualized gravity as the curvature of space around massive objects, he broke through traditional concepts to visualize something new, something unexpected and entirely novel. No neural net, however large and finely tuned, can achieve this kind of creativity. However, once visualized and articulated, this new perspective could be included in a neural net's training, which could help to develop it and extend its implications.

In addition to achieving potentiality (not merely exploring possibilities), human creativity has another qualitative difference as well. It is enormously flexible, operating across all domains of creative endeavors. It flourishes in conceptualizing a new kind of structure for a long bridge, discovering a new kind of subatomic particle, devising a new design for a beautiful fabric, creating a painting in a style never seen before. The creativity of neural nets, by contrast, is extremely specific across very narrow domains. GPT-3 and GPT-4 can detect and capitalize on patterns in verbal or verbal-like sequences, such as short stories and protein folding; DALL-E can create new visual images; LaMDA can craft new dialogue sequences. But they can't design new bridges, offer plans for new scientific discoveries, or point the way to a new mathematical theorem. This doesn't mean, of course, that they can't be useful in aiding human creativity. In fact, it may turn out that is what they do best and where they make their most important contributions: in their symbiotic relations with humans.

## Writing Strategies for Collective Intelligences

Let me turn now to a subject I know well to explore how collective intelligences can work in scenarios of mutual creativity between a human and AI: teaching and practicing the craft of writing. The advent of LLMs such as ChatGPT, GPT-3, and GPT-4 present special challenges to the humanities, where the normal practice is to evaluate student learning and accomplishment by assigning essays. With the demonstrated ability of ChatGPT to compose passable essays on almost any topic, humanities departments must decide how to handle the prospect of a student asking ChatGPT to write an essay and then turning it in as the student's own work. Some departments have already reacted by forbidding students to use ChatGPT at all, declaring that it is a violation of the department's ethical code. As a teacher and parent, I can immediately spot the problem here: there is no reasonable way to determine if this prohibition has been honored or not. One could argue that all it does is train the students to cheat in ways that can't be detected. Other departments, perhaps the majority, have ignored the issue and left the students to try to figure it out on their own, in a kind of "Don't say AI" fashion. Both of these reactions in my view are suboptimal, because they miss the opportunity to enhance the student's learning about algorithmic intelligence and specifically about how LLMs can be used to assist in writing essays and enhance student performance.

A very different approach would *require* students to consult ChatGPT to complete an essay assignment.[9] Then the student could experiment firsthand with the different roles sketched above for humans and AIs. The student would quickly discover both the potential of ChatGPT as a collaborator and its limitations. For example, the student may ask the AI to generate a list of references on topic X, as did a friend of mine, only to discover that although the article titles and journals seemed appropriate, none of the references actually existed. The student might ask the AI to write an essay on topic X but then find that the AI either falls into a recursive loop of endless repetition or else flips into a rant that impossibly exaggerates its powers in the cosmic order (as one of the examples in chapter 6 shows). On the other hand, the student might get lucky and find several ideas to develop further, adding context that would deepen the analysis while also editing the AI's version for relevance and importance. In short, the student would benefit from the AI's wide range of reference and generativity, while using his or her own gifts of consciousness and world experience to separate the wheat from the chaff. The student would learn something about how to craft an essay, and perhaps even more importantly, how to employ collective intelligence to expand his or her own intellectual powers and enhance his or her creativity. The student would also gain experience in what algorithmic intelligence can do and where the limits of its abilities lie, and by implication, what kinds of tasks are best left to human invention and insight—lessons that extend far beyond the college classroom.

In this scenario, what ethical codes should govern these kinds of interactions? Clearly, traditional ideas about avoiding plagiarism would not be sufficient. Even before advanced AIs became widely available to the public, those traditional ideas were being questioned within the field of rhetoric and composition. Specialists in the field worried that university instructions on avoiding plagiarism had been emptied of real intellectual content, so that it was taught simply as a version of academic good manners, on a par with learning which is the correct fork to use for salad. Some argued that the very term "plagiarism" should be avoided, replaced with "unauthorized copying"—a phrase that at least encouraged students to wonder "authorized by whom" and "authorized for what"? One result of this kind of questioning was the copyleft movement and the invention of Creative Commons licenses, which divide the multiple rights usually associated with copyright into several different categories, allowing creators to choose what rights they wanted to grant to the general public and what rights they wanted to reserve.

What would be a "best practices" arrangement for collaborative intelligence productions? Normally such decisions are worked out in collaborative fashion by groups of educators and associated stakeholders, but here are a few suggestions. In my view, one goal should be transparency about who contributed what, with attention given to identifying the contributions of the AI and acknowledging its help. At present, AIs have no legal claim to ownership over their productions, and the general practice for software purveyors, for example Adobe with Photoshop, is explicitly to state that they have no claim to the works created by using the software. However, this could change in the future, especially if AIs become sentient and self-aware. In the 1999 film *Bicentennial Man* (Columbus 1999), Richard Martin (Sam Neil), owner of the robot Andrew (Robin Williams), recognizes the justice of making sure Andrew benefits from his own creativity and sets up a bank account in his name, in which the proceeds for his artwork are deposited. Acknowledging the contributions of AIs could be a preliminary step toward this kind of future, encouraging students to think of AIs as real collaborators and not just useful appliances.

In complex cases where the contributions of the parties is so intermingled that it is difficult to sort out who did what, it may be appropriate to ask students to preface their essays with a paragraph describing their creative process in crafting the work. This would have the advantage of encouraging students to be self-reflexive about their own modes of creativity, helping them to bring to conscious awareness their procedures, which are often semi- or nonconscious, and thus make them available for modification and improvement.

A further consideration, already raised by some creative artists, is what happens to the whole idea of intellectual property when an AI has perused copyrighted works during its training and incorporated them into its general sense of how to produce a picture of X or a poem about Y. Already artists and stakeholders are confronting similar issues in the age of remixing and modifying existing content. How much of a picture, or a song, needs to be altered for it not to count as copyright infringement? As legal cases like these work their way through the courts, collective intelligence will no doubt continue to spread throughout the cultures of developed countries, as more and more people come to rely on ChatGPT and similar models for more and more tasks. Thus our cultures edge toward the realization that the very idea of intellectual property as something owned by an individual who has exclusive rights to it may need to be rethought and reconceptualized on a basis consistent with the reality of collective intelligence and the pervasiveness of cognitive assemblages in producing products of value in the contemporary era.

I do not altogether share the concerns, valid in their own terms, of people who raise red flags about the continuing development of AIs on the grounds that they may get so smart they will pose a danger to humanity.[10] However, one fact seems to me undeniable: the very nature of human intelligence will change as it enters more broadly, and more deeply, into collective intelligence. It will tend to develop along the lines that maximize its synergies with AI, and perhaps become less proficient in the tasks that will be taken over by AIs. But changes in human ways of thinking, and even in the morphology of the human brain, have been happening since the beginning of the species, as the "techno" component of micro/evo/techno recognizes. Acknowledging this is part and parcel of recognizing the interdependencies that lie at the heart of ecological relationalities. The question, from my point of view, is not how to stop it but rather how to forge new visions adequate for understanding, analyzing, and participating in the condition of being human in the contemporary era.

# 10

# Planetary Reversal:
# Ecological Relationality versus Political Liberalism

Planetary reversal: what a concept! Referenced in the title of a 2021 conference in Bremen, Germany, this term highlights a new structure of feeling, a widespread admission that as a global community we humans have been heading in the wrong direction, along with a hope that it might not be too late to reverse course and try other paths. In this chapter, I turn to three powerful resources to interrogate this possibility: Lisa Lowe's *Intimacies of Four Continents* (2015) to diagnose historical contexts for our current problems, Jason Moore's *Capitalism in the Web of Life: Ecology and the Accumulations of Capital* for their continuation into the present, and Kim Stanley Robinson's *Ministry for the Future* (2020) (hereafter *MftF*) to search for possible solutions. Lowe's text highlights the failures of political liberalism by reading across archives to juxtapose it with colonial practices that not only were implemented at the same time but were in fact necessitated by liberalism itself. Jason Moore attacks the still-resonant nature/culture split, arguing that capitalism is best seen not as an economic system but as a way to organize nature, thus crafting a link between the externalities that allow cheap nature to exist and the economic inequalities that Lowe highlights. Robinson's text could easily have been entitled *How to Change the World: A Manual*, because it is less a novel than a compendium of proposed solutions to global inequalities and environmental crises. None of these texts is perfect, and all have at least implicit contradictions, but together they provide much food for thought and perhaps some templates for action.

One version of the many problems we face can be stated simply: a majority of people agree on the causes for environmental pollution, especially rising carbon dioxide levels and other greenhouse gases along with the resultant global warming, yet no one is doing enough to stop it. William Nordhaus,

Sterling Professor of Economics at Yale University and winner of the 2018 Nobel Prize in Economics, states in a recent article that the ambitions of the Paris Accord "are very low and that it will do relatively little to improve the global carbon output ratio at normal economic growth. It is clear that we are nowhere near meeting the goal of zero net emissions by the middle of the century" (Nordhaus 2021, 19). Among the issues he discusses is the free-riding problem, in which "agreements are voluntary, there are no penalties for non-participation, and countries talk loudly but carry no stick" (21). Global warming is already having and will continue to have serious consequences worldwide, from rising ocean levels to heat waves, but each country and region relies on others to pay the price to initiate meaningful change. Too often this takes the form of championing the priorities of one's own country and pointing the finger at others, for example, America at China during the Trump administration. On a regional level, it might be Michigan against Utah, or Detroit against Miami. On an even smaller scale, it might be coal miners versus solar energy companies. Currently there is no institution tasked with caring for the Earth as a whole instead of competing units of kin and clan, regions and nations.[1] In other words, the situation is structural, not merely a personal choice made by some humans, and thus structural analyses are required to understand and deal with it effectively.

## Lisa Lowe's Structural Analyses

Lowe's analysis covers roughly 1750 to 1850, when slavery, labor indenture, and colonial oppressions were pervasive. Against the promises of political liberalism, which she identifies as the "promise of universal rights, emancipation, wage labor, free trade," she points out that repeatedly "such liberties [were] reserved for some and wholly denied to others" (L. Lowe 2015, 3). "By modern liberalism," she clarifies, "I mean broadly the branches of European political philosophy that include the narration of political emancipation through citizenship in the state, the promise of economic freedom in the development of wage labor and exchange markets, and the conferring of civilization to human persons educated in aesthetic and national culture—in each case unifying particularity, difference, or locality through universal concepts of reason and community. I also include in this definition the literary, cultural, and aesthetic genres through which liberal notions of person, civic community, and national society are established and upheld" (3–4). This last sentence, especially, seems to indicate that political liberalism could be part of the solution, because it provides rationales and practices by which citizens can identify themselves and their families with broader national communities, which seems to be moving in the right direction—that is, away from local loyalties and toward global awareness.

In fact, however, Lowe shows that political liberalism did not overcome the problems that underlie our environmental crises but only raised them to a higher scale. "Modern liberalism," she writes, "defined the 'human' and universalized its attributes to European man; it simultaneously differentiated populations in the colonies as less than human" (L. Lowe 2015, 6). Her chapter discussing the writing of John Stuart Mill, one of the principal advocates of British liberal political philosophy, shows this clearly. In the same decade he was writing his classic treatise *On Liberty* (1859) he was also penning defenses of the East Indian Trading Company as an apologist for British imperialism. His reasoning may be characterized as "not yet": the East Indians whose labor, resources, and territories were being exploited to provide goods and commodities for middle-class British homes were denied the benefits of liberty and citizenship, allegedly because they had "not yet" progressed to a level warranting the benefits of liberalism. Thus Lowe dissects the "progressive temporality by education, civilization, and government" through which "normative political theory . . . rationalized the governing of liberty as representative government for some and despotism for others" (L. Lowe 2015, 107).[2]

In my view, Lowe succeeds in demonstrating that the relation between liberalism and oppressive colonialism includes simultaneity and correlation. However, she also implies that it is *causal* because expanding rights and rising levels of living standards in Europe and the United States, including improved education, health care, and access to representative democracy, required that some peoples had to remain or be inducted into hard labor to carry the burdens created by liberating others from these necessities. So stated, her argument is very strong in specific historical cases, such as slavery and the plantation system in the United States; in other contexts, we might think of indentured Chinese in the western United States working to extend railroads across the American continent, thus hugely benefiting the robber barons while exploiting thousands of Chinese men and women as sources of unpaid labor.

However, it is less clear that zero-sum dynamics would remain the same after 1866, after slavery had been outlawed in both the United States and Britain and the industrial revolution increasingly supplied mechanical labor to plow fields, harvest grains, transport goods, and drive factories. It remains an open question whether the ideals of political liberalism are bankrupt or whether they can be modified for more open, equitable, and democratic futures. Is it really the case that there must always be political oppression of some so that others can flourish? If so, then global-scale problems have no solution, for humans everywhere will undoubtedly choose to provide for themselves and their kin (clan, region, nation) against those who have less or nothing at all.

Another voice calling for a change of framework is Jason Moore in *Capitalism in the Web of Life* (2015). He argues that any framework that accepts a nature/culture binary division, including green ones, play into the hands of capitalism's externalization of real costs, because they can be shoved off into the nature side of the binary and thus be insulated from locating them where they belong, on capitalism's side of the ledger. Whereas Lisa Lowe focuses on *human* unpaid and underpaid labor, Moore extends the critique to natural resources as well. Capitalism, in his account, depends on the unpaid labor it extracts from both human and extrahuman sources to create the four cheaps—cheap food, cheap energy, cheap labor, and cheap raw materials (Moore 2015, 1816). Capitalism understood as a way of organizing nature, in contrast to binary nature/society accounts, reveals what he calls the "double internality" (his version of the now-familiar dynamic of reversible internality discussed in chapter 1): nature included inside capitalism, and capitalism inside nature (346). In his account, environmental problems are the unpaid labor that capitalism extracts from the environment, for example by treating the Earth as the "unpaid garage man" for capitalism's toxicities and pollutants (2282). He predicts that within a century, globalized societies will turn away from capitalism simply because it will (or has already) become untenable as a way of organizing human relationships with nature (7204). In this sense, he departs from those who see apocalypse as our inevitable future.

Moore's critique shares similarities with both Robinson's *Ministry for the Future* and my own admittedly utopian schema that focuses on revising political liberal ideals without abandoning the commitment to provide citizenship, liberty, and representative government beyond kin and clan to larger entities. Indeed, my proposed schema of the integrated cognitive framework would expand the circle beyond only humans to enmeshed networks of lifeforms and cognitive media. It would also address the overemphasis on the individual in political liberalism, a stance that puts too much credence in individual effort over social positionality and makes it difficult to see, much less address, systemic factors such as racism and white privilege. It would also revise the image of man (the human individual as has been traditionally conceived) as an economic agent with a recognition that human lives are enmeshed with, and ultimately dependent on, a wide variety of other lifeforms, from viruses and bacteria to insects, animals, and plants, and increasingly in developed societies, on cognitive media as well. It would emphasize collective

TABLE 10.1. Central Aspects: Political Liberalism versus ICF and Ecological Rationalities

| Political Liberalism | ICF and Ecological Rationalities |
|---|---|
| Autonomous self that owns itself | Enmeshed self that has multiple relations |
| Primarily an economic actor | Relations of many kinds, interior and exterior |
| Global capitalism | Planetary awareness and health |
| Free trade | Cooperation among nations on planetary priorities |
| Sovereign citizen with rights | Empathic citizen with responsibilities |
| Rational actor | Participant in cognitive assemblages |
| Free agent with free will | Distributed cognition, distributed agency |
| Brutish nature, social contract | Symbiotic nature, cognitive capacities |
| Only humans | Humans, nonhumans, and technical media |

action and cooperation over economic and political competition, both internally within nation-states and between state actors, a stance that has come closer to common sense in the face of our mounting environmental crises, which require global consensus and global action to be addressed effectively. Instead of the rational actor dear to economic theory, it would recognize that humans in developed societies depend on cognitive media to run much of the country's infrastructure, including airlines, rail lines, electric grids, the internet, the banking system, and almost every other technological system of any complexity, thus creating the hybrid human-techno systems that I call cognitive assemblages. It would endorse programs, procedures, and rules that turn the energies of capitalism toward planetary health and sustainability rather than toward the destruction of environments and habitats. Finally, it would extend the need for care beyond humans to nonhumans and, eventually, to cognitive media. As indicated in chapter 1 and running through all the chapters, the contexts for the ICF are the ecological relationalities now manifesting as second-order emergences. Table 10.1 summarizes central aspects of the ICF and ecological relationalities in contrast to political liberalism.

## The Technosphere and Mediating Mechanisms

Peter Haff's provocative analysis of the technosphere defines it as consisting of the large-scale technologies of the Anthropocene, including transportation networks, power and transmission grids, large-scale energy and resource extraction systems, and other networks including governments and bureaucracies (2014, 126–27). Central to his analysis is the idea of scale, which he calls Stratum I, II, and III, or in my terms, the micro (e.g., bacteria), meso (humans), and macro (e.g., the Federal Reserve) scales.

For my purposes, the critical point from his analysis is the necessity of linking mechanisms that enable parts at the meso level (Stratum II) to interact with systems at the macro scale (Stratum III) and vice versa, illustrated by his example of a well-coordinated navy ship, where all the different scales are in communication with each other and ready to obey commands from the captain. This is in sharp contrast to the technosphere, Haff argues. "The technosphere in general does not offer such simplified structure," Haff comments. "The technosphere is not a giant version of a navy ship. . . . [It] is not an engineered or designed system and during its emergence has not relied on nor required an overall leader. . . . In this regard the technosphere resembles the biosphere—complex and leaderless" (2014, 132).

Our environmental crises can now be seen in a new light, as the challenge of connecting entities at different scale levels to each other through effective linking mechanisms. At the meso scale, kin and clan often dominate (as they do, for example, in failed states such as Somalia and in the United States when the "government" is perceived as intrinsically oppressive to individuals and communities, for example in cyberlibertarianism). As Lisa Lowe points out (2015), one of the effects of liberal political philosophy was to provide linking mechanisms that enabled the idea of kin and clan to expand to the macro level as participation in national citizenship. The problem facing us in the contemporary moment may be understood as a need to develop similar linking mechanisms that enable the meso level (individual humans) to interact with macro systems (governments, banking systems, the technosphere as a whole). Also required is the inclusion of Stratum IV supplementing the divisions of Haff (2014), where the fourth level incorporates all the lower levels and adds a supervening one of the "planetary." Whereas the "global" is currently understood as the flow of things human control (labor, commodities, money), we are now faced with an additional level, understood as phenomena whose flows result from human activities but which humans do not directly control (pollution, global warming, ocean acidity, etc.).[3] Somehow, effects at the planetary level must be made real and palpable to the meso level, so that individuals understand they have a personal stake in how these effects play out and will consequently work through macro-level entities to influence them to change their policies toward the planetary. Envisioning how these various linking mechanisms might operate and become effective (large responsive to small, small accepting a stake in large and focusing their attention on the planetary) is obviously a huge undertaking, and it is perhaps where Robinson's *Ministry for the Future* makes its most important contributions.

## Carbon Coins and Blockchain Verification

One of the principal ways in which *MftF* envisions reorienting capitalist energies toward rescuing the environment instead of bankrolling its destruction is through carbon coins. To understand carbon coins, it is helpful to have some understanding of blockchain verification. Instead of having a bank keep the accounts for transactions (deposits, withdrawals, etc.), blockchain technology uses distributed ledgers and computation to keep track of what currency is created and how it is exchanged. It is designed as a system that self-verifies transactions, thus avoiding the need for a trusted central authority such as a bank facilitating exchanges between entities that do not necessarily trust each other.

Anyone completing a transaction in a digital currency traced through blockchain, for example Bitcoin, announces it through a peer-to-peer network. This transaction is bundled together with others into transaction blocks by computational agents, for example Bitcoin "miners." As Ed Finn explains, the agents then compete to "validate the transactions against the extant communal history of the currency. The outcome of that labor is a new block for the blockchain" (2017, 164). To create this block, the computational agents "must solve an arbitrary and highly complex math problem. The miner who is the first to correctly solve the problem 'wins' that block" (164). The successful agent then earns a fee for the creation of the block, plus a secondary fee for processing the trades. Finn continues, "the first transaction in each new block is a 'generation transaction' that creates a quantity of new Bitcoins" (in this example). Once created, the new block is added to the tail of the blockchain, and the process continues as other blocks are created and verified. Finn emphasizes that "the true radicalism" of blockchain "stems from the fact that the blockchain grounds its authority on *collective computation as an intrinsic form of value*" (165, emphasis added). In the computational era, it is easy to believe that somehow computation itself can create value (compare this with the traditional Marxist stance that labor is the source of value). Here (human) labor is only indirectly involved in producing value; the important component is not human labor as such but rather how much compute power is involved. As Finn points out, this is an attitudinal shift of considerable importance, for it undercuts economic theories of value from Locke to Marx and signals the growing importance of cognitive assemblages in generating wealth. It also anticipates the ways in which the production of economic value is likely to be further revised with ever more pervasive uses of AI.

That money could so easily morph into different forms such as digital currency is made more credible when one realizes that "money is a fiction," albeit a very useful one (Goldstein 2020, xi). In *Money: The True Story of a Made-Up Thing*, Jacob Goldstein emphasizes the point, including on the jacket that "money is a made-up thing, a shared fiction." "Money is fundamentally, unalterably social. The social part of money—the 'shared' in 'shared fiction'—is exactly what makes it money" (Goldstein 2020, xii). Tracing the evolution of money from its first form of IOUs in 3500 BCE through the 1694 establishment of the Bank of England, the 1720 introduction of paper money in France, the 1840 legislation in the United States that allowed each bank to print its own money, and finally the invention in 2008 of Bitcoin, Goldstein shows that at each stage, money succeeds or fails precisely to the extent that it retains credibility. Since 1971, US currency has been backed not by gold or silver but by the "full faith and credit of the US government." The transition into so-called fiat money works because the United States is a superpower, the major factor enabling the US dollar to function as the global reserve currency.[4]

Understanding money as a fiction also makes it easier to comprehend how money can apparently appear and disappear as liquidity increases or decreases. In times of crises, such as 2008 when overfinancing by banks plunged the United States and then the world into recession, central banks intervened to increase liquidity, in this case by buying up lots of government debt, which in turn allowed the government to pump more money into the economy by increasing debt, in effect creating money through a process called quantitative easing. Just as when a company's stock may be diluted when the company creates more shares to sell, so the value of any one person's dollars decreases as more dollars are created. If people see the value of currency decreasing, the result may be a feedback loop resulting in runaway inflation, such as happened in Germany after World War II and more recently in Zimbabwe. In the United States in 2008, this effect was averted because quantitative easing was gradually cut back as the economy recovered.

The idea behind carbon coins, first advanced by Delton B. Chen (2018), proposed that blockchain or distributed ledger technology (DLT) could be used to verify and track carbon offsets. As Chen explains, although carbon offsets are often regarded as commodities, this is not technically correct because they are not physically transported between buyers and sellers. "Carbon markets are usually trading services," he writes, "and the service is recorded as a carbon offset/credit" (Chen 2018, 77). Nevertheless, he proposes that "a single carbon offset represents the service of preventing one metric tonne of carbon dioxide equivalent [written as CO2-e] from entering the atmosphere." Low-carbon projects would then "receive carbon offsets/credits as revenue for reducing carbon emissions or for sequestering carbon," and the carbon amounts "should be measured, reported and verified, and then monitored in case of leakage." He notes that blockchain verification can be used in a similar fashion to the way that diamonds are monitored and traded, with a blockchain ledger "to record each diamond's authenticity and ethical origins." "A blockchain ledger for carbon stocktaking will similarly require that supply chain is monitored and recorded... helping to deliver on the Paris Climate Agreement." In this scheme, carbon coins would function differently than carbon credits, to which they are similar, because they not only could be traded in the carbon market but could be exchanged for dollars, thus acting as a de facto currency.

*Ministry for the Future* takes this idea and runs with it, condensing it into the equation of one carbon coin for each metric ton of $CO_2$-e sequestered, either by restorative agriculture, direct carbon capture from the atmosphere in the form of sequestering dry ice in empty oil wells, or by leaving oil reserves in the ground, pledging not to access them for a century. As the narrative presents it, the challenge is getting enough people to believe in carbon coins. And how they are induced to believe? By having the world's central banks sign on and become the issuers and guarantors of carbon coins. That saving the world from capitalism should depend on central banks, often seen as the epitome of capitalist enterprises, is an irony not lost on the narrator of *MftF*. Left unaddressed is the question of who pays for the carbon coins. If the central banks issue them, this would count as a form of quantitative easing, through which the banks effectively issue money. In that case, their issuance would presumably cause a deflation in the value of national currencies, especially when the idea takes off and more and more carbon coins are issued. *MftF* chooses not to address these concerns, taking at face value the idea that carbon coins can be created without significant effects on traditional currencies.

A less obvious but still important concern is that $CO_2$-e coins, as envisioned by Chen and *MftF*, require blockchain verification. Blockchain has been criticized as being hideously inefficient, gobbling up human and computational resources and increasing global pollution through heat dissipation at server farms, extensive use of rare Earth metals in computers that are detrimental to human and environmental health, and its association with regressive cyberlibertarian ideologies.[5] *MftF* does not address these concern, nor does it clarify why blockchain verification would be required (presumably to give each metric ton of sequestered carbon an identity, since in nature, one molecule of carbon dioxide is indistinguishable from another). Since carbon coins require endorsement by central banks to succeed, why not link their credibility directly to global banking systems rather than blockchain? Surely the identity question could be settled in a much more efficient way, for example by including a marker molecule in the mix.

Leaving these questions unanswered, *MftF* accepts the blockchain connection as necessary without further examining it, just as it does not address the global economic consequences of issuing carbon coins. This example illustrates Robinson's problem in proposing specific solutions; some may not work or may have negative effects that exacerbate problems rather than solve them. In my view, the enduring contribution of *MftF* lies not so much in its proposed solutions as in its robust faith that there *are* solutions and ways to achieve them on a global scale.

### Confronting Climate Catastrophes: Frank Lay and Mary Murphy

Insofar as *MftF* has a plot in a conventional sense, it revolves around Frank Lay, Mary Murphy, and their interactions. Frank is introduced in the opening chapters as an aid worker in India when it is hit by a massive heat wave later estimated to have killed twenty million people. At first Frank's office has a generator to supply electricity for the air conditioner when the power goes out, but soon a group of armed young men seize it, telling him, "We need this more than you do" (Robinson 2020, 9). Then Frank leads a group of people who have taken shelter in his building down to the local lake, where he stays immersed that day and night. When morning comes, he is surrounded by dead bodies; he is the sole survivor, an experience that scars him for life. Possibly he was able to survive because he had a hidden flask of fresh water. When he is pulled out by rescuers, looking like something far from human, he asks them why he survived. His rescuer suggests, "Maybe you were better hydrated?"—a suggestion too close to the truth for Frank's conscience. Answering his own question, the rescuer says, "It's just fate." As Frank tells it, "Fate, I agreed" (28). Thereafter "fate" will reappear in the text as a cop-out, an abdication of ethical responsibility.

Mary Murphy, "an Irish woman of forty-five years of age, ex-minister of foreign affairs in the government of the Irish Republic" (Robinson 2020, 18), has the awesome responsibility of leading the Ministry for the Future, established by the Paris Accords to serve as the voice for future generations and those who have no voice, that is, the nonhumans of the Earth.[6] Janus-Athena, an AI expert of indeterminate gender on her staff, suggests the idea of a carbon coin and tells Mary that for it to succeed, she must win the backing of key central banks, especially those in the United States, the UK, the European Union, and China. She recognizes that this would be a new form of quantitative easing without going into the economic implications, and (following Chen 2018) suggests that one carbon coin (or carboni) should be issued for each metric ton of $CO_2$-e sequestered for a century, authenticated through blockchain verification. The central banks would establish the coins' interest rates and designate a floor beneath which the carbon coins could not fall. The carbon coins (or fractions thereof) would trade against other currencies and on their own market, just as do government bonds at present.

The proposal's genius is to redirect the capitalist drive for profits away from burning carbon to sequestering it, or as Mary puts it, to "go long on civilization" (Robinson 2020, 240). Pitching the idea at a meeting of the central banks, Mary is met with unyielding resistance. This is not in their purview; their job is to stabilize national currencies, keep inflation under control, and establish prime interest rates for banking systems. Slowly, however, the proposal starts to gain traction, first with the central bank of Germany, then China, whose female minister of finance proves a key ally for Mary. Once the central banks get on board, the oil companies follow, convinced not because they are saving the planet but because it makes economic sense. As Mary points out to them, "You can short civilization if you want. Not a bad bet really. But no one to pay you if you win. Whereas if you go long on civilization, and civilization (therefore) survives, you win big. So the smart move is to go long" (240).[7] A revolution in Saudi Arabia hastens the trend. With control wrested away from the Saudis, the government renames its country Arabia and announces that its oil will be used solely for the manufacture of plastics and other nonburning purposes, thus making the government eligible for a huge amount of carbon coins. Thereafter energy company lawyers descend on Zurich and the ministry to bargain for their clients' also winning carbon coins, either by leaving their oil in the ground or by using their empty wells to draw carbon dioxide directly from the atmosphere in the form of dry ice.

This is fiction, of course, and the author is in control of how events proceed. He narrates enough sources of resistance for the representation to be credible, but it is scarcely a surprise when the wheel begins to turn in a positive direction. Mary's dark thoughts in the midst of the struggle can be taken as a plan for the book: "Lose, lose, lose, lose, lose, lose, fuck it—win" (Robinson 2020, 453). The narrative's utopian force derives not only (or even primarily) from its positive outcomes, however, but also from the vignettes employed to illustrate how the turn is achieved. These serve to craft for readers the linking mechanisms that allow movement across scales, from meso to macro and on to the planetary.

### Geoengineering at Multiple Scales

An important way that *MftF* achieves an aesthetic effect of operating at multiple scales is through the myriad of voices it crafts, ranging from an enormous entity such as the Sun to the subatomic scale of a photon. This can be seen as a literary strategy equivalent to (or at least in tune with) Jason

Moore's insistence that nature is internal to capitalism, just as capitalism is internal to nature. When natural entities (the Sun, a photon) are given voices as individual speakers on a par with human narrators, nature is represented as internal to the human world, while the human world is simultaneously represented as internal to nature, for example when humans are trapped in the Indian heat wave mentioned earlier. Most narrators, understandably, are located at the meso scale of humans, but even here, there is enormous diversity in the occupation, mission, and background of the different speakers. The uber-narrator, if I may call him that, speaks in the third person with thoughts focalized through Mary and Frank, but many other speakers narrate their own first-person accounts. Often these are not identified by name, so they are remembered primarily by associating them with the role they play in the text. Frequently these concern geoengineering projects at various scales: the anonymous Indian pilot flying missions to pump sulfate particles into the upper atmosphere to create an effect similar to a volcanic eruption and thus temporarily bring down global temperatures by a degree or two; the scientists who are attempting to slow seawater rise by pumping water from underneath glaciers in the Arctic and Antarctica, slowing their slides into the ocean. One of the most memorable, because of her comically ironic tone, is the wife of a farmer (perhaps in India?), haranguing him to work fertilizer into their small plot of inherited land to increase its fertility. At her insistence, they register their land and establish a low benchmark figure for its percentage of compost. Over several seasons, they labor to work into the soil manure and vegetable compost and then call for another soil assessment; their efforts are rewarded by more than twenty-three carbon coins, amounting to over seventy thousand (dollars?). It is through vignettes such as this that the large-scale and rather abstract schema of carbon coins is linked to the human reality of a couple trying to make a better life for themselves and their family.

Perhaps no contemporary writer is fonder of geoengineering projects than Robinson, whose Martian trilogy (*Red Mars*, *Green Mars*, *Blue Mars*; 1990, 1995, 1997) is an epic narrative of exo-engineering on a planetary scale. In *MftF*, certain passages suggest that he is keenly aware of recent criticisms of geoengineering projects and nevertheless remains unconvinced that the criticisms are valid.[8] The Indian project to inject sulfate particles into the atmosphere, for example, is a violation of the Paris Accords but justified (according to the Indians) by their experience of losing twenty million people in a heat wave. The most explicit rejection comes from Pete Griffen, an "icehead" scientist leading the Antarctic glacier project. "So someone asked tonight in the mess tent, is what we're doing down here geoengineering? Who the hell knows! What's in a word? Call it Glacier Elevation Operations. . . .

There are some things man was not meant to know—my ass! We are meant to know everything we can find out. So get over that whole wimpy line of objection. And I'll tell you what the unintentional side effects of slowing down the glaciers of Antarctica will be: nothing. Nada. No side effects whatsoever, and the beaches and coastal cities of the world will stay out of the drink" (Robinson 2020, 265).

Later a member of Mary's staff comments that "geoengineering [is] no longer a useful word or concept. Everything people do at scale is geoengineering. Glacier slowdown, direct air capture, soil projects like 4 per 1000, they're all geoengineering. . . . Meanwhile, we're working with CCCB to list all [the] ways carbon drawdown could be quantified and confirmed, in ways that would allow for carbon coins to be created and paid to individuals.[9] All geoengineering, all good. The word itself needs to be rehabilitated" (Robinson 2020, 356–57). The important point seems to be that all avenues should be open, all possibilities explored to combat climate change. Standing on principle holds little water in this text—as little as the Arctic pumps that go dry because they have succeeded in excavating all the melt water beneath the glaciers.

### War for the Earth

The palpable urgency to save the environment before the damage becomes irreversible is deeply connected with another linking mechanism: a covert war against those who continue to pollute, burn carbon, and exhaust Earth's resources. The figures carrying out the so-called War for the Earth are always shadowy figures in the text, for example the people who take over a fishing boat, free the slaves who served as crew, and set the boat on fire with the comment "No more fishing." The same is true of the Children of Kali, the Indian terrorist group who give Frank a message to take to the world when he tries, unsuccessfully, to join them: "You can tell them that they must change their ways. If they don't, we will kill them" (Robinson 2020, 49). Thus individuals at the meso level band together to affect change at the macro scale, targeting capitalists grown rich by burning carbon, companies that continue to use fossil fuels for their huge container ships, and airlines that build, buy, and operate commercial aircraft.

The linking mechanisms operate not only in the (fictional) world but also in the text's narratives, for example in this passage analyzing the nineteen largest energy companies and the five hundred or so people who run them. By focusing on a few hundred decision makers, the text crafts a link between the macro-scale global energy companies and those who decide policy and direction; further, it extends the link down to individual families, thus focusing the reader's attention on the meso scale of individual actors. "They will be good people. Patriotic politicians, concerned for the fate of their beloved nation's citizens; conscientious, hard-working corporate executives, fulfilling their obligations to their board and their shareholders. Men, for the most part; family men for the most part; well-educated, well-meaning. Pillars of the community. Givers to charity . . . they will want the best for their children" (Robinson 2020, 30). The implication is that the people in charge are not necessarily immoral or unethical; in their own eyes, they are "pillars of the community" who "want the best for their children." How then to reach them, persuade them to do otherwise? One highly controversial solution the text proposes is to bring the War for the Earth out of the boardroom into the bedroom, that is, into the spaces where the meso scale resonates most intensely.

That and similar strategies are illuminated by an anonymous narrator who reports retrospectively on the war after a decade of fighting. "They killed us so we killed them. . . . Methods were worked up over many iterations. We took a lot of losses at first. . . . Most of us didn't want to do [suicide bombing]. . . . We weren't that crazy, and we wanted to be more effective than that. Much better to kill and disappear. Then you can do it again. . . . Drones are best. Much of the job becomes intelligence, finding the guilty, finding their moments of exposure. . . . A swarm of incoming drones the size of sparrows, moving at hundreds or thousands of meters per second. . . . The guilty died by the dozens in those years [2030s–2040s]" (Robinson 2020, 135).

As the passage suggests, the advent of armed drones and "pebble mob missiles" proves to be a game changer, because it puts long-range deadly weapons into the hands of terrorists, or anyone who could afford a few hundreds or thousands of dollars. In one swoop, drone warfare renders traditional large-scale weapons such as tanks, destroyer ships, and fighter planes obsolete; any one drone can be taken out, of course, but masses of them attacking at once are invincible, because some will succeed in hitting their targets. The narrator comments, "It was mutual assured destruction, not of civilian populations, but of war machinery" (Robinson 2020, 348). Noting that "the world was trembling on the brink, something had to be done," the narrator supplies a rationale for the terrorists. "The state monopoly on violence had probably been a good idea while it lasted, but no one could believe it would ever come back. . . . It was like eating beef: some things were just too dangerous to continue doing" (369). (Spreading mad cow disease through cattle herds is one of the strategies the environmental terrorists adopted.)

To reinforce the point, the narrator makes explicit why the meso scale and its corollaries—favoring one's kin, clan or nation—has been rendered obsolete by terrorism.

> Either everyone's happy or no one is safe. But we're never happy. So we'll never be safe. Or put it this way: Either every culture is respected, or no one is safe. Either everyone has dignity, or no one has it.
>
> Because why? Because this:
>
> A private jet owned by a rich man—boom.
>
> A coal-fired plant in China—boom. . . .
>
> No place on Earth was safe. (Robinson 2020, 347)

Thus terrorism in the War for the Earth has the effect of creating linking mechanisms—albeit deadly ones—between the planetary and meso scales. From a private jet carrying a rich capitalist to a polluting coal plant in China—all are subject to disruption and annihilation. Although the values of liberal philosophy in one sense are forced to expand beyond kin, clan, and country to include all humans, and (somewhat later) nonhuman animals as well, a major contradiction looms. This result was possible only because of the oxymoronic concept of ethical terrorism. As a linking mechanism, this is a terrorism not devoted to a specific religion, region, or country but seen by its practitioners as a battle waged on behalf of the Earth itself.

The linkages are not always between scales; they also include intrascale attacks by individuals on individuals deemed to be criminals guilty of mass pollution or exploitation. The individual nature of the attacks is emphasized by an anonymous narrator (presumably a member of the Children of Kali) when he talks about one killing. "Leap up in the air, come down on the vent, crash down into room right on bed, trailing rope ladder. Stab the guilty one in the torso quickly four times, then the neck, several times. . . . Back up the rope ladder. . . . Out onto compound wall, up to roof, drone waiting to carry me up and away like packaged goods" (Robinson 2020, 136). The graphic detail of this passage is an exception. More frequently, the killings are described through mass numbers that convey little of the horror that would be involved in such scenes (for example, the seven thousand killed on Crash Day when sixty planes go down), or the events are focalized through innocents who benefit from the murders, as with the freed slaves on the fishing boat or miners relieved of indentured labor when the mine is taken over by dissidents. In these ways, the text damps down the horror and creates an atmosphere in which terrorism can be seen not only as a viable option but indeed as an ethical one.

As a result, a strain of schizophrenia runs through the text, a kind of doublethink in which the text recognizes two conflicting realities simultaneously. In stark contrast to the terrorism is the faith that Mary and her team put in laws and legal remedies. Noting that "there were crucial clauses in the [Paris] Agreement," the narrator points out that "the text of these articles and their clauses had been fought over sentence by sentence, phrase by phrase, word by word" (Robinson 2020, 351). "Were they foolish to have tried so hard for words," the narrator asks, but then argues for the importance of words (and implicitly of laws). "Words are gossamer in a world of granite.... If you give up on sentences you end up in a world of gangsters and thieves and naked force, hauled into the street at night to be clubbed or shot or jailed" (352). Apparently supporting the necessity for laws is Tatiana, an expert in Russian international treaty law working at the Ministry for the Future: "'Rule of law is all we've got,' she said darkly. 'We tell people that and then try to make them believe it'" (36). Ironically, Tatiana herself is later shot dead by bullets in a terrorist attack, a recognition that two (and more) can play at the terrorist game. Despite her apparent argument in favor of law, there may be another irony lurking in the phrase, "try to make them believe it." One way to read the passage is that people need to be convinced that laws are effective; but a second way is that the attempt is to make people believe the Ministry for the Future is itself a law-abiding institution. As we will see, this is depicted in the text as correct only within certain boundaries, with major covert exceptions.

## Black Wing at the Ministry for the Future

Frank's and Mary's paths intersect when, walking home alone, Mary is kidnapped by Frank, who snaps a handcuff around her wrist and shows her a gun. Agreeing they can talk in her apartment, Mary succeeds in somewhat normalizing the situation, although Frank remains very much on edge and threatening throughout. The charge he makes to Mary is that she is not doing enough to forestall environmental collapse. "We're doing what we can," she replies, but he is adamant. "No you're not. You're not doing everything you can, and what are you are doing is not going to be enough" (Robinson 2020, 96). "They're killing the world," he continues. "People, animals, everything. We're in a mass extinction event, and there are people trying to do something about it. You call them terrorists, but it's the people you work for who are the terrorists" (97). When she asks what more she can do, he has a ready answer. "Like identifying the worst criminals in the extinction event and going after them" (99). When she objects that the ministry does bring lawsuits, he asks, "What about targeted assassinations?" "Of course not," she answers. He is not having it. "Why of course . . . the violence of carbon burning kills many more people than any punishment for capital crimes ever would. So really your morality is just a kind of surrender." She objects, "I believe in the rule of law," but he points out that would be fine "if the laws were just. But in fact they're allowing the very violence you're so opposed to!" There is a hidden irony in this scene that readers may realize, although Mary does not: Frank himself had earlier stolen a rifle and picked out a target, but when push came to shove he was unable to pull the trigger, so in effect he is trying to coerce Mary into doing what he could not do himself. Nevertheless, this passage makes perhaps the text's most explicit case for an ethical terrorism, positioning it not as an oxymoron but as a realistic view of the situation.

Although Mary escapes from her kidnapper when her doorbell rings, she is shaken by the encounter and soon brings up the topic with Badim Bahadur, her East Indian chief of staff. "I think maybe we need a black wing," she confides. "Actions that are maybe illegal, or in some sense ill-advised. . . . We could consider them in secret, on a case-by-case basis" (Robinson 2020, 109). But Badim is way ahead of her, confessing that such a black wing already exists: "That even I myself might have started it after you hired me as your chief of staff" (110). However, the requirement for such a black wing is that the principal officer, to preserve deniability, remain ignorant of the details and even of its existence; so even Badim's admission is cast in a hypothetical or conditional mode ("might have"). Here again, because of this constraint, the text gives no details of how the black wing operates, the facilities it has sabotaged, or the people it has killed.

The exception arrives when Tatiana, a beloved member of Mary's staff, is killed. Mary demands to talk with Badim, and whatever reservations she may have had fly out the window. "And listen, your black wing—sic them on those bastards!" "I already have," Badim answers. "When we find them, we'll kill them." "Good," Mary replies. "I hope you can find them" (Robinson 2020, 453). This is the last word we hear on the subject, so readers never learn if the offenders (probably Russians hurt by the changes Tatiana was working to bring about, Badim surmises) were in fact killed and if so, how. Once again, narrative techniques are introduced that acknowledge the terrorism of the ministry itself, but still tamp down its full realization.

The War for the Earth includes adversaries working on behalf of the polluters, so it is no surprise when Mary herself becomes a target. Since the Ministry for the Future is located in Zurich, the Swiss police and secret service have the job of protecting her, which soon includes around-the-clock surveillance. At one point she narrowly escapes assassination by an exhausting midnight climb over a towering Alpine path, accompanied by her Swiss guards.

Her close call draws attention to how easily the asymmetric warfare of terrorism can get out of control and itself become a new regime of lawlessness.

This may be why Robinson, alert to this possibility, includes a chapter in which an anonymous narrator, an East Indian self-made businessman, decides to meet with the Children of Kali, who look at him "if calculating where to insert the knives" (Robinson 2020, 389). "We see what's happening," a female member tells him. "We brought you here to tell you to do more." However, he has a message for them as well, replying, "I came when you asked, to tell you the time has come to change tactics. That's good thing, and it's partly because of what you did. You were doing the needful, I know that" (389–90). The woman asserts that "we are still doing the needful," but he ripostes: "It's a question of what's needful now" (390). "Conditions have changed," he continues. "Together we helped to change them. . . . I'm telling you, it's time to change. The big criminals are dead or in jail, or in hiding and rendered powerless. So now if you keep killing, it's just to kill. Even Kali didn't kill just to kill, and certainly no human should" (391). Still resisting, the woman tells him they listen to Kali, not to him, whereupon he suddenly says, "I am Kali." The narrator comments, "Suddenly he felt the enormous weight of that, the truth of it. They stared at him and saw it crushing him. The War for the Earth had lasted years, his hands were bloody to the elbow. For a moment he couldn't speak, and there was nothing more to say." Although the East Indian narrator is not identified in the text, Robinson in an interview has said that he intends it to be Badim, Mary's chief of staff. The text's coyness in not explicitly revealing Badim's identity is another way in which environmental terrorism is acknowledged but not fully accounted for in the text.

This powerful passage marks a pivot in the text, a movement away from the War for the Earth to a celebration of what has been accomplished. One could summarize the text's stance toward terrorism by saying it was "needful" but must stop at some point. People now avoid jet planes and racing cars; diesel-powered ships have been transformed to solar-electric sailing vessels; air travel is now mostly accomplished by airships such as blimps; and increasingly, carbon coins are having an impact as more and more tons of $CO_2$-e are sequestered. Without the drone attacks, without Crash Day, without mad cow disease, without assassinations: would the world have changed so dramatically? The text implies that might have happened, but not fast enough to forestall environmental catastrophe.

### Novel as Linking Mechanism

As the denouement approaches, it becomes clear that Robinson is determined to write a utopia. Every major global problem is addressed in the text, and an appropriate solution (or solutions) is suggested. Crushing student debt leads to a national strike in which students refuse to pay, which sends the banks into crisis; when they demand a bailout, government officials agree but on the condition that the government then becomes a major shareholder, in effect nationalizing them. YourLock, a technology sponsored by the ministry, solves the problem of data privacy and collection by offering consumers an alternative to social media in which citizens own their data and can decide for themselves if and on what terms they want to sell it. When the banks are reluctant to create carbon coins, YourLock offers an alternative banking system authenticated by blockchain (again without the text acknowledging its problems), thus undermining the banks' hegemonic hold on the economy. Refugees worldwide are given global citizenship and allowed to emigrate to the country of their choice, or to repatriate to their native countries, now made more stable by global interventions and consensus. The citizens of Hong Kong are successful in resisting the autocratic demands of mainland China, having demonstrated every Saturday for thirty straight years (!).[10] This narrator makes an interesting observation when she (or he) notes that among other oppressive measures, Beijing had tried to suppress the Cantonese language. But the southern part of China speaks this language, including the city of Guangzhou and "a hundred million of us" (Robinson 2020, 517). That "meant all of Guangzhou didn't believe in Beijing either—they were more with Hong Kong than with Beijing, even if they never did that much to show it. But language is family. Language is the real family," a reorientation of liberal values that redefines kin as linguistic community. Even nonhuman animals benefit when wildlife corridors are established from Yukon to Yellowstone, from the Rockies down into Central America, and in east-west corridors from California to Colorado. In these ways, the text suggests the move away from capitalism that Jason Moore had predicted would happen within this century.

## Summary of Linking Mechanisms

Table 10.2 revisits the proposed schema for ecological relationalities, adding the relevant linking mechanisms.

Of course, it is extremely unlikely that all such measures would have succeeded. Realism is less a concern for this text, however, than is imagining how the world could change at every scale level, from bedrooms to boardrooms. If you have a pet problem you think is important, chances are this text offers one or more solutions to it. In the process, *MftF* gives a local habitation and a name to a large variety of linking mechanisms that can contest Haff's

TABLE 10.2. Central Aspects: Political Liberalism, Ecological Rationalities, and Linking Mechanisms

| Political Liberalism | Ecological Relationality | Linking Mechanisms |
| --- | --- | --- |
| Autonomous self, owns itself | Enmeshed, embodied self | Cognitive assemblages |
| Primarily an economic actor | Relations of many kinds | Climate disasters |
| Global capitalism | Planetary awareness | Environmental terrorism |
| Free trade | Cooperation on global priorities | Paris Accords, Ministry for the Future |
| Sovereign citizen with rights | Empathic citizen, responsibilities | War for the Earth |
| Rational actor | Participant in cognitive assemblages | YourLock |
| Free agent with free will | Distributed cognition, agency | Blockchain |
| Exploitive capitalism | Capitalism tied to planet health | Carbon coins |
| Brutish nature, social contract | Symbiotic nature, cognition | Assemblages |
| Only humans | Humans, nonhumans, cognitive media | Wildlife corridors |

argument that "one of the principal paradigms of the Anthropocene world [is that] . . . humans are components of a larger sphere they did not design, do not understand, do not control and from which they cannot escape" (2014). On the contrary, Robinson would reply, it is possible to turn the technosphere's trajectory toward a more biophilic, just, and sustainable direction. You just need to invent or create the necessary linking mechanisms.

Life coaches know that one of the secrets to achieving one's goals is to be able to visualize them, to visualize yourself as a character in the life you want to live. At its best, that is the contribution that *MftF* offers its readers. Through its myriad of voices, hundreds of narrators, dozens of scenes and adventures, it catalyzes the imagination of readers to invite them into this hard-won and always fragile future in which it is possible to solve problems, even at the planetary level. In this sense, the novel itself functions as a giant linking mechanism operating across multiple scales to join the present of us, its readers, to futures in which we may want to live.

*Acknowledgments*

Behind any single-authored book stands a chorus of collaborators, colleagues, and fellow travelers whose efforts undergird, support, and extend the insights of the supposed single author. Much of the material in the preceding chapters was developed through personal and long-distance relationships with other scholars from whose ideas and thoughts I have greatly benefited. In no particular order, I want to thank Leif Weatherby, who organized a conference at New York University that put me and Terrence Deacon, on whose work I draw extensively in chapter 2, in direct conversation. Working from different premises, he and I had very different views about the texts produced by AIs such as GPTs, and his skepticism encouraged me to think more critically about my own position. Leif himself, along with Brian Justie, was an important interlocutor for his work on the importance of indexical pointers with GPTs.

Rita Raley's laser-sharp mind helped me think more about the arguments in chapter 6 on critical responses to GPT texts at the conference "Transformations of Attention" at the University of California, Santa Barbara, organized by Alan Liu. Isaac Mackey in the Computer Science Department at UC Santa Barbara generously shared with me his knowledge of neural nets and how they work. Katherine Boda of Australia National University generously read a draft of chapter 6 and solicited her graduate student Galen Cuthbertson for comments; I thank both of them for their many helpful thoughts.

Mark Hansen from Duke University has for many years been both a colleague and a treasured interlocutor, and he continued to be so for this book. Beatrice Fazi from the University of Sussex inspired me to think more deeply about what exactly computation is, a question posed to me by Stuart Moulthrop when I visited his seminar at the University of Wisconsin–Milwaukee. Louise

Amoore, during my visit to Durham University, UK, kindly organized a conference in my honor there and remains an inspiration for her important books *Cloud Ethics: Algorithms and the Attributes of Ourselves and Others* and the *Politics of Possibility: Risk and Security beyond Probability*.

During a visit to Cambridge, MA, I was privileged to meet Michael Levin from Tufts University and hear him speak about the latest results from his lab on xenobots and related topics. I shared a cab ride with his colleague Douglas Blackiston and heard with great interest his (somewhat skeptical) views on cellular cognition. Benjamin Bratton's work on the "Stack," collective intelligence, and "artificialization" continues to challenge and enlighten me. At the conference "Patterns" at Birkbeck College, London, arranged by Scott Lash and Joel McKim, I was privileged to meet Guiseppe Longo and engage with him in a friendly disagreement about the prerequisites for emergence. Stuart Kauffman, whose collaboration with Longo in a seminal article on how biological evolution is fundamentally different from material processes, has long been a figure of admiration and collegiality for me.

Sherryl Vint of the University of California, Riverside, had an important conversation with me at a conference in South Korea that gave me important insights into the nature of what I came to call ecological relationality. Wendy Chun's work on algorithmic discrimination continues to be an important inspiration for my own thinking on the matter, as does Luciana Parisi's challenging but always enlightening work on recursivity and incomputability.

Bruce Clarke's two books on the work and correspondence of Lynn Margulis and her collaborator James Lovelock were invaluable resources for chapter 3. Cary Wolfe of Rice University (and soon Arizona State University) made me aware of how much our current research runs along the same lines when he gave a lecture at a conference in Fribourg, Switzerland, on his book-in-progress on "jagged ontologies." He and I later shared the stage at a symposium arranged by Claire Webb of the Berggruen Institute in Los Angeles and debated our different perspectives, he from a deep interest in philosophy and critical theory, and I from a more technically oriented perspective. I also thank Claire Webb for her marvelous work in arranging all manner of interesting events and her continued interest in my research.

In addition to all the ways in which I have benefited from the work of individual scholars, I am also grateful for the institutional support I have received. The English Department at the University of California, Los Angeles, has provided valuable IT support and office space; the Literature Program at Duke University has contributed access to software packages essential for my work; and the Berggruen Institute in Los Angeles has given me an honorary fellowship in their "Future Humans" project.

My greatest debt is to my companion and partner, Nicholas Gessler, whose vast technical expertise is a source of continuing wonder and inspiration, and whose world-class collections inspire my curiosity and provide material support for many of my ideas.

I am also pleased to acknowledge here permissions from journals, publishers, and individuals to reprint materials. I thank Utkarsh Ankit and the Code Emporium for reprinting his "heat map" of a Transformer word sequence that appears in chapter 6. In addition, I thank *New Literary History* for permission to reprint "Inside the Mind of an AI: Materiality and the Crisis of Representation," which appeared in *New Literary History* 53, no. 4 (Winter 2022)/54, no. 1 (Spring 2023):635–66 and formed the basis for chapter 6. I thank Duke University Press and *American Literature* for permission to reprint "Subversion of the Human Aura: A Crisis of Representation," which was published in vol. 95., no. 2 (June 2023): 255–79 and appears here in slightly revised form as chapter 8. I thank *Critical Inquiry* for permission to reprint "Can Computers Create Meanings? A Cyber/Bio/Semiotic Perspective," vol. 46, no. 1 (2019): 32–55, which forms the basis for chapter 2, and for permission to reprint "Microbiomimesis: Bacteria, Our Cognitive Collaborators," vol. 47, no. 4 (Summer 2021):777–87, which appears here in slightly revised form as part of chapter 4.

# Notes

## Chapter One

1. I mean no disrespect to clams in saying they do not have brains. What they have instead is a decentralized nervous system consisting of ganglia and a nerve cord, which enables them to receive and process information from their environments. Like every other lifeform, they therefore have cognitive capacities.

2. I understand this usage is changing, and that some standard dictionaries do list "human" as a noun—thank goodness!

3. I am indebted to Nicholas Gessler for this example.

4. In contrast to digital computers, analogue computers do not necessarily rely on electronics; nor do older versions of computers that relied on core memory, etc. But of course the overwhelming majority of contemporary digital computers do use electronics.

5. For a summary of the possibility that information is more fundamental even than matter and energy, see Ananthaswamy 2017.

6. In its original 1945 version, von Neumann architecture included a processing unit with an arithmetic logic unit plus processing registers, a control unit, internal memory, and external mass data storage as well as input/output mechanisms. In contemporary usage, it denotes a computer with a stored program and shared bus for instruction fetch and data operations.

7. The locus classicus for this claim is philosopher John Searle's "Chinese Room." For a summary of Searle's thought experiment, along with the many rebuttals to his argument and Searle's subsequent revisions, see Searle 1999 and Searle 1984.

8. The idea of a symbiosis between humans and computational media is not a new thought. As early as 1960, Licklider wrote about what was then only a future possibility of human-computer symbiosis. "In the anticipated symbiotic partnership, men will set the goals, formulate the hypotheses, determine the criteria, and perform the evaluations. Computing machines will do the routinizable work that must be done to prepare the way for insights and decisions in technical and scientific thinking" (1960, 4). What has changed between then and now is the extent to which agency and decision making has been bestowed on computational media.

9. Geoffrey Hinton was the coinventor of backpropagation, catalyzing a crucial leap forward in AI research. Widely esteemed by AI experts, he was employed by Google in a research capacity. To his credit, he resigned from Google so that he could issue warnings about GAI without compromising his employers.

10. An example of reflexive approaches is M. C. Escher's *Drawing Hands* (1948), in which the left hand draws the right while simultaneously being drawn by the right.

11. There is a strong similarity here with Donna Haraway's (1988) notion of "situated knowledges," which argues that there is no God's-eye point of view that would guarantee objective reality.

12. Popper ([1934] 1959) emphasized the power of disconfirmation in his book *The Logic of Scientific Discovery*, in which he argued that scientific experiments cannot verify a result as true, but they can falsify a hypothesis.

13. DiCaglio (2021) discusses the importance of scale and how attention to it transforms one's worldview. Horton (2021) shows how attention to scale can be the basis for a theory of media.

14. Barad's more recent work on quantum field theory does include such case studies, including possible quantum effects in the phylogenetic development of tadpoles by electrical/quantum fields.

15. At the point when she wrote *Meeting the Universe Halfway* (2007), Barad was apparently unaware of Simondon's work; his name does not appear in the index to *Meeting the Universe Halfway*.

16. "D. Haraway" is cited in the acknowledgements of McFall-Ngai et al. (2013, 3235) for "helpful discussion and comments on the manuscript."

17. The caveat here would be group selection, for example in the sociobiological study of seagulls that shows the birds act to protect the young of their relatives in close approximation to the genetic inheritances they share with the young (Pierotti 1980).

18. "SF" as Haraway uses it, means science fiction, speculative fiction, string figures.

19. For an indication of his concerns, see Stiegler 2010.

20. The critical literature on algorithmic culture is vast, but particularly noteworthy are Amoore 2021 and Hörl 2017.

21. Gessler 2010 usefully distinguishes between historical diachronic emergence, how we got to where we are in the present, and contemporaneous synchronic emergence, the interactions that contribute to the "now" in which we live. While the diachronic elements that created historical emergence may have faded away or no longer be present, the synchronic elements creating our "now" must continue to interact for the world to exist in its present form.

22. The field of artificial life, popular in the 1990s and 2000s, proposed that computer simulations capable of creating reproducing and mutating patterns, such as Conway's Game of Life (n.d.), were actually alive, a speculation that roiled received notions about what life was and whether there could be some essence of life that transcended the medium in which it was instantiated. These discussions have faded away, but they have returned in another guise with artificial intelligence and arguments that sufficiently advanced AIs should be regarded as alive and have rights similar to those of living beings, especially humans. In addition, a recent article by Abramov and colleagues (2021) argued that blockchain-based distributed systems possess what they call "bioanalogous properties," including response to environment, growth and change, replication, and homeostasis.

### Chapter Two

1. This chapter was first published as Hayles 2019, © 2019 by the University of Chicago. 00093-1896/19/4601-0007$10.00. All rights reserved.

2. In two recent articles, Deacon relates three versions of information (those of Ludwig Boltzmann, Claude Shannon, and Charles Darwin) to absential phenomena: Boltzmann's work in thermodynamics gave a statistical definition of entropy, allowing entropy to serve as a background relative to which an absence could be detected (i.e., systems that should have been running down but were not). Shannon's theory of information defined informational content relative to messages that could have been sent but were not. The implicit theory of (genetic) information in Darwin relates the presence of existent species to species that went extinct and so were absent. Deacon claims that such a strategy of negative logic (my term, not his) would provide a firm basis for defining information, which in turn would allow biosemiosis to be put on a firm foundation as well. See Deacon 2007 and 2008.

3. Several of M. C. Escher's black-and-white prints specialize in encouraging just this kind of flip; see for example, *Sky and Water* (June 1938).

4. There is an obvious parallel here with Maturana and Varela's idea of autopoiesis ([1972] 1980), which Deacon cites but does not in my view acknowledge sufficiently.

5. Gerald Edelman, the Nobel Prize–winning neuroscientist, has argued that in humans, recursive dynamics are largely responsible for the emergence of consciousness through what he calls "reentrant signaling" (Edelman and Gally 2013).

6. A notable attempt in this regard is the Danish theoretician Søren Brier's concept of cybersemiosis, which unites four different frameworks into an integrated framework: cosmology/physics; evolution/biology; historical/sociology and linguistics; and personal life history/phenomenology (see Brier 2008; Brier 2011).

7. For an extended analysis of these works, please see Hayles (2022b). For a discussion of Chun's (2021) *Discriminating Data*, see Hayles (2022a).

## Chapter Three

1. As Margulis emphasized, her theories of endosymbiosis and symbiogenesis have predecessors in the nineteenth century. She was quick to give credit to these researchers and personally oversaw the first English translation of Boris Kozo-Polyansky's *Symbiogenesis: A New Principle of Evolution*, published in 2010, a year before Margulis's death.

2. In *Symbiotic Planet*, Margulis (1998, 123) writes that "Lovelock admits . . . that he gave up original notions that Gaia is 'teleological.' He no longer asserts that the living planetary system behaves together to optimize conditions for all its members." Margulis insists, "My Gaia is not vague, quaint notion of a mother Earth that nurtures us. The Gaia hypothesis is science" (123). Her disclaimer notwithstanding, the designation of Earth as Gaia has allowed nonscientists, especially people concerned about environmental issues, readily to identify with Lovelock's ideas. In *The Revenge of Gaia*, he declared himself unrepentant precisely for this reason: "I know that to personalize the Earth System as Gaia, as I have often done and continue to do in this book, irritates the scientifically correct, but I am unrepentant because metaphors are more than ever needed for a widespread comprehension of the true nature of the Earth and an understanding of the lethal dangers that lie ahead" (Lovelock 2006, 147). Moreover, Margulis may have spoken too soon when she asserted that Lovelock had given up on Gaia being teleological. In *Revenge*, the end glossary says of Gaia theory, "The theory sees this system as having a goal—the regulation of surface conditions so as always to be as favourable as possible for contemporary life" (Lovelock 2006, 162). Even if there were no other contraindications, the Great Oxidation Event alone would seem to contradict this view.

3. In *Facing Gaia*, Latour analyzes the dilemma Lovelock faced when trying to describe a system dense with interactions without implying a whole. Calling this an "impossible question," Latour articulates it as "how to obtain effects of connection among agencies without relying on an untenable conception of the whole" (2017, 123). Latour's own solution is to argue that "the interaction between a neighbor who is actively manipulating his neighbors and all the others who are manipulating the first one defines what could be called *waves of action*, which respect no borders and, even more importantly, never respect any fixed scale" (2017, 126). Bruce Clarke cites this passage and argues that the idea there are no boundaries is counterfactual, since each organism has well-defined boundaries delimited by the covering (membrane, skin) that allows it to exist (Clarke 2020, 78–80).

4. We may hear here echoes of the special issue of the *American Naturalist* dedicated to finding an "evo-eco" explanation for this "immunitarian" phenomenon; see Reznick and Bornstein 2013.

5. Maturana (1980) wrote that he had developed the central concept of autopoiesis before he began his collaboration with Francisco Varela. Varela, for his part, believed that autopoietic theory was the collaborative result of his work with Maturana.

6. The importance of the paper may be indicated by the fact that when I was a graduate student at Caltech and took a course in neurophysiology from Roger Sperry in 1968, nearly a decade after its first publication, Sperry assigned this paper for the class—at a time when a decade was an eon in a fast-moving field such as neuroscience.

7. Ananthaswamy (2017) reports on scientific research that speculates information is the fundamental entity that underlies everything else.

8. This article is not the only place that Margulis and Sagan speculate about the future of humans and machines, although their speculation here is one of the most sustained.

9. This leaves aside Stephen Jay Gould's argument about spandrels, that is, phenotypic behaviors or structures that are not selected for directly but coevolve because they are coupled genetically or otherwise with traits that do give selective advantage. See Gould and Lewontin 1979.

## Chapter Four

1. Parts of this chapter were first published as Hayles 2021, © 2021 by the University of Chicago. 00093-1896/21/4704-0007$10.00. All rights reserved.

2. The classic treatment of mimesis is of course Erich Auerbach's magisterial *Mimesis* (1968), which established the tradition of contextualizing mimetic strategies in terms of their relation to contemporary social, economic, and intellectual conditions. It thus reinforces the human-centric focus of mimesis. Notable exceptions to this focus are the mimetic theories of Roger Caillois. In two early essays first published in 1934–35 in the surrealist journal *Minotaure*, Caillois displaced the act of mimesis from humans to insects (Caillois 2003a and 2003b). These essays exhibited his interest in nonhuman forms of creativity, which he saw as freeing creative acts from rational thought and agency. I am grateful to Bill Brown for calling my attention to the Caillois references.

3. This is Fyodor Urnov's accolade in "Human Nature"; he acknowledges it is a "strong word, so I am going to use it carefully" (see Bolt 2019).

4. A visit to the Addgene website, a company that offer CRISPR repair kits, shows a range of $75 to $300, depending on functionality and gene targeted. The price for scientific laboratories in 2016 was just $65 per plasmid. In 2015, Addgene had shipped "some sixty thousand CRISPR-related plasmids to researchers in over eighty different countries" (Jinek et al. 2012, 112).

5. The technosymbiosis here also typically includes computer algorithms, used to design the RNA sequence quickly and efficiently. Here too, nonconscious cognitions are employed to aid in the CRISP constructions, expanding the reach of technosymbiosis beyond living systems into cognitive media.

6. Haraway (2016, 28) makes similar points about what she calls the human respons-ability.

7. Modularity has long been recognized as a crucial aspect of design for artificial life, as well as for computer design (discussed in chapter 2 as "intermediation"). See, for example, Koza 2010.

## Chapter Five

1. Marij van Strien (2014) argues that Laplace was not responding directly to Newtonian mechanics but rather to "general philosophical principles, namely the principle of sufficient reason and the law of continuity," which derived from Leibniz. The point is a useful clarification, but the effect is not to deny the relation of Laplace's Demon to Newtonian mechanics but to make it more indirect than direct.

2. This point has been made by many scholars in the STS community, including Bruno Latour, Simon Schaffer, and Steven Shapin. It was often misunderstood by scientists such as Alan Sokal, who thought the STS scholars were denying the existence of "the law of gravity," for example (Sokal 1996). Rather, the STS writers were pointing to the distinction between underlying regularities and their abstract representations as the "laws of nature."

3. In legal terms, an "entailing law" limits the successors that can inherit a property. By analogy, an "entailing law" in physics and chemistry refers to the underlying regularities that limit (or determine) the states the system can occupy.

4. "Preadaption" occurs when a new biological function develops from an existing adaptation without a significant change in structure. For example, bird feathers may have evolved to keep birds warm but, once there, found new uses in flight and mating displays.

5. Kauffman (1996 and 2000, 22) has independently developed the idea of an "adjacent possible."

6. David Hand at the Jet Propulsion Laboratory / NASA in Pasadena, California, is among the scientists there looking at the possibility of a "second origin of life" in the solar system in the moons that encircle Saturn and Jupiter. For example Titan, a moon of Saturn, has environmental conditions at the "triple point" of methane, similar to Earth's environment at the "triple point" of water, where $H2O$ exists as a gas, liquid, and solid. If life tends to arise in liquid environments, it is possible that life may have arisen in liquid methane lakes.

## Chapter Six

1. This chapter was first published as Hayles (2022c). I am pleased to acknowledge the help of Isaac Mackey, Computer Science Department, University of California, Santa Barbara; and Rita Raley, English Department, University of California, Santa Barbara. Mackey contributed comments and corrections to a manuscript draft. Raley offered comments after my presentation at the "Transformations of Attention" conference, March 4, 2022, Santa Barbara. Also helpful were comments by Katherine Boda, along with remarks by PhD student Galen Cuthbertson at Australian National University. Any errors that remain are of course my sole responsibility.

2. Other large programs using Transformer architectures include Google's LaMDA (Language Model for Dialogue Applications) and BERT (Bidirectional Encoder Representations from Transformers); and Megatron-Turing NLG 530B from Microsoft and Nvidia.

3. Recently, OpenAI released an API (Application Programming Interface) for Instruct-GPT, which was then released on the web as ChatGPT. OpenAI claims InstructGPT (and then ChatGPT) are better than GPT-3 at following English instructions. It trained these models with humans in the loop, making the models "more truthful and less toxic," according to the OpenAI website. (In their concern for misuse of GPT, which they acknowledge can "generate outputs that are untruthful, toxic, or reflect harmful sentiments," they initially had refused, in February 2019, to release the code for a fully trained GPT-2.) Now, using a technique called "reinforcement learning from human feedback (RLHF)," they used prompts submitted by their customers to the API to rank outputs from several models, and then fine-tuned GPT-3 with this data. The resulting InstructGPT models "are much better at following instructions than GPT-3. They also make up facts less often, and show small decreases in toxic output generation" (R. Lowe and Leike 2022). Obviously, LLMs are a moving target. This chapter will focus on GPT-3, the dominant model at the time this chapter was written; the following chapter will focus on the newer ChatGPT, of great interest because it has been released to the general public, and GPT-4.

4. "Human-competitive" is defined by Koza (2010) using a set of eight criteria, considering models that are publishable in peer-reviewed journals or that would be eligible to receive patents. Koza focuses on technical fields such as circuit design and protein folding; for the humanities and specifically for literary studies, "human-competitive" may be taken to indicate a text sufficiently complex in concept, rhetoric, style, etc. to be considered interesting or worthy of critical commentary.

5. There has been a long tradition of electronic poetry, dating back to the mid-twentieth century (Rettberg 2019). With its paratactic techniques, oblique references, and allusive structures, poetry is well suited to a variety of coded forms, from slot algorithms (where the computer randomly generates a word from a list of options) to the compound slices that put parts of different words together, as in Nick Montfort's and Stephanie Strickland's *Sea and Spar Between*, a mash-up of Melville's *Moby Dick* and Emily Dickinson's poetry. There have been far fewer examples of computer-generated prose, and none that match GPT-3's ability to generate syntactically correct and semantically coherent paragraphs.

6. My addition of the phrase "some aspects of" is intended to clarify that computer models are always radically incomplete relative to the complexities of the real world.

7. Danny Hillis, for example, once made a computer from Tinker Toys pieces. For details, see Computer History, n.d.

8. These forms can include a diagram, a verbal description such as Alan Turing (1936) used in his foundational paper on computation, or the vacuum tubes of ENIAC before transistors were invented.

9. The same is true for the laws of physics, as Cartwright (1983) argues. The so-called laws of nature are also abstractions that ignore small deviations due to very weak interactions, quantum fluctuations, and other small causes.

10. "Fudge factors" is a technical term of art for chemistry graduate students at Caltech.

11. An early instantiation of a neural net was the Mark I Perceptron, invented in 1958 by Frank Rosenblatt, a psychologist at Cornell University. The Perceptron was an image classifier but in practice proved unable to recognize many classes of patterns. For a detailed description of the Mark 1 and the surrounding controversy, see Olazaran (1996). By contemporary standards, the Perceptron was a relatively crude device that was primarily electromechanical (rather than electronic). It had an array of 20×20 photocells, randomly connected to the neurons, that could

produce a four-hundred-pixel image. Potentiometers encoded the weights, and updating the weights during successive inputs was performed by electric motors.

12. "Convolution" in mathematics means a function that is shifted over another function to create a third function that is a blend of the two. As applied in convolutional neural networks, it means a filter (vector expressed as a matrix) that is multiplied by an existing vector to yield a third vector/matrix, typically to sharpen distinctions in an image.

13. Tokens separate a section of text into smaller units, typically either words, characters, or subwords. They are the building blocks of natural language that neural networks use to process a natural language text. Typically tokens are about four characters. "Hamburger," according to the OpenAI website, would be broken into "ham" "bur" "ger," while "pear" would be a single token. For OpenAI applications, the combined text length of prompt and response cannot be more than the model's maximum content length, which for most OpenAI models is 2,048 tokens, or about fifteen hundred words (see Open AI website, https://beta.openai.com/docs/introduction/key-concepts, accessed March 25, 2024).

14. Two of the authors, Ailan N. Gomez and Illia Polosukhin, list other affiliations, but a footnote (Vaswani et al. 2017, n.) explains that they did the research while at Google Brain and Google Research, respectively. Rita Raley asked me if "attention" here is no more than a metaphor for or analogy to human attention. There are some functional similarities between the ways in which the attention mechanism works in Transformer architecture and in humans; for instance, research has shown that human attention is a multilevel process, and the same is true of Transformer attention. I therefore consider it a homology, which has more constraints than an analogy and therefore signals a stronger resemblance.

15. Forty-five terabytes equal forty-five thousand gigabytes or forty-five million megabytes. Common Crawl is a nonprofit organization that crawls the web and provides its archives and data sets free to the public. Common Crawl's web archive consists of petabytes of data collected since 2011. WebText is an OpenAI corpus created by scraping web pages, specifically outbound links from Reddit, based on "whether other users found the link interesting, educational, or just funny" (Radford et al. 2018, 3). (Note the relevance of "funny" when GPT-3 tries to tell jokes.)

16. O'Giebly (2021) asks a similar question. She arrives at no certain answer but explores the possibility through her own experiences with writer's block, when she spent several sessions writing under hypnosis, a practice that seems to bypass consciousness and produce writing from unconscious or nonconscious processes.

17. Noam Chomsky, Ian Roberts, and Jeffrey Watumull also compare how ChatGPT uses language with how a child uses language, arguing that the difference mean that ChatGPT generates only the appearance of a competent language user, lacking the contexts characteristic of actual language use (Chomsky et al. 2023). Although I am sympathetic to their argument, my own analysis leads to a different conclusion, that programs like ChatGPT are opening new ways to think about machine language that makes its study worthwhile in its own context.

18. Iconic representations work through morphological resemblance, for example, a woodcut of a priest related through similarity of form to an actual priest. Symbolic representations rely on an arbitrary relation between the sign vehicle and the representamen, mediated through the interpretant.

19. Recognizing the problem of a large language program lacking real-world context, the "Say-Can" project links a language program's output with a robot's embodied actions (Ahn et al. 2022). The linking works by first having the language program generate a suggested action to solve an everyday problem (e.g., spilling a drink on a desk) and then evaluates it via a value

function geared to the robot's repertoire of available actions, which provides a contextualized real-life solution. Although I applaud the project's aim to provide contextual grounding for a language program's discourse, it seems a shame to waste GPT-3's prodigious verbal gifts on such mundane texts as "find a sponge." The program's real talent lies in generating complex narratives, which is why it is suitable for literary analysis.

20. Another indication of this kind of fragility occurs with adversarial attacks on programs intended to classify images. Changing on a few pixels in an image can cause a wildly incongruous classification, for example, mistaking a library for a lion (Su et al. 2019). These are the kinds of mistakes that no human would be likely to make, and they stem from the program's absolute lack of embodied experience of the human lifeworld.

21. Elkins and Chun (2020), teaching at Kenyon College, arranged for GPT-3's responses to be graded as if they were written by an undergraduate. Both of their samples received high grades (A and A−), with laudatory comments by the instructor.

22. This aspect of GPT-3's productions is why Tobias Rees calls its texts "a structural analysis of *and a structuralist production* of language (Rees 2022, 178, emphasis in original)."

23. The human author, K. Allado-McDowell (who has expressed a preference for the pronoun "they"), says that the responses of the machine, indicated by sans serif type, are direct transcripts of GPT-3's responses. I find this implausible, since the kind of fragility of reference exhibited in other GPT-3 responses is nowhere to be seen. In addition, certain repetitions in the text suggest that the human author either wrote these passages or manipulated the prompts until the desired response was achieved.

24. I note the implication here that parrot vocalizations are merely stochastic nonsense, which is unfair to the intelligence of these remarkable birds. As researchers in parrot communication have discovered, parrots in the wild vocalize with each other as a way of fitting into the flock and bonding with others. Alex, a famous African gray parrot, was able not only to use words appropriately but also to generate new combinations appropriate to the situation (Irene Pepperberg, a cognitive ornithologist at Harvard, has made several videos about Alex; see Pepperberg 2009). He liked crackers and bananas and would ask for them by name; when presented with dried banana chips, he appropriately named them "banacker" (DePasquale 2016). Although parrots may not know the human meaning of the words they mimic, they have a contextual understanding of their appropriateness. When the owner enters the room and a parrot asks "How are you?," the utterance likely is not an inquiry about health but a recognition that the phrase is appropriate to the owner's appearance after an absence (see Davis and Roy 2020 for further evidence of the cognitive capabilities of parrots).

25. Proposing that the question of whether neural nets can create meaning is a literary genre in itself, Bender and Koller (2020, 5185) refer to such articles as "BERTology papers," after the large language model from Jacob Devlin's team at Google.

26. This example is from gwern.net, created by Gwern Branwen (2022), testing the capabilities of GPT-3 to respond to various kinds of prompts.

27. "All your base are belong to us" circulates as humor meme on the web; the badly translated phrase is from the opening cutscene of the video game *Zero Wing*.

28. In the spring 2022 issue of *Daedalus*, dedicated to Artificial Intelligence, inside the front and back covers are examples of DALL-E's productions responding to such prompts as "an illustration of a happy turnip walking a dog-robot" and "an artist painting the future of humans cooperating with AI."

## Chapter Seven

1. I was not sure if the proof was correct, so I asked ChatGPT to clarify. The proof works by positing the number N + 1, where N is the product of all the primes. Since it adds 1 to this product, N is not divisible by any of the primes other than 1, which means that it must also be a prime. If N is a prime, then the procedure can be repeated to generate an infinite number of more primes, which shows that there must be no limit to the amount of possible primes. This query illustrates how people (like me) can use ChatGPT, available free on the web, to clear up life's little mysteries. And GPT-4 is more powerful than ChatGPT.

2. Asked to perform this task, GPT-4 writes as the first line "I heard his voice across the crowd," and then gives as the last line, "Crowd the across voice his heard I," which sounds like a very bad translation, by a person with minimal knowledge of English, of the first line (Bubeck et al. 2023, 79).

3. Sacks (2021, 34) describes Jimmie as a "lost soul," a "man without a past (or future), stuck in a constantly changing, meaningless moment."

4. For 2023–24, I am an honorary fellow of the "Future Humans" project initiated by the Berggruen Institute in Los Angeles. As such, I was asked to contribute to the *Multispecies Lexicon* they are developing to make the case not just for animal rights, but for determining how nonhuman forces and actors should be regarded in ethical terms. My contribution is on the rights that could be claimed for AIs. In Canada, Australia, and elsewhere, rivers have been declared to be legal persons, when they are shown to be essential to certain indigenous communities. A similar legal claim could be made for AIs who have capacities crucial to the communities they serve, whether or not they are presumed to have a sense of selfhood.

## Chapter Eight

1. This chapter was first published as Hayles 2023. Republished by permission of Duke University Press.

2. For an analysis of the full context of Benjamin's "aura," see Hansen 2008.

3. For an example of a deep fake, see Jordan Peele's imitation of Barak Obama (Peele 2018).

4. For a "gentle introduction" to machine vision, see Jason Brownlee, n.d.

5. Ishiguro comments in an interview with *Wired* (Knight 2021) that in 2017 he met Jennifer Doudna, who was awarded the 2020 Nobel Prize in Chemistry for her development of the gene-editing tool. When he first heard about her work, he recounts, he thought "it's going to make a meritocracy something quite savage."

6. For a summary of the importance of kinship relations in human societies, see Hasty et al. 2022.

## Chapter Nine

1. The caveat here is the possibility that post-Transformer architectures, if modified in the ways that the Microsoft team suggests, might have the technical capability to generate a sense of self, as explored in chapter 7.

2. Here it should be noted that the robot Adam appears to be conscious, so he is a counterexample to the decoupling of intelligence and consciousness.

3. Dennett (2020) offers a few friendly amendments; and Panagiotaropoulos et al. (2020) are dubious that attention was a better overall umbrella than the global workspace.

4. Typical is the work of Cynthia Breazeal at the MIT Media Lab on emotional robots. Simone (2021, 7) makes the convincing case that AI, since the beginning, has relied on practices that aim to convince users that the software agents are more intelligent than they actually are, a phenomenon that he calls "banal deception." His point is that even when there is no malicious intent, these ordinary, everyday deceptions influence how we think about ourselves as humans, our relations with other humans, and our relations with machines.

5. This is why attempts to create algorithms that can construct stories have failed so spectacularly.

6. I am grateful to Evens for sharing with me a pdf of his manuscript before its publication, while it was in press; it is to this pdf that page numbers in the following citations refer.

7. It should go without saying that many kinds of knowledges resist or elude verbal formulations. Indeed, wordsmiths, whose life endeavors are precisely putting things into words, tend to be as aware of this limitation as anyone, if not more so.

8. Amoore (2013) brilliantly analyzes the shift toward a space of possibility in contrast to the statistics of probability in the context of politics and security. She writes, "The contemporary politics of possibility marks a change in emphasis from the statistical calculation of probability to the algorithmic arraying of possibilities such that they can be acted upon. Redeploying the data analytics first devised for retail and consumer products, the 'association rules' of new security software establish possible projected futures of guilt by association" (23).

9. Hoffman (2023, 343–673) gives an example of this approach.

10. The latest person to voice such concerns is esteemed researcher Geoffrey Hinton, coauthor of a seminal paper on backpropagation (Rumelhart et al. 1986), who, as mentioned in chapter 1, resigned his position at Google so that he could critique advanced AI without compromising Google's research efforts (see Metz 2023).

## Chapter Ten

1. The United Nations might be a candidate, but it has no ceded powers to care for the global environment as a whole. The Paris Accords constitute an attempt in this direction, but as Nordhaus (2021) observes, they have little to no enforcement powers.

2. For a contrary and more nuanced view, see Tunick (2006).

3. Morton (2016) writes about entities too large for humans easily to conceptualize, naming them hyperobjects. In this way he usefully draws attention to issues of scale and offers constructive suggestions for dealing with them.

4. The fact that the credit rating of the United States had been downgraded and the continuing dramas in the US Congress over raising the debt ceiling have dealt significant blows to the status of the dollar as the reserve currency, amid signs it is shifting to Chinese currency instead.

5. Marr (2023) identifies the five most important problems with blockchain as scalability, energy consumption, security, complexity, and interoperability. Golumbia (2016) discusses its connection with cyberlibertarianism and similar ideologies, in which "government" is seen by definition as oppressive and subversive of freedom and liberty, an Alice-in-Wonderland inversion of the functions of democratic government usually perceived as securing individual freedom and liberty.

6. Latour (2017 and 2018) suggests a similar protector of the Earth.

7. The reasoning is similar to strategy employed in recent success of the pro-environment investment company Engine No. 1 to convince other investors to elect Exxon board members sympathetic to developing plans for pivoting to clean energy—not that it is good for the environment but that it makes sense from an economic point of view, as a hedge against a global movement away from fossil fuels.

8. There are two main lines of objections: that such projects may have unintended consequences, and that they will delay or sabotage the hard work of changing people's attitudes and the political will to address the climate crisis. See, for example, Hamilton (2015), who argues that proposals to shield the Earth from the Sun by pumping sulfate particles into the atmosphere are misguided because the real problem is bringing about a shift in attitudes and political systems.

9. CCCB is the Center for Contemporary Culture in Barcelona. Working with the CCCB lab, Timothy Morton has published "performed talks," narrated as if he were the (fictional) Minister for the Future.

10. Given the present Chinese crackdown on Hong Kong and the success of the government's crusade against protestors, readers are likely to regard this prophecy as unintentionally ironic.

# Bibliography

Abramov, Oleg, Kirstin L. Bebell, and Stephen J. Mojzsis. 2021. "Emergent Bioanalogous Properties of Blockchain-Based Distributed Systems." *Origins of Life and Evolution of Biospheres* 51:131–65. https://doi.org/10.1007/s11084-021-09608-1.

Agüera y Arcas, Blaise. 2022. "Artificial Neural Nets Are Making Strides toward Consciousness." *Economist*, September 2. https://www.economist.com/by-invitation/2022/09/02/artificial-neural-networks-are-making-strides-towards-consciousness-according-to-blaise-aguera-y-arcas.

Ahn, Michael, et al. 2022. "Do as I Can, Not as I Say: Grounding Language in Robotic Affordances." April 4. arXiv:2204.01691v1.

Allado-McDowell, K. 2021. *Pharmako-AI*. Peru: Ignota Books.

Amoore, L. A. 2013. *The Politics of Possibility: Risk and Security beyond Probability*. Durham, NC: Duke University Press.

———. 2016. "Cloud Geographies: Computing, Data, Sovereignty." *Progress in Human Geography* 42, no. 1:4–24.

———. 2021. *Cloud Ethics: Algorithms and the Attributes of Ourselves and Others*. Durham, NC: Duke University Press.

Ananthaswamy, Anil. 2017. "Inside Knowledge: Is Information the Only Thing That Exists?" *New Scientist*, March 29. https://www.newscientist.com/article/mg23431191-500-inside-knowledge-is-information-the-only-thing-that-exists/.

Ankit, Utkarsh. 2020. "Transformer Neural Network: Step-by-Step Breakdown of the Beast." *Towards Data Science*, April 24. https://towardsdatascience.com/transformer-neural-network-step-by-step-breakdown-of-the-beast-b3e096dc857f.

Auerbach, Erich. 1968. *Mimesis: The Representation of Reality in Western Literature*. Translated by Willard R. Trask. Princeton, NJ: Princeton University Press.

Bakker, Karen. 2022. *The Sounds of Life: How Digital Technology Is Bringing Us Closer to the Worlds of Animals and Plants*. Princeton, NJ: Princeton University Press. Kindle.

Barad, Karen. 2007. *Meeting the Universe Halfway: Quantum Physics and the Entanglement of Matter and Meaning*. Durham, NC: Duke University Press.

Barsalou, Lawrence W. 1999. "Perceptual Symbol Systems." *Behavioral Brain Science* 22:577–660.

———. 2008. "Grounded Cognition." *Annual Review of Psychology* 59:617–45.

Barthes, Roland. 1975. *S/Z: An Essay*. New York: Hill and Wang.
———. 1977. "The Death of the Author." In *Image, Music, Text*, translated by Stephen Heath, 142–48. New York: Hill and Wang.
Beer, Gillian. 2009. *Darwin's Plots: Evolutionary Narrative in Darwin, George Eliot and Nineteenth-Century Fiction*. 3rd ed. Cambridge: Cambridge University Press.
Bell, Gordon, and Jim Gemmell. 2009. *Total Recall*. New York: Dutton Adult Library.
Bender, Emily M., Timnit Gebru, Angelina McMillan-Major, and Shmargaret Schmitchell. 2021. "On the Dangers of Stochastic Parrots: Can Language Models Be Too Big?" *FaccT '21: Proceedings of the 2021 ACM Conference on Fairness, Accountability, and Transparency*, 610–23.
Bender, Emily, and Alexander Koller. 2020. "Climbing towards NLU: On Meaning, Form, and Understanding in the Age of Data." *Proceedings of the 58th Annual Meeting of the Association for Computational Linguistics*, 5185–87.
Benjamin, Walter. 2006. "The Work of Art in the Age of Its Technological Reproducibility, Second Version." In *Selected Writings*, vol. 3, *1935–1938*, edited by Howard Eiland and Michael Jennings, 101–33. Cambridge, MA: Belknap Press of Harvard University Press.
Bennett, Jane. 2010. *Vibrant Matter: A Political Ecology of Things*. Durham, NC: Duke University Press.
Biden Administration / White House. 2023. "Fact Sheet: Biden-Harris Administration Secures Voluntary Commitments from Leading Artificial Intelligence Companies to Manage Risks Posed by AI." July 21. https://www.whitehouse.gov/briefing-room/statements-releases/2023/07/21/fact-sheet-biden-harris-administration-secures-voluntary-commitments-from-leading-artificial-intelligence-companies-to-manage-the-risks-posed-by-ai/.
Blackiston, Douglas, Sam Kriegman, Josh Bongard, and Michael Levin. 2023. "Biological Robots: Perspectives on an Emerging Interdisciplinary Field." *Soft Robotics* 10, no. 4:674–86. https://doi.org/10.1089/2022.0142.
Blackiston, Douglas J., Emma Lederer, Sam Kriegman, Simon Garnier, Josh Bongard, and Michael Levin. 2021. "A Cellular Platform for the Development of Synthetic Living Machines." *Science Robotics* 6, no. 52. https://www.science.org/doi/10.1126/scirobotics.abf1571.
Blackiston, Douglas J., Khanh Vien, and Michael Levin. 2017. "Serotonergic Stimulation Induces Nerve Growth and Promotes Visual Learning via Posterior Eye Grafts in a Vertebrate Model of Induced Sensory Plasticity." *Nature, npj Regenerative Medicine* 2, no. 8. https://www.nature.com/articles/s41536-017-0012-5.
Bolt, Adam, dir. 2019. "Human Nature." Episode of *Nova*. PBS.
Bonazzi, Mauro. 2023. "Protagoras." In *The Stanford Encyclopedia of Philosophy* (Fall edition), edited by Edward N. Zolta and Uri Noddman. https://plato.stanford.edu/archies/fall 2023/entries/protagoras/.
Bongard, Josh, and Michael Levin. 2021. "A Cellular Platform for the Development of Synthetic Living Machines." *Science Robotics* 6, no. 52. https://www.science.org/doi/10.1126/scirobotics.abf1571.
Boston Dynamics. 2023. "Automate Your Inbound: Automate Loading with Stretch." Webinar presented by Mike Fair and Spencer Brouwer. https://resources.bostondynamics.com/webinar-impact-your-inbound.
Bostrom, Nick. 2016. *Superintelligence: Paths, Dangers, Strategies*. Oxford: Oxford University Press.
Braidotti, Rosi. 2013. *The Posthuman*. London: Polity.
Branwen, Gwern. 2022. "GPT-3: Creative Fiction." February 10. https://www.gwern.net/GPT-3.

Brier, Søren. 2008. *Cybersemiosis: Why Information Is Not Enough!* Toronto: University of Toronto Press.
———. 2011. "Cybersemiosis." In *Glossarium Bitrun*. http://gloarrarium.bitrun.unileon.es/glossary/cybersemiotics.
Britton, Sheilah, and Dan Collins. 2003. *The Eighth Day: The Transgenic Art of Eduardo Kac*. Tempe: Institute for Studies in the Arts, Arizona State University.
Brownlee, Jason. n.d. Machine Learning Mastery. https://machinelearningmastery.com/object-recognition-with-deep-learning/. Accessed March 24, 2021.
Bubeck, Sébastien, Varum Chandrasekaran, Ronen Eldan, Johannes Gehrke, Eric Horvitz, Ece Kamar, Peter Lee, Yin Tat Lee, Yuanzhi Li, Scott Lundberg, Harsha Nori, Hamid Palangi, Marco Tulio Ribeiro, and Yi Zhang. 2023. *Sparks of Artificial General Intelligence: Early Experiments with GPT-4*. April 13. arXiv:2023.2712c5 [cs.CL].
Bush, Vannevar. 1945. "As We May Think." *Atlantic* 176, no. 1 (July): 101–8.
Butler, Judith. 2006. *Gender Trouble: Feminism and the Subversion of Identity*. New York: Routledge.
Caillois, Roger. 2003a. "Mimicry and Legendary Psychasthenia." In *The Edge of Surrealism: A Roger Caillois Reader*, translated by Claudine Frank and Camille Naish, edited by Claudine Frank, 89–103. Durham, NC: Duke University Press.
———. 2003b. "The Praying Mantis: From Biology to Psychoanalysis." In *The Edge of Surrealism: A Roger Caillois Reader*, trans. Claudine Frank and Camille Naish, edited by Claudine Frank, 66–81. Durham, NC: Duke University Press.
Cartwright, Nancy. 1983. *How the Laws of Physics Lie*. Oxford: Oxford University Press.
Chen, Delton B. 2018. "Utility of the Blockchain for Climate Mitigation." *Journal of British Blockchain Association* 1, no. 1:75–80.
Chomsky, Noam, Ian Roberts, and Jeffrey Watumull. 2023. "Noam Chomsky: The False Promise of ChatGPT." *New York Times*, March 8. https://www.nytimes.com/2023/03/08/opinion/noam-chomsky-chatgpt-ai.html.
Chun, Wendy Hui Kyong. 2021. *Discriminating Data: Correlation, Neighborhoods, and the New Politics of Recognition*. Cambridge: Cambridge University Press.
Clarke, Bruce. 2020. *Gaian Systems: Lynn Margulis, Neocybernetics, and the End of the Anthropocene*. Minneapolis: University of Minnesota Press.
———. 2022. *Writing Gaia: The Scientific Correspondence of James Lovelock and Lynn Margulis*. Cambridge: Cambridge University Press.
Cole, David. 2015. "The Chinese Room Argument." In *The Stanford Encyclopedia of Philosophy* (Winter ed.), edited by Edward N. Zalta. https://plato.stanford.edu/archives/win2015/chinese-room.
Columbus, Chris, dir. 1999. *Bicentennial Man*.
Computer History. n.d. "Danny Hillis." https://www.computerhistory.org/collections/catalog/102630799. Accessed March 24, 2024.
Conway's Game of Life. n.d. https://playgameoflife.com/. Accessed March 24, 2024.
Cross Labs AI. 2022. "Michael Levin and Josh Bongard on Xenobots." https://www.youtube.com/watch?v=Aiwq7sAmwg.
Damasio, Antonio. 2000. *The Feeling of What Happens: Body and Emotion in the Making of Consciousness*. Boston: Mariner Books.
———. 2010. *Self Comes to Mind: Constructing the Conscious Brain*. New York: Vintage.

Davis, Nicola, and Eleanor Ainge Roy. 2020. "Study Find Parrots Weigh Up Probabilities to Make Decisions." *Guardian*, March 3. https://www.theguardian.com/science/2020/mar/03/study-finds-parrots-weigh-up-probabilities-to-make-decisions#:~:text=Some%20parrots%20weigh%20up%20probabilities,to%20identify%0colours%20and%20count.

Deacon, Terrence W. 1998. *The Symbolic Species: The Co-evolution of Language and the Brain*. New York: W. W. Norton.

———. 2007. "Shannon-Boltzmann-Darwin: Redefining Information, Part 1." *Cognitive Semiotics* 1, no. S1 (Fall): 123–48.

———. 2008. "Shannon-Boltzmann-Darwin: Redefining Information, Part 2." *Cognitive Semiotics* 2 (September): 169–96.

———. 2011. *Incomplete Nature: How Mind Emerged from Matter*. New York: W. W. Norton.

———. 2023. "In the Shadow of Descartes: The Difference between Computing and Thinking Is Both Physical and Semiotic." Signs of Artificial Life Symposium, March 30, New York University.

DeepMind. 2017. "Learning from Scratch." https://deepmind.com/blog/alphago-zero-learning-scratch/.

Dehaene, Stanislas. 2014. *Consciousness and the Brain: How the Brain Codes Our Thoughts*. New York: Penguin.

Dennett, Daniel C. 1992. *Consciousness Explained*. New York: Back Bay Books.

———. 2020. "On Track to a Standard Model." *Cognitive Neuropsychology* 37, nos. 3–4:173–75.

DePasquale, Christine. 2016. "Conservation Can Be Enhanced by Human Study of Animal Cognition." American Association for the Advancement of Science. October 19. https://www.aaas.org/news/conservation-can-be-enhanced-human-study-animal-cognition.

DiCaglio, Joshua. 2021. *Scale Theory: A Nondisciplinary Inquiry*. Minneapolis: University of Minnesota Press.

Doudna, Jennifer A., and Samuel H. Sternberg. 2017. *A Crack in Creation: Gene Editing and the Unthinkable Power to Control Evolution*. New York: Mariner.

Dreyfus, Hubert. 1972. *What Computers Can't Do: The Limits of Artificial Intelligence*. New York: Harper and Row.

———. 1992. *What Computers Still Can't Do: A Critique of Artificial Reason*. Cambridge, MA: MIT Press.

Dunbar, Robin I. M. 2009. "The Social Brain Hypothesis and Its Implications for Social Evolution." *Annals of Human Biology* 36, no. 5:562–72.

Edelman, Gerald M. 1987. *Neural Darwinism: The Theory of Neuronal Group Selection*. New York: Basic Books.

Edelman, Gerald, and Joseph A. Gally. 2013. "Reentry: A Key Mechanism for the Integration of Brain Function." *Frontiers in Integrative Neuroscience* 7, no. 63. https://www.ncbi.nlm.nih.gov/pmc/articles/PMC3753453/.

Edelman, Gerald, and Giulio Tononi. 2001. *A Universe of Consciousness: How Matter Becomes Imagination*. New York: Basic Books.

Edwards, Chris. 2011. "Memory Points Way to SyNAPSE Chip." *Engineering and Technology*. January 10. https://eandt.theiet.org/content/articles/2011/06/memory-points-way-to-synapse-chip/.

Eldredge, Niles, and Steven Jay Gould. 1972. "Punctuated Equilibria: An Alternative to Phyletic Gradualism." In *Models in Paleobiology*, edited by T. M. Schopf, 82–115. San Francisco: Freeman Cooper.

Elkins, Katherine, and Jon Chun. 2020. "Can GPT-3 Pass a Writer's Turing Test?" *Journal of Cultural Analytics* 5, no. 2:1–16.

Emmerich, Roland, dir. 1996. *Independence Day*.

European Parliament. 2022. "EU AI Act: First Regulation on Artificial Intelligence." December 19. https://www.europarl.europa.eu/news/en/headlines/society/20230601STO93804/eu-ai-act-first-regulation-on-artificial-intelligence.

Evens, Aden. Forthcoming. *Discontents of the Digital*. Minneapolis: University of Minnesota Press.

Fazi, M. Beatrice. 2018. *Contingent Computation: Abstraction, Experience and Indeterminacy in Computational Aesthetics*. London: Rowman and Littlefield.

Felski, Rita. 2008. *Uses of Literature*. Hoboken, NJ. Wiley-Blackwell. Kindle.

Field, Hayden. 2021. "Inside a Hot-Button Research Paper: Dr. Emily M. Bender Talks Large Language Models and the Future of AI Ethics." *Emerging Tech Brew*, February 1. https://www.emergingtechbrew.com/stories/2021/02/01/inside-hotbutton-research-paper-dr-emily-m-bender-talks-large-language-models-future-ai-ethics.

Fields, Chris, and Michael Levin. 2020. "How Do Living Systems Create Meaning?" *Philosophies* 5, no. 36. https://www.mdpi.com/2409-9287/5/4/36.

Finn, Ed. 2017. *What Algorithms Want: Imagination in the Age of Computing*. Cambridge, MA: MIT Press.

Fodor, Jerry. 1983. *Modularity of Mind: An Essay on Faculty Psychology*. Cambridge, MA: MIT Press.

Folbarth, Anja, Jutta Jahnel, Jascha Bareis, Carsten Orwat, and Christian Wadephul. 2020. "Tackling Problems, Harvesting Benefits: A Systematic Review of the Regulatory Debate around AI." KIT Scientific Working Papers 197. September 7. arXiv.cs.arXiv:2209.05468.

Ford, Martin. 2015. *The Rise of the Robots: Technology and the Threat of a Jobless Future*. New York: Basic Books.

Foucault, Michel. (1969) 1973. *The Order of Things: An Archaeology of the Human Sciences*. New York: Vintage.

Foucault, Michel. (1969) 1979. "What Is an Author?" *Screen* 20, no. 1 (Spring): 13–34.

Frankish, Keith. 2016. "Illusionism as a Theory of Consciousness." *Journal of Consciousness Studies* 23:1–39.

Fredkin, Edward. 2003. "Introduction to Digital Philosophy." *International Journal of Theoretical Physics* 42, no. 2 (February): 189–247. https://link.springer.com/article/10.1023/A:1024443232206.

Freedman, Walter J. 1992. "Tutorial in Neurobiology: From Single Neurons to Brain Chaos." *International Journal of Bifurcation and Chaos* 2:451–82.

Gabrys, Jennifer. 2016. *Program Earth: Environmental Sensing Technology and the Making of a Computational Planet*. Minnesota: University of Minnesota Press.

———. 2022. "Smart Forests Atlas." www.jennifergabrys.net/2022/11/smart-forests-atlas.

———. 2023. "Forests That Compute." Zoom lecture given at the Bergen Centre for Electronisk Kunst, Bergen Norway, October 18. https://bek.no/en/jennifer-gabrys-forests-that-compute.

Gantz, Valentino M., and Ethan Bier. 2015. "The Mutagenic Chain Reaction: A Method for Converting Heterozygous to Homozygous Mutations." *Science* 348, no. 6233 (April 25): 442–44. sciencemag.org/content/348/6233/442.full.

Gantz, Valentino M., Nijole Jasinskiene, Olga Tatarenkova, and Aniko Jazeka. 2015. "Highly Efficient Cas9-Mediated Gene Drive for Population Modification of the Malaria Vector

Mosquito *Anopheles stephensi*." *Proceedings of the National Academy of Sciences*, December 8. www.pnas.org/content/112/49/E6736.
Gennaro, Rocco J. 2012. *The Consciousness Paradox: Consciousness, Concepts, and Higher-Order Thoughts*. Cambridge, MA: Cambridge University Press.
Gessler, Nicholas. 2010. "Fostering Creative Emergences in Artificial Cultures." www.academia.edu/70941595/.
Gilbert, S. F, J. Sapp, and A. I. Tauber 2012. "A Symbiotic View of Life: We Have Never Been Individuals." *Quarterly Review of Biology* 87, no. 4 (December): 325–41. https://doi.org/10.1086/668166. PMID 23397797.
Gleick, James. 2008. *Chaos: Making a New Science*. New York: Penguin.
Goldberg, Kenneth. 1995–2004. *The Telegarden*. https://goldberg.berkeley.edu/garden/Ars/.
———, ed. 2000. *The Robot in the Garden: Telerobotics and Telepistemology in the Age of the Internet*. Cambridge: MIT Press.
Goldstein, Jacob. 2020. *Money: The True Story of a Made-Up Thing*. New York: Hachette Books.
Golumbia, David. 2016. *The Politics of Bitcoin: Software as Right-Wing Extremism*. Minneapolis: University of Minnesota Press.
Gough, Michael Paul. 2022. "Information Dark Energy Can Resolve the Hubble Tension and Is Falsifiable by Experiment." *Entropy* 24, no. 3:385. https://doi.org/10,3390/e24030385. https://www.mdpi.com/1099-4300/24/3/385. https://www.mdpi.com/1099-4300/24/3/385.
Gould, Stephen Jay, and Richard C. Lewontin. 1979. "The Spandrels of San Marco and the Panglossian Paradigm: A Critique of the Adaptationist Programme." *Proceedings of the Royal Society*, ser. 21, 205, no. 1161 (September): 581–98.
Graziano, Michael S., Arvid Guterstam, Branden J. Bio, and Andrew I. Wilterson. 2019. "Toward a Standard Model of Consciousness: Reconciling the Attention Schema, Global Workspace, Higher-Order Thought, and Illusionist Theories." *Cognitive Neuropsychology* 37, nos. 3–4:155–72.
Grosz, Elizabeth. 2011. *Becoming Undone: Darwinian Reflections on Life, Politics, and Art*. Durham, NC: Duke University Press.
Guillory, John. 2010. "Genesis of the Media Concept." *Critical Inquiry* 36, no. 2. https://www.journals.uchicago.edu/doi/pdfplus/10.1086/648528.
Haff, Peter. 2014. "Humans and Technology in the Anthropocene: Six Rules." *Anthropocene Review* 1, no. 2:126–36. https://journals-sagepub-com.proxy.lib.duke.edu/doi/full/10.1177/2053019614530575.
Hamilton, Clive. 2015. "Geoengineering Is Not a Solution to Climate Change." *Scientific American*, March 10. https://www.scientificamerican.com/article/geoengineering-is-not-a-solution-to-climate-change/.
Hammond, Andrew, et al. 2016. "A CRISPR-Cas9 Gene Drive System Targeting Female Reproduction in Malaria Mosquito Vector *Anopheles gambia*." *Nature*, January 1. www.nature.com/articles/nbt.3439?foxtrotcallback=true.
Hannah, Eric C., and Intel Corp. 2007. "Cosmic Ray Detectors for Integrated Chips." US Patent, December 18. http://patft.uspto.gov/netacgi/nph-Parser?Sect1=PTO1&Sect2=HITOFF&d=PALL&p=1&u=%2Fnetahtml%2FPTO%2Fsrchnum.htm&r=1&f=G&l=50&s1=7,309,866.PN.&OS=PN/7,309,866& RS=PN/7,309,866.
Hansen, Miriam Bratu. 2008. "Benjamin's Aura." *Critical Inquiry* 34 (Winter): 336–75.
Harari, Yuval Noah. 2015. *Homo Deus: A Brief History of Tomorrow*. New York: Vintage.
Haraway, Donna J. 1988. "Situated Knowledges: The Science Question in Feminism and the Privilege of Partial Perspective." *Feminist Studies* 14, no. 3 (Autumn): 575–99.

———. 2012. "Awash in Urine: DES and Premarin in Multispecies Respons-ability." Women's Studies Quarterly, vol. 20, no. 1/2 (Spring/Summer): 301–16.

———. 2016. *Staying with the Trouble: Making Kin in the Chthulucene*. Durham, NC: Duke University Press.

Harding, Sandra. 2005. "Rethinking Standpoint Epistemology: What Is 'Strong Objectivity'?" In *Feminist Theory: A Philosophical Anthology*, edited by Ann E. Cudd and Robin O. Andreasen, 218–36. Oxford: Blackwell.

Hasty, Jennifer, David. G. Lewis, and Marjorie M. Snipes, senior eds. 2022. *Introduction to Anthropology*. OpenStax.org. https://openstax.org/books/introduction-anthropology/pages/11-1-what-is-kinship#:~:text=Through%20kinship%20systems%2C%20humans%20create,and%20how%20closely%2C%20vary%20widely.

Hayles, N. Katherine. 1990. *Chaos Bound: Orderly Disorder in Literature and Science*. Ithaca, NY: Cornell University Press.

———. 1993. "Constrained Constructivism: Locating Scientific Inquiry in the Theater of Representation." In *Realism and Representation: Essays on the Problem of Realism in Relation to Science, Literature, and Culture*, edited by George Levine, 27–43. Madison: University of Wisconsin Press.

———. 1995. "Making the Cut: The Interplay of Narrative and System, or What System Theory Can't See." *Cultural Critique*, no. 30 (Spring): 71–100.

———. 1999. *How We Became Posthuman: Virtual Bodies in Cybernetics, Literature, and Informatics*. Chicago: University of Chicago Press.

———. 2005. *My Mother Was a Computer: Digital Subjects and Literary Texts*. Chicago: University of Chicago Press.

———. 2006. "Unfinished Work: From Cyborg to Cognisphere." *Theory, Culture and Society* 23:159–66.

———. 2007a. "Hyper and Deep Attention: The Generational Divide in Cognitive Modes." *Profession*, 187–99.

———. 2007b. "Intermediation: In Pursuit of a Vision." *New Literary History* 38, no. 1:99–125.

———. 2017. *Unthought: The Power of the Cognitive Nonconscious*. Chicago: University of Chicago Press.

———. 2015. *How We Think: Digital Media and Contemporary Technogenesis*. Chicago: University of Chicago Press.

———. 2018. "Literary Texts as Cognitive Assemblages: The Case of Electronic Literature." *Electronic Book Review*. http://electronicbookreview.com/essay/literary-texts-as-cognitive-assemblages-the-case-of-electronic-literature/.

———. 2019. "Can Computers Create Meanings? A Cyber/Bio/Semiotic Perspective." *Critical Inquiry* 46, no. 1:32–55.

———. 2021. "Microbiomimesis: Bacteria, Our Cognitive Collaborators." *Critical Inquiry* 47, no. 4: (Summer): 777–87. https://doi.org/10.1086/714511.

———. 2022a. "Approximating Algorithms: From *Discriminating Data* to Talking with an AI." *History and Theory* 61, no. 4 (December): 162–65. https://onlinelibrary.wiley.com/toc/14682303/2022/61/4.

———. 2022b. "Ethics for Cognitive Assemblages: Who's in Charge Here?" In *The Palgrave Handbook of Critical Posthumanism*, edited by Stefan Herbrechter, Ivan Callus, Manuela Rossini, Marija Grach, Megen de Bruin-Molé, and Christopher John Müller. London: Palgrave

Macmillan. https://doi.org/10.1007/978-3-030-42681-1_11-1. https://link.springer.com/referenceworkentry/10.1007/978-3-030-42681-1_11-1.

———. 2022c. "Inside the Mind of an AI: Materiality and the Crisis of Representation." *New Literary History* 54, no. 1:635–66. https://doi.org/10.1353/nlh.2022.a898324.

———. 2023. "Subversion of the Human Aura: A Crisis in Representation." *American Literature* 95, no. 2 (June): 255–79.

Hazen, Robert M. 2010. "Evolution of Minerals." *Scientific American* 302, no. 3 (March 1). https://www.scientificamerican.com/article/evolution-of-minerals/.

Hazen, Robert M., Dominic Papineau, Wouter Bleeker, Robert T. Downs, John M. Ferry, Timothy J. McCoy, Dimitri A. Sverjensky, and Hexiong Yang. 2008. "Mineral Evolution." *American Mineralogist* 93:1693–720.

Heldén, Johannes, and Håkan Jonson. 2013. "Evolution." https://www.johanneshelden.com/evolution/.

———. 2014. *Evolution*. Stockholm: OEI.

Higgins, Jacob S., Lawson T. Lloyd, Sara H. Sohail, and Gregory S. Engel. 2021. "Photosynthesis Tunes Quantum-Mechanical Mixing of Electronic and Vibrational States to Steer Excitation Energy Transfer." *PNAS* 118, no. 11 (March 9). https://doi.org/10.1073/pnas.202824011. https://www.pnas.org/doi/10.1073/pnas.2018240118.

Hoffman, Reid. 2023. *Impromptu: Amplifying Our Humanity through AI*. Anacortes, WA: Dallapedia. Kindle.

Hoffmeyer, Jesper. 1997. *Signs of Meaning in the Universe*. Bloomington: Indiana University Press.

———. 2009. *Biosemiotics: An Examination into the Signs of Life and the Life of Signs*. Rpt. ed. Scranton, PA: University of Scranton Press.

Hörl, E., ed. 2017. *General Ecology: The New Ecological Paradigm*. London: Bloomsbury Academic.

Horton, Zachary. 2021. *The Cosmic Zoom: Scale, Knowledge and Mediation*. Chicago: University of Chicago Press.

Ishiguro, Kazuo. 2006. *Never Let Me Go*. New York: Vintage.

———. 2021. *Klara and the Sun*. New York: Alfred A. Knopf.

Jinek, Martin, Krzysztof Chylinski, Ines Fonfara, Michael Hauer, Jennifer A. Doudna, and Emmanuelle Charpentier. 2012. "A Programmable Dual-RNA-Guided DNA Endonuclease in Adaptive Bacterial Immunity." *Science*, August 17. science.sciencemag.org/content/337/6096/816.full.

Kac, Eduardo. 2001. *The Eighth Day, a Transgenic Artwork*. https://www.ekac.org/8thday.html.

Kauffman, Stuart. 1996. *At Home in the Universe*. Rpt. ed. Oxford: Oxford University Press.

———. 2000. *Investigations*. Oxford: Oxford University Press.

Kirschenbaum, Matthew. 2021. "Spec Acts: Reading Form in Recurrent Neural Networks." *ELH* 83, no. 2:361.

Knight, Will. 2021. "*Klara and the Sun* Imagines a Social Schism Driven by AI." *Wired Magazine*, March 8. https://www.wired.com/story/kazuo-ishiguro-interview.

Kotler, Shlomi, Gabriel A. Peterson, Ezad Shojaee, Florent Lecocq, Katarina Cicak, Alex Kwiatkowski, Shawn Geller, Scott Glancy, Emanuel Knill, Raymond W. Simmonds, José Aumentado, and John D. Teufel. 2021. "Direct Observation of Deterministic Macroscopic Entanglement." Science 372, no. 6542 (May 7): 622–25. https://doi.org/10.1126/science.abf2998.

Koza, John R. 2010. "Human-Competitive Results Produced by Genetic Programming." *Genetic Programming and Evolvable Machines* 11 (May 25): 251–84. https://doi.org/10.1007/s10710-010-9112-3.

Kozo-Polyansky, Boris M. 2010. *Symbiogenesis: A New Principle of Evolution*, annotated ed. Edited by Lynn Margulis. Translated by Victor Fet. Cambridge, MA: Harvard University Press.

Kriegman, Sam, Douglas Blackiston, Michael Levin, and Josh Bongard. 2021. "Kinematic Self-Replication in Reconfigurable Organisms." *PNAS* 118, no. 49. https://doi.org/10.1073/pnas.2112672118.

Lanphier, Edward, et al. 2015. "Don't Edit the Human Germ Line." *Nature*, March 12. www.nature.com/news/don-t-edit-the-human-germ-line-1.17111.

Laplace, Pierre Simon. (1814) 1951. *A Philosophical Essay on Probabilities*. Translated from the original French 6th ed. by F. W. Truscott and F. L. Emory. Mineola, NY: Dover.

Latour, Bruno. 1993. *We Have Never Been Modern*. Translated by Catherine Porter. Cambridge, MA: Harvard University Press.

———. 2002. "Morality and Technology: The End of the Means." Translated by Couze Venn. *Theory, Culture and Society* 19, nos. 5–6:247–56.

———. 2017. *Facing Gaia: Eight Lectures on the New Climactic Regime*. Translated by Catherine Porter. Cambridge: Polity.

———. 2018. *Down to Earth: Politics in the New Climactic Regime*. Cambridge: Polity, 2018.

Latour, Bruno, and Timothy M. Lenton. 2019. "Extending the Domain of Freedom, or Why Gaia Is So Hard to Understand." *Critical Inquiry* 45, no. 3 (March). https://doi.org/10.1086/702611.

LeCain, Timothy James. 2015. "Against the Anthropocene: A Neo-materialist Perspective." *International Journal of History, Culture and Modernity* 3, no. 1 (April 23): 1–28.

Lettvin, J. Y., Humberto R. Maturana, Warren S. McCulloch, and Water H. Pitts. 1959. "What the Frog's Eye Tells the Frog's Brain." *Proceedings of the Institute for Radio Engineers* 47, no. 11 (November): 1940–51.

Levin, Michael, and Daniel Dennett. 2020. "Cognition All the Way Down: How to Understand Cells, Tissues and Organisms as Agents with Agendas." *Aeon*, October 14. https://aeon.co/essays/how-to-understand-cells-tissues-and-organisms-as-agents-with-agendas.

Levin, Michael, and Rafael Yuste. 2022. "How Evolution 'Hacked' Its Way to Intelligence from the Bottom Up." *Aeon Magazine*, March 8. https://aeon.co/essays/how-evolution-hacked-its-way-to-intelligence-from-the-bottom-up.

Levine, George, ed. 1993. *Realism and Representation: Essays on the Problem of Realism in Relation to Science, Literature, and Culture*. Madison: University of Wisconsin Press.

Liang, Puping, et al. 2015. "CRISPR/Cas9-Mediated Gene Editing in Human Tripronuclear Zygotes." *Protein and Cell* 6 (May). link.springer.com/article/10.1007/s13238-015-0153-5.

Libet, Benjamin. 1993. *Neurophysiology of Consciousness: Selected Papers and New Essays*. Basel, Switzerland: Birkhäuser.

Libet, Benjamin, and Stephen M. Kosslyn. 2005. *Mind Time: The Temporal Factor in Consciousness*. Cambridge, MA: Harvard University Press.

Licklider, J. C. R. 1960. "Man-Computer Symbiosis." *IRE Transactions on Human Factors in Electronics* HFE-1 (March): 4–11.

Liévane, Daniel. 2021. "Remarkable Progress Has Been Made in Understanding the Folding of Proteins." *Economist*, July 31. https://www.economist.com/leaders/2021/07/31/remarkable-progress-has-been-made-in-understanding-the-folding-of-proteins?utm_medium=cpc.adword.pd&utm_source=google&ppccampaignID=17210591673&ppcadID

=&utm_campaign=a.22brand_pmax&utm_content=conversion.direct-response.anonymous&gad_source=1&gclid=Cj0KCQiA1rSsBhDHARIsANB4EJZdhHCwE-IT1b8DVRDABOJRDzGDORGTZuMvym5JnS0IcR5gkwRYpE8aAgyDEALw_wcB&gclsrc=aw.ds.

Longo, Giuseppe, and Maël Montévil. 2011. "The Inert vs. the Living State of Matter: Extended Criticality, Time Geometry, Anti-entropy; An Overview." *Frontiers in Physiology* vol. 3, article 39 (1–7). https://hal.archives-ouvertes.fr/hal-01192906.

Longo, Giuseppe, Maël Montévil, and Stuart Kauffman. 2012. "No Entailing Laws, but Enablement in the Evolution of the Biosphere." *Proceedings of the 14th Annual Conference on Genetic and Evolutionary Computation*, January 10 (Philadelphia: American Computing Machinery). arXiv:1201.2069v1.

Lovecraft, H. P. 1971. *At the Mountains of Madness, and Other Tales of Terror*. New York: Ballantine Books. https://www.hplovecraft.com/writings/texts/fiction/mm.aspx.

Lovelock, James. 1988. *The Ages of Gaia: A Biography of Our Living Earth*. New York: W. W. Norton.

———. 2006. *The Revenge of Gaia: Earth's Climate in Crisis and the Fate of Humanity*. New York: Basic Books.

———. 2009. *The Vanishing Face of Gaia: A Final Warning*. New York: Basic Books.

———. 2016. *Gaia: A New Look at Life on Earth*. Rpt. ed. Oxford: Oxford University Press.

———. 2020. *Novacene: The Coming Age of Hyperintelligence*. Cambridge, MA: MIT Press.

Lowe, Lisa. 2015. *Intimacies of Four Continents*. Durham, NC: Duke University Press.

Lowe, Ryan, and Jan Leike. 2022. "Aligning Language Models to Follow Instructions." *OpenAI.com*, January 27. openai.com/blog/instruction-following/.

Luhmann, Niklas 1995. *Social Systems*. Translated by John Bednarz Jr. with Dirk Baecker. Stanford, CA: Stanford University Press.

"Lynn Margulis: The Woman Who Defied Darwin." 2021. *Attic*, January 8. https://www.theattic.space/home-page-blogs/2021/1/8/the-woman-who-defied-darwin#:~:text=%E2%80%9CScience's%20Unruly%20Earth%20Mother%E2%80%9D%20remained,Margulis%20reveled%20in%20her%20role.

Manning, Christopher D. 2022. "Human Language Understanding and Reasoning." *Daedalus* 151, no. 3:127–38.

Margulis, Lynn. 1967. "On the Origin of Mitosing Cells." *Journal of Theoretical Biology* 14, no. 3:255–74. https://doi.org/10.1016/0022-5193(67)90079-3.

———. 1997. "Big Trouble in Biology: Physiological Autopoeises versus Mechanistic Neo-Darwinism." In *Slanted Truths: Essay on Saia, Symbiosis, and Evolution*, by Dorion Sagan and Lynn Margulis, 265–82. New York: Springer-Verlag.

———. 1998. *Symbiotic Planet: A New Look at Evolution*. New York: Basic Books.

Margulis, Lynn, and Dorion Sagan. 1995. *What Is Life?* New York: Simon and Schuster.

———. 1997. *Microcosm: Four Billion Years of Microbial Evolution*. Berkeley: University of California Press.

———. 2002. *Acquiring Genomes: A Theory of the Origins of Species*. New York: Basic Books.

Marr, Bernard. 2023. "The Five Biggest Problems with Blockchain Technology That Everyone Must Know." *Forbes*, April 23. https://www.forbes.com/sites/bernardmarr/2023/04/14/the-5-biggest-problems-with-blockchain-technology-everyone-must-know-about/?sh=765d5f4a55d2.

Maturana, Humberto R. 1980. "Autopoiesis: Reproduction, Heredity, and Evolution." In *Autopoiesis, Dissipative Structures and Spontaneous Social Orders*, edited by M. Zeleny, 45–79. AAAS Selected Symposium 55 (AAAS National Annual Meeting, Houston TX, January 3–8, 1979). Boulder, CO: Westview. https://cepa.info.55.

Maturana, Humberto R., and Franciso J. Varela. (1972) 1980. *Autopoiesis and Cognition: The Realization of the Living*. Dordrecht: D. Reidel.

———. 1987. *The Tree of Knowledge: The Biological Roots of Human Understanding*. Boston: New Science Library.

Mayr, Ernst. 2002. "Foreword." In *Acquiring Genomes: A Theory of the Origin of Species*, by Lynn Margulis and Dorion Sagan, xi–xvi. New York: Basic Books.

Macpherson, C. B. (1962) 2011. *The Political Theory of Possessive Individualism: Hobbes to Locke*. Rpt. ed. Oxford: Oxford University Press.

McEwan, Ian. 2019. *Machines Like Me*. New York: Doubleday.

McFall-Ngai, M, M. G. Hadfield, T. C. Bosch, H. V. Carey, T. Domazet-Lošo, A. E. Douglas, N. Dubilier, G. Eberl, T. Fukami, S. F. Gilbert, U. Hentschel, N. King, S. Kjelleberg, A. H. Knoll, N. Kremer, S. K. Mazmanian, J. L. Metcalf, K. Nealson, N. E. Pierce, J. F. Rawls, A. Reid, E. G. Ruby, M. Rumpho, J. G. Sanders, D. Tautz, and J. J. Wernegreen. 2013. "Animals in a Bacterial World, a New Imperative for the Life Sciences." *Proceedings of the National Academy of Sciences, U.S.A.* 110, no. 9 (February 26): 3229–36. https://doi.org/10.1073/pnas.1218525110.

Metz, Cade. 2023. "'The Godfather of A.I.' Leaves Google and Warns of Danger Ahead." *New York Times*, May 1. https://www.nytimes.com/2023/05/01/technology/ai-google-chatbot-engineer-quits-hinton.html.

Mill, John Stuart. 1859. *On Liberty*. London: John W. Parker and Son.

"Mind." 2010. *Oxford Languages*. Google. https://www.google.com/search?q=definition+mind&oq=definition+mind&gs_lcrp=EgZjaHJvbWUyBggAEEUYOTIHCAEQABiABDIHCAIQABiABDIHCAMQABiABDIHCAQQABiABDIHCAUQABiABDIHCAYQABiABDIGCAcQRRg80gEINTgxN2owajeoAgCwAgA&sourceid=chrome&ie=UTF-8.

Modha, Dharmendra S. 2015. "Introducing a Brain-Inspired Computer." https://www.research.ibm.com/articles/brain-chip.shtml.

Montfort, Nick, and Stephanie Strickland. 2010. *Sea and Spar Between*. https://nickm.com/montfort_strickland/sea_and_spar_between/.

Moore, Jason. 2015. *Capitalism in the Web of Life: Ecology and the Accumulations of Capital*. New York: Verso. Kindle.

Morowitz, Harold. 2002. *The Emergence of Everything: How the World Became Complex*. Oxford: Oxford University Press.

Morton, Timothy. 2016. *Hyperobjects: Philosophy and Ecology after the End of the World*. Minneapolis: University of Minnesota Press.

Nagel, Thomas. 1974. "What Is It Like to Be a Bat?" *Philosophical Review* 83, no. 4 (October): 435–50.

National Counterproliferation Center. 2016. "Biological Warfare." *Worldwide Threat Assessment*, report to Senate Armed Serviced Committee, Office of the Director of National Intelligence. www.dni.gov/index.php/ncpc-features/1548-features-2.

National Intelligence Council. 2021. "Unclassified Summary of the Assessment of COVID-19 Origins." https://www.dni.gov/files/ODNI/documents/assessments/Unclassified-Summary-of-Assessment-on-COVID-19-Origins.pdf.

Newitz, Annalee. 2017. *Autonomous*. New York: Tor.

Nordhaus, William A. 2021. "Steps toward International Climate Governance." *Bulletin of the American Academy of Arts and Sciences*, Spring, 18–23.

Nuñez, Rafael, and Walter J. Freeman, eds. 2000. *Reclaiming Cognition: The Primacy of Action, Intention and Emotion*. Exeter, UK: Imprint Academic.

O'Giebly, Meghan. 2021. "Babel: Could a Machine Have an Unconscious?" n +1, no. 40. https://www.nplusonemag.com/issue-40/essays/babel-4/.

Okojie, Irenosen. 2021. "Introduction." In *Pharmako-AI*, by K. Allado-McDowell, i–vii. Peru: Ignota Books.

Olazaran, Mikel. 1996. "A Sociological Study of the Official History of the Perceptrons Controversy." *Social Studies of Science* 26, no. 3:611–59.

OpenAI. 2023. *GPT-4 Technical Report*. arXiv:2303.09774. https://arxiv.org/abs/2303.08774.

Panagiotaropoulos, Theofanis, Liping Wang, and Stanislas Dehaene. 2020. "Hierarchical Architecture of Conscious Processing and Subjective Experience." *Cognitive Neuropsychology* 37, no. 3–4:180–83.

Peele, Jordan. 2018. "Jordan Peele Uses AI, Barack Obama in Fake News." *Good Morning America*. https://www.youtube.com/watch?v=bE1KWpoX9Hk.

Peirce, C. S. 1958. *Collected Papers*. Vols. 7–8. Edited by Arthur W. Burks. Cambridge, MA: Harvard University Press.

———. 1998. *The Essential Peirce: Selected Philosophical Writings*. Vol. 1. Edited by Nathan Houser and Christian Kloesel. Bloomington: Indiana University Press.

Pepperberg, Irene. 2009. *Alex and Me: How a Scientist and a Parrot Discovered a Hidden World of Animal Intelligence—and Formed a Deep Bond in the Process*. Illustrated ed. New York: Harper Perennial.

Petzold, Charles. 2000. *Code: The Hidden Language of Computer Hardware and Software*. Sebastopol, CA: Microsoft Press.

Pierotti, R. 1980. "Spite and Altruism in Gulls." *American Naturalist* 115, no. 282.29. https://kuscholarworks.ku.edu/bitstream/handle/1808/17547/Pierotti_AN_115%282%29_290.pdf?sequence=1&isAllowed=y.

Plaue, Ethan, and William Morgan, with GPT-3. 2021. "Secrets and Machines: A Conversation with GPT-3." *e-flux*, no. 123 (December). https://www.e-flux.com/journal/123/437472/secrets-and-machines-a-conversation-with-gpt-3/.

Popper, Karl. (1934) 1959. *The Logic of Scientific Discovery*. Translation of *Logik der Forschung*. Abingdon-on-Thames: Routledge.

Powers, Richard. 2019. *The Overstory*. New York: W. W. Norton.

Prigogine, Ilya, and Isabelle Stengers. 1984. *Order Out of Chaos*. New York: Bantam.

Qureshi, Faisal Z., Demetri Terzopoulos, and Piotr Jasiobedzki. 2005. "Cognitive Vision for Autonomous Satellite Rendezvous and Docking." *Proceedings of the 9th IAPR (Journal of Document Analysis and Recognition) Conference on Machine Vision Applications*. http://web.cs.ucla.edu/~dt/papers/mva05/mva05.pdf.

Radford, Alex, Karthik Narasimhan, Tim Salimans, and Ilya Sutkever. 2018. "Improving Language Understanding by Generative Pre-Training." https://s3-us-west-2.amazonaws.com/openai-assets/research-covers/language-unsupervised/language_understanding_paper.pdf.

Raley, Rita. 2022. "Attention." Transformations of Attention Conference at the University of Santa Barbara, March 4. Santa Barbara, CA.

Rees, Tobias. 2022. "Non-human Words: On GPT-3 as a Philosophical Laboratory." *Daedalus* 151, no. 2:165–82.

Rettberg, Scott. 2019. *Electronic Literature*. Cambridge: Polity.

Reznick, David N., and editor Judith L. Bornstein. 2013. "A Critical Look at Reciprocity in Ecology and Evolution: An Introduction to the Symposium." *American Naturalist* 181, no. S1: A1–8. https://doi.org/10.1086/670030.

Rizzolatti, Giacomo, and Laila Craighero. 2004. "The Mirror-Neuron System." *Annual Review of Neuroscience* 27:169–92. https://doi.org/10.1146annurev.neuro.27.070203.144230.

Robinson, Kim Stanley. 1990. *Red Mars*. New York: Bantam Spectra.

———. 1995. *Green Mars*. New York: Bantam Spectra.

———. 1997. *Blue Mars*. New York: Bantam Spectra.

———. 2020. *Ministry for the Future*. London: Orbit Books.

Rosen, Robert. 1991. *Life Itself: A Comprehensive Inquiry into the Nature, Origin and Fabrication of Life*. New York: Columbia University Press.

Rosenthal, David. 1991. *The Nature of Mind*. New York: Oxford University Press.

———. 2005. *Consciousness and Mind*. New York: Oxford University Press.

Rouvroy, Antoinette. 2012. "The End(s) of Critique: Data-Behaviourism vs. Due Process." In *Privacy, Due Process and the Computational Turn: Philosophers of Law Meet Philosophers of Technology*, edited by M. Hildebrant and E. De Vries 165–88. New York: Routledge.

Rumelhart, David E., Geoffrey E. Hinton, and Ronald J. Williams. 1986. "Learning Representations by Back-Propagating Errors." *Nature* 323 (October): 533–36.

Sacks, Oliver. 2021. *The Man Who Mistook His Wife for a Hat*. New York: Vintage.

Sagan, Dorion, and Lynn Margulis. 1987. "Gaia and the Evolution of Machines." *Whole Earth Review* 65 (Summer): 15–21.

———. 1997. "Futures." In *Slanted Truths: Essays on Gaia, Symbiosis, and Evolution*, by Lynn Margulis and Dorion Sagan, 235–46. Berlin: Springer-Verlag.

Savulescu, Julian, et al. 2015. "The Moral Imperative to Continue Gene Editing Research on Human Embryos." *Protein and Cell* 6 (July). link.springer.com/article/10.1007/s13238-015-0184-y.

Scott, Kevin. 2022. "I Do Not Think It Means What You Think It Means: Artificial Intelligence, Cognitive Work and Scale." *Daedalus* 151, no. 2:75–84.

Searle, John. 1984. *Minds, Brains, and Science*. Cambridge, MA: Harvard University Press.

———. 1999. "The Chinese Room." In *The MIT Encyclopedia of Cognitive Science*, edited by R. A. Wilson and F. Keil. Cambridge, MA: MIT Press. https://web.mit.edu/morrishalle/pubworks/papers/1999_Halle_MIT_Encyclopedia_Cognitive_Sciences-paper.pdf.

Shapin, Steven, and Simon Schaffer. 1985. *Leviathan and the Air-Pump: Hobbes, Boyle, and the Experimental Life*. Princeton, NJ: Princeton University Press.

Shaw, Helen. 2021. "In *Klara and the Sun*, Artificial Intelligence Meets Real Sacrifice." *Vulture.com*, March 6. https://www.vulture.com/article/review-klara-and-the-sun-kazuo-ishiguro.html.

Siddarth, Divya, Daron Acemoglu, Danielle Allen, Kate Crawford, James Evans, Michael Jordan, and E. Glen Weyl. 2021. "How AI Fails Us." Justice, Health, and Democracy Impact Initiative and Carr Center for Human Rights Policy. Harvard University, Cambridge, MA. https://ethics.harvard.edu/files/center-for-ethics/files/howai_fails_us_2.pdf.

Simondon, Gilbert. 2017. *On the Mode of Existence of Technical Objects*. Translated by Cecile Malaspina and John Rogove. Minneapolis: University of Minnesota Press.

———. 2020. *Individuation in Light of Notions of Form and Information*. Translated by Taylor Atkins. Minneapolis: University of Minnesota Press.

Simone, Natal. 2021. *Deceitful Media: Artificial Intelligence and Social Life after the Turing Test*. New York: Oxford University Press.

Simonite, Tom. 2017. "Google's Learning Software Learns to Write Learning Software." *Wired Magazine*, October 13. https://www.wired.com/story/googles-learning-software-learns-to-write-learning-software/onAutoML.

Slater, Avery. 2020. "Automating Origination: Perspectives from the Humanities." In *The Oxford Handbook of Ethics in AI*, edited by Markus D. Dubber, Frank Pasquale, and Sunit Das, 521–38. Oxford: Oxford University Press.

Snow, Michael. (1975) 2020. *Cover to Cover*. Facsimile ed. New York: Primary Information.

Sokal, Alan D. 1996. "Transgressing the Boundaries: Towards a Transformative Hermeneutics of Quantum Gravity." *Social Text* 46/47 (Spring Summer): 217–52.

Spivak, Gayatri C. 1988. "Can the Subaltern Speak?" *Die Philosophin* 14, no. 27:42–58.

StackOverflow.com. n.d. "Cosmic Rays: What Is the Probability They Will Affect a Program?" https://stackoverflow.com/questions/2580933/cosmic-rays-what-is-the-probability-they-will-affect-a-program. Accessed October 24, 2018.

"Starter Motors." n.d. AutoElectro. https://www.autoelectro.co.uk/starter-motors. Accessed March 26, 2023.

Stiegler, Bernard. 1998. *Technics and Time, 1: The Fault of Epimetheus*. Translated by Richard Beardsworth and George Collins. Stanford, CA: Stanford University Press.

———. 2010. *Taking Care of Youth and the Generations*. Stanford, CA: Stanford University Press.

Su, Jiawei, Danilo Vasconcellos Vargas, and Sakurai Kouichi. 2019. "One Pixel Attack for Fooling Deep Neural Networks." October 17. arXiv:1710.08864.

Traweek, Sharon. 1992. *Beamtimes and Lifetimes: The World of High Energy Physics*. Cambridge, MA: Harvard University Press.

Tunick, Mark. 2006. "Tolerant Imperialism: J. S. Mill's Defense of British Rule in India." *Review of Politics* 68, no. 4 (Fall): 586–611.

Turing, Alan. 1936. "On Computable Numbers, with an Application to the Entscheidungs Problem." *Proceedings of the London Mathematical Society*, 42, no. 1. https://www.cs.virginia.edu/~robins/Turing_Paper_1936.pdf.

Ullrich, George. 1997. "Threat Posed by Electromagnetic Pulse (EMP) to Military Systems and Civil Infrastructure." Committee on National Security, House of Representatives, July 16. https://commdocs.house.gov/committees/security/has197010.000/has197010_1.htm.

*The Universe Is Hostile to Computers*. 2021. Veritasium, August 31. https://www.youtube.com/watch?v=AaZ_RSt0KP8.

Unsworth, John. 1997. "Creating Digital Resources: The Work of Many Hands." Talk delivered at the Digital Resources for the Humanities, September 14, Oxford, England. https://johnunsworth.name/drh97.html.

van Strien, Marij. 2014. "On the Origins and Foundations of Laplacian Determinism." *Studies in History and Philosophy of Science*, part A, 45:24–31. https://doi.org/10.1016/shpsa.20131.2003.

Varela, Francisco J. 1981. "Describing the Logic of the Living: The Adequacy and Limitations of the Idea of Autopoiesis." In *Autopoiesis: A Theory of Living Organization*, edited by Milan Zeleny, 35–79. North Holland Series in General Systems Research 3. New York: North Holland.

Varela, Francisco J., Evan Thompson, and Eleanor Rosch. 1991. *The Embodied Mind: Cognitive Science and Human Experience*. Cambridge, MA: MIT Press.

Vaswani, Ashish, et al. 2017. "Attention Is All You Need." Proceedings of the Thirty-First Conference on Neural Information Processing Systems (NIPS), Long Beach, CA, December. August 2. ArXiv:1706.03762:1.

Verbeek, Peter-Paul. 2011. *Moralizing Technology: Understanding and Designing the Morality of Things*. Chicago: University of Chicago Press.

Villeneuve, Denis, dir. 2017. *Blade Runner 2049*.

von Foerster, Heinz. 1984. *Observing Systems*. Cambridge, MA: Intersystems, 1984.

von Uexküll, Jakob. 2010. *A Foray into the Worlds of Animals and Humans with a Theory of Meaning*. Translated by Joseph D. O'Neil. Minneapolis: University Of Minnesota Press.

Vopson, Melvin. 2019. "The *Mass-Energy-Information* Equivalence Principle." *AIP [American Institute of Physics] Advances* 9, no. 9. aip.scitation.org/doi/10.1063/1.5123794.

Weatherby, Leif, and Brian Justie. 2022. "Indexical AI." *Critical Inquiry* 48, no. 2:381–415.

Weaver, Warren. 1967. *US Philanthropic Foundations: Their History, Structure, Management, and Record*. New York: Harper and Row.

Westerlaken, Michelle, Jennifer Gabrys, Danilo Urzed, and Max Ritts. 2023. "Unsettling Participation by Foregrounding More-Than-Human Relations in Digital Forests." *Environmental Humanities* 15, no. 1:87–108.

"What You Need to Know about Keyless Ignition Systems." n.d. *Edmunds Car Guides*. https://www.edmunds.com/car-technology/going-keyless.htm. Accessed March 24, 2024.

Wheeler, Wendy. 2006. *The Whole Creature: Complexity, Biosemiotics and the Evolution of Culture*. London: Lawrence and Wishart.

———. 2016a. *Expecting the Earth: Life, Culture, Biosemiotics*. London: Lawrence and Wishart.

———. 2016b. "In Other Tongues: Ecologies of Meaning and Loss." *Modern Forms*, November 17. modernforms.org/blog/colourful-speculation/.

Wolfe, Cary. 2024. "Embodiment, Enaction, Representation." Lecture at Berggruen Institute, Los Angeles. March 14.

Wolf, Gary, and Steven Jonas. n.d. "Quantified Self: Self-Knowledge through Numbers." https://quantifiedself.com/about/team/. Accessed March 27, 2024.

Wolfram, Stephen. 2002. *A New Kind of Science*. New York: Wolfram Media.

Woods, Derek. 2022. "Prosthetic Symbiosis." *Centennial Review* 22, no. 1 (Spring): 157–86.

Woolf, Max. n.d. "In Response to Philosophical Comments on Tech Forum Hacker News . . . , the Model Itself Has Written a Rebuttal." GitHub. https://gist.github.com/minimaxir/f4998c20f2520ad5969b03c9590f16ce. March 24, 2024.

Ziegler, J. F. 1996. "Terrestrial Cosmic Rays." *IBM Journal of Research and Development* 14, no. 1 (January): 19–39. https://ieeexplore.ieee.org/document/538944314.1.

Ziegler, J. F., and W. A. Lanford. 1979. "Effects of Cosmic Rays on Computers." *Science* 206 (4420): 776–88. https://10.1126/science 206.4420.776.

Zuse, Konrad. 1969. *Rechnender Raum, Schriften zur Datenverarbeitung*. Vol. 1. Braunschweig: Freidrich Vieweg and Sohn.

# Index

Abramov, Oleg, 254n22
absential phenomena, 56, 58–59, 66–67, 255n2 (chap. 2)
*Acquiring Genomes* (Margulis and Sagan), 76, 88
agential realism, 17, 21–25
*Ages of Gaia, The* (Lovelock), 78
Agüera y Arcas, Blaise, 163–64
algorithms, 33, 47–48, 65–71, 161–64, 170, 191–92, 207–19, 262n8
Allado-McDowell, K., 149, 260n23
AlphaGo, 72–73
Amazon, 71, 178
Amoore, Louise, 70–71, 262n8
analogue computers, 37, 253n4
Analytical Engine, 174
Ananthaswamy, Anil, 256n7
"Animals in a Bacterial World, a New Imperative for the Life Sciences" (McFall-Ngai), 26
Ankit, Utkarsh, 144
anthropocentrism, 1–4, 8–12, 15–17, 38, 40, 48, 59, 68–70, 88, 99–100, 109, 165, 247
apocalypse. *See* dystopia
artificial intelligence, 13, 46–47, 50, 72, 151–52, 158–59, 166, 213. *See also* algorithms; ChatGPT; collaboration; collective intelligence; conscious robots; creativity; Digital Reasoning; general artificial intelligence (GRI); GPT-3; GPT-4; large language models (LLMs); neural nets; OpenAI GPT; regulation
Artificial Intelligence Act (AIA), 45, 177. *See also* regulation
artificial life, 134, 254n22
"Attention Is All You Need" (Vaswani et al.), 138, 142–44
attention span, 19, 32–33
"At the Mountains of Madness" (Lovecraft), 157

Auerbach, Erich, 256n2
authenticity, 139
autocatalytic networks, 57
"Automating Origination" (Slater), 173–74
AutoML, 15
*Autonomous* (Newitz), 45–46, 182–90
autopoiesis, 18, 21, 26–30, 41–42, 74, 81–93, 96–97, 255n4, 256n5
"Autopoiesis" (Maturana), 86
*Autopoiesis and Cognition* (Maturana and Varela), 29, 81, 85–86
autopoietic Earth, 88

Babbage, Charles, 174
backpropagation, 140–41, 253n9, 262n10
bacteria, 21, 27, 41–42, 59, 75–76, 88–91, 101–4. *See also* micromimesis; microorganisms
Bakker, Karen, 7
Balzac, Honoré de, 148
Barad, Karen, 17–18, 21–25, 254nn14–15
Barthes, Roland, 147–48
*Beamtimes and Lifetimes* (Traweek), 176
Beer, Gillian, 114
Bell, Gordon, 218
Bender, Emily, 153–56
Benjamin, Walter, 45, 180, 192
Bennett, Jane, 10–11, 125
BERT, 257n2 (chap. 6), 260n25
*Bicentennial Man*, 226
Biden, Joseph, 178
Bier, Ethan, 106
"Big Trouble in Biology" (Margulis), 77, 87, 89–90
biological cognition, 102
biological mandate, 10, 13. *See also* evolution
biologism, 53
biomineralization, 129, 131

biosemiotics, 9, 40, 51–54, 59–64, 67–68, 74, 91, 94–97. *See also* signs
blockchain technology, 234–37, 254n22
Bohr, Niels, 17, 21
Boltzmann, Ludwig, 255n2 (chap. 2)
Bongard, Josh, 101–2, 115
Bostrom, Nick, 16
Braidotti, Rosi, 10–11, 17
Branwen, Gwern, 157–60, 260n26
Breazeal, Cynthia, 262n4 (chap. 9)
Brier, Søren, 255n6
Bush, Vannevar, 223

Caillois, Roger, 256n2
"Can the Subaltern Speak?" (Spivak), 159
*Capitalism in the Web of Life* (Moore), 49, 228, 231
carbon coins, 234–39
Cartwright, Nancy, 122, 258n9
causality, 28, 44–48, 64–66, 81–83, 149, 166–77, 208, 230
celestial mechanics, 119–20, 130
cellular cognition, 41–42, 101–2, 117–18
Cetacean Translation Initiative (CETI), 7
*Chaos Bound* (Hayles), 123
Charpentier, Emmanuelle, 101, 104–5
ChatGPT, 33, 44, 144, 168–69, 177, 204–5, 222, 224–27, 258n3, 259n17, 261n1 (chap. 7)
Chen, Delton B., 235–36
chiasmus, 13, 34
"Chinese Room" thought experiment, 67–68, 73, 253n7
Chomsky, Noam, 16, 259n17
Chun, Jon, 260n21
Chun, Wendy Hui Kyong, 71
Clarke, Bruce, 74, 77, 79–80, 92–93, 98
*Cloud Ethics* (Amoore), 71
cognition, 1–4, 8–9, 11–12, 90–94, 96–99, 103, 118. *See also* autopoiesis; bacteria; biological cognition; cellular cognition; computer cognition; consciousness; embodied cognition; grounded cognition; integrated cognitive framework (ICF); nonconscious cognition; planetary cognition; plant cognition
"Cognition All the Way Down" (Levin and Dennet), 102
cognitive assemblages, 3–8, 12–15, 33–34, 39–41, 49–54, 69–70, 92, 100–102, 117–18, 207, 212, 219, 232–34. *See also* ecological relationality; technosymbiosis
cognitive horizon, 102, 118
Cognitive Revolution, 11–12
collaboration, 205–7, 226
collective computation, 234
collective intelligence, 50, 79, 204–5, 224–27. *See also* intelligence
colonialism, 161–62, 230

Common Crawl texts, 153, 259n15
computation, definition of, 139
computational media, 38–39. *See also* cognitive assemblages; general artificial intelligence (GRI); symbiosis; technosymbiosis
computational neuroscience, 208–9
computational theory of mind, 61–62
computer cognition, 52–53, 60–64, 67–68, 71–72, 91
consciousness, 1–4, 40, 45–47, 164, 188, 193–97, 202, 206–12. *See also* nonconscious cognition
conscious robots, 173, 181–82, 193–203
constrained constructivism, 18, 23–24
constraints, 53–58, 63
constructivism, 22–23
convolutional neural nets (CNNs), 138, 141–42, 259n12. *See also* neural nets
Conway's Game of Life, 134, 254n22
copyrights, 226–27
cosmic rays, 64–65
*Cover to Cover* (Snow), 36
Covid-19, 16, 105–6, 110–11
Craighero, Laila, 209
"Creating Digital Resources" (Unsworth), 204
creativity, 47, 103, 125, 132, 173–76, 222–25
CRISPR-Cas9, 41–42, 101, 103–11, 256–57nn4–5 (chap. 4). *See also* gene editing
critical race theory, 17
Csíkszentmihályi, Mihály, 174–75
cybernetics, 41, 81, 165
cybersemiosis, 255n6
cybersemiotics, 51

Damasio, Antonio, 173, 216
dark energy, 85
dark matter, 85
DARPA, 182
Darwin, Charles, 10, 31, 51, 114, 255n2 (chap. 2). *See also* neodarwinism
*Darwin's Plots* (Beer), 114
dataism, 48, 208, 218–20
Dawkins, Richard, 86
Deacon, Terrence W., 9, 40, 51–63, 66, 145–46, 206, 255n2 (chap. 2), 255n4
"Death of the Author, The" (Barthes), 147
deep fakes, 138, 177, 180–81
deep learning algorithms, 11, 72
DeepMind, 72–73
Defense Special Weapons Agency, 7
Dehaene, Stanislas, 210–11
Dempster, M. Beth, 28–29
Dennett, Daniel, 10, 42, 102, 117, 210–11, 262n3 (chap. 9)
"Derp Learning" (Branwen), 158–59
Derrida, Jacques, 148
Devlin, Jacob, 260n25
DiCaglio, Joshua, 254n13

diffractive methodology, 24
Digital Reasoning, 71
digital simulations, 222–23
*Discontents of the Digital* (Evens), 47, 222–23
*Discriminating Data* (Chun), 71
distributed ledger technology (DLT), 235
"Don't Edit the Human Germ Line" (Lanphier et al.), 107
double internality, 34, 49. *See also* reversible internality
Doudna, Jennifer, 101, 104–5, 261n5
*Drawing Hands* (Escher), 254n10
Dreyfus, Hubert, 151–52
Dunbar, Robin, 164
dynamic heterarchy, 54, 102
dystopia, 47–49, 73, 220–21

ecological relationality, 19–20, 40–42, 48–50, 73, 111, 118, 219–20, 232, 246–47. *See also* cognitive assemblages; relationality; technosymbiosis
ecotheory, 31
Edelman, Gerald, 206, 209, 255n5
*Eighth Day, a Transgenic Artwork* (Kac), 7
Einstein, Albert, 85, 224
Eldredge, Niles, 135
electromagnetic pulses (EMPs), 7–8
Elkins, Katherine, 260n21
embodied cognition, 209
*Embodied Mind, The* (Varela), 83–84
*Emergence of Everything, The* (Morowitz), 126
emotions, 213, 215, 262n4 (chap. 9)
enablements, 124
enaction, 84, 95
endosymbiosis, 255n1
"End(s) of Critique, The" (Rouvroy), 71
energy, 84–85
entailing laws, 80, 121–24, 257n3
environmental crises, 228–29, 233, 240–44
environmental humanities, 39–40
epiphylogenetic memory, 31, 33
Escher, M. C., 254n10, 255n3
Esposito, Roberto, 80
Evens, Aden, 47, 222–23
evolution, 41, 101–2, 114–17, 131, 134–36; and adjacent possibles, 10, 42–43, 72, 124, 257n5 (chap. 5); and autopoiesis, 83, 85–87, 90, 92; biological, 2, 18–19, 89, 101, 119–20, 123–25, 131–32, 134, 137, 211; bio-techno, 51–54, 58, 60–62, 69, 73, 92; comparative, 120; and constraints, 54–57; directing, 15–16, 42, 110; and forests, 37–38; and integrated cognitive framework (ICF), 12–13; and machines, 90–93, 99–100, 138; mineral, 43, 120, 125–37; of money, 235; technological, 33, 134–37. *See also* biological mandate; symbiosis
explicit knowledge, 223–24
extinction, 136–37

*Facing Gaia* (Latour), 256n3 (chap. 3)
Fazi, M. Beatrice, 139
Felski, Rita, 156
feminist theory, 17, 125
Fields, Chris, 74, 80, 94–96
Finn, Ed, 234
First Great Inversion, 13
Fodor, Jerry, 208
Ford, Martin, 191, 213
Foucault, Michel, 147–48
Freedman, Walter J., 83
free will, 215–19

Gabrys, Jennifer, 7, 37
*Gaia* (Lovelock), 78, 80–81
"Gaia and the Evolution of Machines" (Sagan and Margulis), 91–93, 98–99
*Gaian Systems* (Clarke), 77, 79, 92–93
Gaia theory, 21, 74–75, 77–81, 93, 255n2 (chap. 3)
Gantz, Valentino, 106
Gardner, Howard, 176
Gebru, Timnit, 153–54, 156
Gemmell, Jim, 218
gender, 186–89
gene drives, 106
gene editing, 6, 15–16, 20–21, 41–42, 101, 104–10, 112, 117, 136, 261n5. *See also* CRISPR-Cas9
general artificial intelligence (GRI), 16
*General Ecology* (Hörl), 71
General Protection of Data Regulation (GPDR), 177
generative adversarial networks (GANs), 12
generative uninspiration, 175
genetic drift, 86, 89
geoengineering, 239–40
germ-line editing, 107–8, 111
Gessler, Nicholas, 127, 254n21
*Gesture and Speech* (Leroi-Gourhan), 31
Gibbs free energy, 129, 132
Gilbert, Scott, 26–28
global warming, 229
global workspace model, 210
Goldberg, Kenneth, 7
Golding, William, 78
Goldstein, Jacob, 234–35
Gomez, Ailan N., 259n14
Google, 72, 163–64, 178, 259n14, 260n25, 262n10
Gough, Michael Paul, 85
Gould, Stephen Jay, 135, 256n9
GPT-3, 33, 39, 50, 138–40, 144–63, 168, 182, 204, 223–25, 260nn20-23
GPT-4, 33, 39, 45, 166–73, 182, 204, 222, 224, 261n2 (chap. 7)
gradient descent, 140–41, 171
Graham, Paul, 159
Graziano, Michael, 210–11

Great Oxidation Event, 127–28, 130–31, 255n2 (chap. 3)
Gross, Oskar, 175
Grosz, Elizabeth, 10–11, 17
grounded cognition, 209
Guillory, John, 54

Haff, Peter, 232–33, 246–47
Hand, David, 257n6 (chap. 5)
Harari, Yuval Noah, 47–48, 207–21
Haraway, Donna, 17–18, 26–30, 39, 50, 93, 100, 254n11, 257n6
Harding, Sandra, 23
Hazen, Robert, 120, 125–36
heat maps, 143–44
Heidegger, Martin, 151
Heisenberg, Werner, 21
Heldén, Johannes, 66
Hewlett-Packard, 182
Hillis, Danny, 258n7
Hinton, Geoffrey, 16, 140–41, 253n9, 262n10
Hoffman, Reid, 206
Hoffmeyer, Jesper, 9, 40, 53, 63, 66
Hofstadter, Douglas, 158–59
homeodynamics, 55–58
*Homo Deus* (Harari), 47–48, 207–21
Hörl, Erich, 71
Horton, Zachary, 254n13
"How AI Fails Us" (Siddarth), 205–6
"How Evolution 'Hacked' Its Way to Intelligence from the Bottom Up" (Levin and Yuste), 102
*How the Laws of Physics Lie* (Cartwright), 122
"How to Write Usefully" (Graham), 159–60
*How We Became Posthuman* (Hayles), 47–48, 86–87, 217
*How We Think* (Hayles), 32
HRL Laboratories, 182
Huang, Junjiu, 107
human aura, 45–46, 180–81, 183–84, 189, 192–93, 196–97, 202–3
human-competitive productions, 142, 206, 258n4
HumanEval data set, 167
humanism, 207–9, 212, 217–19
"Human Nature" (*Nova* episode), 104, 256n3 (chap. 4)
human rights, 48, 179, 219
humor, 157–58, 260n27

IBM, 12, 64–65, 182
ICITE, 71
imperialism, 230
"Indexical AI" (Weatherby and Justie), 145
individuation, theory of, 18, 25–26
"Inert vs. the Living State of Matter, The" (Longo and Montévil), 125
inferences, 78, 82, 89, 124, 145–59, 163–64, 223

informational closure, 27–28
information theory, 84–85, 145–46, 255n2 (chap. 2), 256n7
integrated cognitive framework (ICF), 2, 8–13, 16–17, 20, 46–47, 74–75, 96–97, 101, 118, 231–32
intelligence, 115–17, 172, 207, 212, 217. *See also* collective intelligence
intermediation, 51–54, 57–58
interpretation, 64–70, 99, 109
*Intimacies of Four Continents* (Lowe), 48, 219, 228–31
Ishiguro, Kazuo, 45–46, 182–83, 189–93, 202, 208, 261n5

Jackson, Zakkiyyah Iman, 17
jagged ontologies, 2
James, William, 115, 117
Jonson, Håkan, 66
Justie, Brian, 145

Kac, Eduardo, 7
Kant, Immanuel, 123–24
Karlsruhe Institute of Technology, 178–79
Kauffman, Stuart, 10, 42–43, 57, 72, 80, 119–25, 129–30, 133–34, 257n5 (chap. 5)
kinship, 112, 197–201, 246
Kirschenbaum, Matthew, 149–50, 170
*Klara and the Sun* (Ishiguro), 45–46, 182–83, 189–93, 202, 208
Koller, Alexander, 154–55
Koza, John R., 258n4
Kozo-Polyansky, Boris, 255n1

LaMDA, 163–64, 224, 257n2 (chap. 6)
Lanphier, Edward, 107–8
Laplace, Pierre Simon, 119, 257n1 (chap. 5)
large language models (LLMs), 2, 16, 43–45, 138–45, 153–54, 163–65, 169, 181, 205–6, 223–27. *See also* ChatGPT; GPT-3; GPT-4; OpenAI GPT
latent knowledge, 223–24
Latour, Bruno, 70, 78–79, 84, 256n3 (chap. 3), 257n2 (chap. 5)
*Laws of Form, The* (Spenser-Brown), 28
laws of motion, 122–24
LeCain, Timothy James, 14
Leibniz, Gottfried Wilhelm, 257n1 (chap. 5)
Leroi-Gourhan, André, 31
Lettvin, J. Y. (Jerry), 82–83
Levin, Michael, 10, 42, 74, 80, 94–96, 101–2, 112–18
Levine, George, 22
liberalism, 195, 219–20, 228–33
Libet, Benjamin, 216
Licklider, J. C. R., 253n8
life-logging movement, 218–19
literary criticism, 2, 40, 43–44, 139, 147–50, 153–57, 163, 181

Locke, John, 184
lock-in effects, 40, 57–58
logic gates, 58, 63–65
*Logic of Scientific Discovery, The* (Popper), 254n12
Longo, Giuseppe, 10, 42, 80, 119–25, 129–30, 133–34
long short-term memory (LSTM), 141–42. *See also* neural nets
Lovecraft, H. P., 157
Lovelace, Ada, 174
Lovelock, James, 41, 74, 77–81, 83–84, 87, 93, 97–99, 129, 255–56nn2–3 (chap. 3)
Lowe, Lisa, 48, 219, 228–31, 233
Luhmann, Niklas, 27–28, 79, 95, 148

*Machines Like Me* (McEwan), 45–46, 182–84, 193–202, 208
Macpherson, C. B., 184
Manning, Christopher D., 154
Marcus, Gary, 158
Margulis, Lynn, 29, 41, 74–81, 87–93, 97–99, 123, 134, 136, 255nn1–2 (chap. 3)
Mark I Perceptron, 258n11
Martian trilogy (Robinson), 239
mass-energy-information equivalence principle, 85
materiality, 54
material processes, 9–11, 40, 42, 79–80
Maturana, Humberto, 18, 27–28, 41, 74, 81–87, 89–90, 93–98, 255n4, 256n5
Mayr, Ernst, 76–77
McCulloch, Warren, 82–83
McEwan, Ian, 45–46, 182–84, 193–202, 208
McFall-Ngai, Margaret, 26–27, 29
meaning-making practices, 51–52, 58–62, 66–73, 93, 100, 103, 109, 138
media, 52–55, 58
mediation, 54
*Meeting the Universe Halfway* (Barad), 22, 254n15
Megatron-Turing NLG 530B, 257n2 (chap. 6)
Menabrea, Luigi Federico, 174
Meta, 178
metastable, 19–20
*Microcosm* (Margulis and Sagan), 88
micro/evo/techno relationality. *See* relationality
micromimesis, 42, 104–5, 108, 110–12. *See also* mimesis
microorganisms, 1, 6, 11, 41, 43, 77, 103, 137. *See also* bacteria
Microsoft, 44, 168–73
Mill, John Stuart, 230
mimesis, 42, 101, 103–4, 109–11, 256n2. *See also* micromimesis
*Mimesis* (Auerbach), 256n2
"Mineral Evolution" (Hazen), 126
mineralization, 132–33
mineralogy, 43

*Ministry for the Future* (Robinson), 49, 228, 231, 233–34, 236–47
mitochondria, 14, 75
modularity, 115–16, 257n7 (chap. 4)
Mojica, Francisco, 103–4
*Money* (Goldstein), 234–35
Montévil, Maël, 10, 42, 80, 119–25, 129–30, 133–34
Moore, Jason, 34, 49–50, 228, 231, 238–39, 246
Morgan, William, 160–61
Morowitz, Harold, 54, 125–26, 128, 131
morphodynamics, 55–58
*My Mother Was a Computer* (Hayles), 52

Nagel, Thomas, 152–53
natural language processing (NLP), 141
natural selection, 58–59
negative logic, 53, 255n2
neocybernetic systems theory (NTS), 74, 77, 93, 97. *See also* systems theory
neodarwinism, 27, 30, 76, 85–86, 89–90, 97. *See also* Darwin, Charles
neomaterialism, 10–11, 14
neural nets, 7, 11, 19–21, 39, 43–44, 72, 138–42, 151–53, 163–64, 181, 223–24, 258n11. *See also* BERT; ChatGPT; convolutional neural nets (CNNs); DeepMind; GPT-3; GPT-4; long short-term memory (LSTM); OpenAI GPT; recurrent neural nets (RNNs); Transformer architecture
neuromorphic chips, 12, 21
*Never Let Me Go* (Ishiguro), 192
Newitz, Annalee, 45–46, 182–90
New Materialism, 125
Newton, Isaac, 84, 119–20, 257n1 (chap. 5)
noise of materiality, 140–41
nonconscious cognition, 1–2, 4, 33–34, 67–68, 94, 101–3, 109, 215–17, 259n16. *See also* consciousness
Nordhaus, William, 228–29
*Nova*, 104
*Novacene* (Lovelock), 81, 84, 99
null strategy, 44, 139, 147–50

*Observing Systems* (von Foerster), 35
O'Giebly, Meghan, 259n16
O'Gorman, Marcel, 32
Okojie, Irenosen, 149
*On Liberty* (Mill), 230
"On the Dangers of Stochastic Parrots" (Bender et al.), 153–56, 260n24
*On the Mode of Existence of Technical Objects* (Simondon), 25
"On the Origin of Mitosing Cells" (Margulis), 75
OpenAI, 178, 258n3
OpenAI DALL-E, 155, 163, 222, 224, 260n28
OpenAI GPT, 2, 33, 43–44, 138, 153, 166–68
operational closure, 27–28, 81–82, 90

*Order of Things, The* (Foucault), 147
origin stories, 2, 18–21, 27, 31. *See also* evolution; quantum mechanics; technics
overfitting, 140–41
*Overstory, The* (Powers), 145
*Oxford Handbook of Ethics in AI, The*, 173

Panagiotaropoulos, Theofanis, 262n3 (chap. 9)
Paris Accord, 229, 236
pattern completion, 116–17
Peirce, C. S., 9, 52–53, 59, 62–63, 66, 94, 109, 145
*Pharmako-AI* (Allado-McDowell), 149
phase spaces, 10, 121–25, 127, 133
photosynthesis, 75–76, 128, 130
Pitts, Walter, 82–83
plagiarism, 225–26
planetary cognition, 79, 93–94, 98
planetary reversal, 228
plant cognition, 91
Plaue, Ethan, 160–61
political liberalism. *See* liberalism
Pollock, Jackson, 222
Polosukhin, Illia, 259n14
Popper, Karl, 23, 254n12
possessive individualism, 184
postcolonial/decolonial theory, 17
posthuman, the, 47–48
Powers, Richard, 145
preadaption, 257n4
preimplantation genetic diagnosis (PGD), 108
Prigogine, Ilya, 56
*Program Earth* (Gabrys), 7
propaganda, 169, 217
"Prosthetic Symbiosis" (Woods), 29–31, 33–34
Protagoras, 3
protein foldings, 21
pseudorandomness, 65–66
punctuated equilibria, 135, 137

quantitative easing, 235–36
quantum field theory, 254n14
quantum mechanics, 17–19, 21, 24–25. *See also* origin stories

racism, 161–62, 231
Raley, Rita, 148
randomness, 65–66, 76, 124, 131
Rasch, William, 28
*Realism and Representation* (Levine), 22
RealToxicityPrompts data set, 167
recurrent neural nets (RNNs), 138, 141–42, 149. *See also* neural nets
recursive dynamics, 29–30, 54, 57, 65, 81, 255n5
reentry signaling, 206, 209, 255n5
Rees, Tobias, 155, 163, 260n22
reference frame theory (RFT), 74, 80, 95–98

reflexivity, 22, 27, 41, 254n10
regulation, 45, 169, 177–79. *See also* Artificial Intelligence Act (AIA)
reinforcement learning with human feedback (RLHF), 167, 258n3
relationality, 2–3, 17–21, 25–26, 31, 34–38, 71, 227
religion, 23
representationalism, 22, 45, 139, 165, 181, 208
*Revenge of Gaia, The* (Lovelock), 80, 255n2 (chap. 3)
reversible internality, 34–38, 49, 87–88, 111–12, 129, 166, 174–76, 179. *See also* double internality
Rizzolatti, Giacomo, 209
Roberts, Ian, 259n17
Robinson, Kim Stanley, 49, 228, 231, 233–34, 236–47
"Rocks from Space" (Gessler), 127
Rosch, Eleanor, 84
Rosen, Robert, 134
Rosenblatt, Frank, 258n11
Rouvroy, Antoinette, 71
rule-based reward models (RBRM), 167
"ruliness," 222
Rumelhart, David, 140–41

Sacks, Oliver, 173, 261n3 (chap. 7)
Sagan, Dorion, 74–77, 80, 88, 90–93, 97–99
*Sarrasine* (Balzac), 148
Savulescu, Julian, 107–8
"Say-Can" project, 259n19
scale, 17–18, 24, 26, 30–31, 52, 232–33, 238–42, 247, 254n13
Schaffer, Simon, 257n2 (chap. 5)
science and technology studies (STS), 22–23, 257n2 (chap. 5)
Scott, Kevin, 154
Searle, John, 67–68, 73, 253n7
Second Great Inversion, 13
"Secrets and Machines" (Plaue and Morgan), 160–61
semiosphere, 9, 66, 69–70
sexuality, 186–89, 194–95
Shannon, Claude, 255n2 (chap. 2)
Shapin, Steven, 257n2 (chap. 5)
Shaw, Helen, 190
Siddarth, Divya, 205–6
Sidney, Philip, 153
signs, 9–11, 40, 52–53, 58–59, 66–69, 94–95, 109. *See also* biosemiotics
Simondon, Gilbert, 18–20, 25–26, 35, 254n15
Simone, Natal, 262n4 (chap. 9)
situated knowledges, 254n11
"Sketch of the Analytical Engine, A" (Lovelace), 174
Slater, Avery, 44–45, 166, 173–75, 179
slavery, 230–31
"Smart Forests Atlas" (Gabrys), 37
Snow, Michael, 36
social brain theory, 164
social cohesion, 215, 220

social systems, 27–28
Sokal, Alan, 257n2 (chap. 5)
*Sounds of Life, The* (Bakker), 7
spandrels, 256n9
*Sparks of Artificial General Intelligence* (Microsoft report), 168–73
"Spec Acts" (Kirschenbaum), 149–50
Spenser-Brown, George, 28
Sperry, Roger, 256n6
Spivak, Gayatri, 159
standard model of consciousness, 210–11
*Staying with the Trouble* (Haraway), 18, 26
Stengers, Isabelle, 56
Sternberg, Samuel, 105
Stiegler, Bernard, 18–19, 31–34
stochastic parrot view, 44
strong objectivity, 23–24
superhumans, 208
symbiogenesis, 18, 27, 29–30, 74–77, 88–90, 97–98, 255n1
*Symbiogenesis* (Kozo-Polyansky), 255n1
symbiosis, 14, 18, 26–34, 38–41, 74–77, 98, 117, 253n8. *See also* evolution; technosymbiosis
*Symbiotic Planet* (Margulis), 88, 255n2 (chap. 3)
"Symbiotic View of Life, A" (Gilbert et al.), 26–27
sympoiesis, 18, 28–30
SyNAPSE chip, 12, 21, 182
systems theory, 28, 30, 74, 93, 95, 148. *See also* neocybernetic systems theory (NTS)
*S/Z* (Barthes), 148

technics, 2, 18–20, 31–38. *See also* origin stories; Transformer architecture
*Technics and Time* (Stiegler), 18–19, 31–32
technological change, 20
technological reproducibility, 45
technosphere, 232–33
technosymbiosis, 34, 38–41, 47, 50, 74, 77, 97–100, 102, 109, 117, 257n5 (chap. 4). *See also* cognitive assemblages; ecological relationality; symbiosis
*Telegarden, The* (Goldberg), 7
teleodynamics, 40, 53, 55, 57–58, 69–70
theory of consciousness, 210
theory of information. *See* information theory
theory of mind, 170, 172
Thompson, Evan, 84
Thompson, William Irwin, 89
Toivonen, Hannu, 175
Tononi, Guilio, 206, 209
tools, 20, 34, 108–9
*Total Recall* (Bell and Gemmell), 218
Transformer architecture, 16, 43, 138–39, 142–44, 157, 168, 170, 172–73, 181, 223, 257n2 (chap. 6), 259n14. *See also* neural nets
Traweek, Sharon, 176
*Tree of Knowledge, The* (Maturana and Varela), 86

Trump, Donald, 229
Turing, Alan, 258n8

Ullrich, George, 7
umwelten, 67–70, 73, 94, 117, 150–52, 154–55, 163
*Universe Is Hostile to Computers, The*, 65
unmediated flux, 23, 25
"Unsettling Participation by Foregrounding More-Than-Human Relations in Digital Forests" (Westerlaken et al.), 37
Unsworth, John, 204
*Unthought* (Hayles), 1, 3–4, 8, 202, 216
Urnov, Fyodor, 256n3 (chap. 4)
*Uses of Literature* (Felski), 156
utopia, 73, 231, 238, 245–46

vaccines, 6, 105–6
Van der Waals corrections, 140
*Vanishing Face of Gaia, The* (Lovelock), 80
vanishing gradient problem, 141–42
van Strien, Marij, 257n1 (chap. 5)
Varela, Francisco, 18, 27–29, 41, 74, 81–87, 89–90, 94–98, 255n4 (chap. 2), 256n5
Vaswani, Ashish, 142–44
Verbeek, Peter-Paul, 70
Vindevogel, Marie, 65
von Foerster, Heinz, 35
von Neumann architectures, 11, 19, 63, 84, 140, 150–51, 253n6
von Uexküll, Jakob, 67, 69, 150
Vopson, Melvin, 85

Watumull, Jeffrey, 259n17
Weatherby, Leif, 145
Weaver, Warren, 135
*We Have Never Been Modern* (Latour), 84
*What Computers Can't Do* (Dreyfus), 151
*What Computers Still Can't Do* (Dreyfus), 151
"What Is an Author" (Foucault), 147
"What Is It Like to Be a Bat?" (Nagel), 152–53
"What the Frog's Eye Tells the Frog's Brain" (Lettvin et al.), 82–83, 95
Wheeler, Wendy, 9, 40, 53, 59–62, 66, 68, 109
Williams, Ronald, 140–41
Wolfe, Cary, 2, 28
Woods, Derek, 18, 29–31, 33–34
world horizons, 3, 16, 67–70, 73, 94, 117, 150–52, 154–55, 163
*Worldwide Threat Assessment*, 106–7
*Writing Gaia* (Clarke), 77

xenobots, 21, 42, 101–2, 112–14, 117

Yuste, Rafael, 102, 114–18

Zuse, Konrad, 11